Information Refinement Technologies for Crisis Informatics

Marc-André Kaufhold

Information Refinement Technologies for Crisis Informatics

User Expectations and Design Principles for Social Media and Mobile Apps

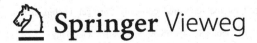
Springer Vieweg

Marc-André Kaufhold
Darmstadt, Germany

Darmstadt, Technische Universität Darmstadt, doctoral thesis

ISBN 978-3-658-33343-0   ISBN 978-3-658-33341-6 (eBook)
https://doi.org/10.1007/978-3-658-33341-6

Responsible Editor: Stefanie Eggert
This Springer Vieweg imprint is published by the registered company Springer Fachmedien Wiesbaden GmbH part of Springer Nature.
The registered company address is: Abraham-Lincoln-Str. 46, 65189 Wiesbaden, Germany

Foreword

Social media and mobile apps have been part of our lives for almost 20 years now and are used by large parts of the population; however, not only in everyday situations, but also in emergencies, crises, and disasters. These technologies offer many opportunities for crisis management, including a fast flow of information among involved actors, direct organization of civil self-help, or situational awareness for emergency services. Still, the potential information overload, increased complexity for emergency managers, or spread of misinformation pose substantial risks, as demonstrated recently by the COVID-19 pandemic. Hence, by combining knowledge, methods, and theory from computing and social sciences, the research area of *crisis informatics* has been investigating the effective use and design of information and communication technology (ICT) to prepare for, respond to, and recover from critical situations.

The dissertation of Marc-André Kaufhold addresses this research area and explores the process of *information refinement*, comprising the applied practices and used technologies to refine obtained information according to event-related, organizational, societal, and technological boundary conditions in order to improve crisis management. The dissertation provides *empirical and technical findings* for computer science, including contributions to the disciplines of *human-computer interaction (HCI)*, *computer supported cooperative work (CSCW)*, and *information systems (IS)*. The work addresses (1) the research of expectations, perceptions, and usage patterns of the population regarding ICT usage, focussing on social media and crisis apps, and derives (2) implications for the design and implementation of these applications to support effective information refinement.

Marc-André Kaufhold's dissertation presents outstanding work. The methodological *spectrum*, from qualitative and quantitative empirical studies to technical contributions in computer science, and the quality of the work, in particular visible by the publication performance of the candidate in quantity (63 publications), quality (9x CORE-A/A* until submission of the dissertation), and the already achieved scientific impact (about 650 citations), are in my opinion excellent for a dissertation. Marc-André Kaufhold has clearly demonstrated that he is capable of independent scientific work and his dissertation was graded summa cum laude. I wish Marc and his research all the best in his future career! For the readers, I wish a pleasant and instructive reading.

Prof. Dr. Christian Reuter
Full Professor for Science and Technology for Peace and Security
(PEASEC)
Technical University of Darmstadt
Darmstadt, Germany

Acknowledgements

My dissertation would not have been possible without the support of many people, projects, and organizations. I am incredibly grateful to my doctoral advisor *Prof. Dr. Christian Reuter* for our discussions and his advice during the entire research process. I also thank my co-advisor *Prof. Dr. Gunnar Stevens* for his valuable support during my research activities as well as *Prof. Dr. Matthias Hollick, Prof. Dr. Max Mühlhäuser*, and *Prof. Dr. Arjan Kuijper* who kindly agreed to join the doctoral committee. Furthermore, I would like to thank *Prof. Dr. Volkmar Pipek, Prof. Dr. Thomas Ludwig,* and *Dr. Marén Schorch* who supported my research during my time at the Institute for Information Systems at the University of Siegen.

I am grateful to my colleagues of the research group Science and Technology for Peace and Security (PEASEC) at the Department of Computer Science at the Technical University of Darmstadt for their valuable feedback to my concepts and ideas. Here, I would especially thank my colleagues *Jasmin Haunschild* and *Thea Riebe*, who worked together with me in research projects and co-authored some of my conference papers and journal articles. Moreover, I would like to thank (former) student assistants, especially *Lukas Dratwa, Margarita Grinko, Sabrina Neuhäusel, Leo Bauer*, and *Helene Pleil*, who have contributed in several ways to my thesis, including literature research, implementation of ICT and proofreading of publications.

Furthermore, my special thanks to all co-authors of my conference papers and journal articles, including *Markus Bayer, Nicola Rupp,* and *Matthias Habdank*, as well as participants in my empirical studies for the insightful collaboration and for their contribution to my work. I am also obliged to the supporting and underlying research centres, programmes, and projects *EmerGent* (project funded by the European Union Framework Programme for Research and Technological Development, FP7 No. 608352) and *KontiKat* (junior research group funded by the German Federal Ministry for Education and Research, BMBF No. 13N14351) at the University of Siegen, as well as *ATHENE* (national research centre funded by the German Federal Ministry of Education and Research and the Hessen State Ministry for Higher Education, Research and Arts) and *MAKI* (collaborative research centre funded by the German Research Foundation, DFG SFB 1053) at the Technical University of Darmstadt.

Finally, I also feel deep gratitude to my family, friends and in particular my girlfriend *Thea*, who have supported me through the years of my research project. Thank you very much!

Marc-André Kaufhold

Abstract

In the past 20 years, mobile technologies and social media have not only been established in everyday life, but also in crises, disasters, and emergencies. Especially large-scale events, such as 2012 Hurricane Sandy or the 2013 European Floods, showed that citizens are not passive victims but active participants utilizing mobile and social information and communication technologies (ICT) for crisis response (Reuter, Hughes, et al., 2018). Accordingly, the research field of *crisis informatics* emerged as a multidisciplinary field which combines computing and social science knowledge of disasters and is rooted in disciplines such as human-computer interaction (HCI), computer science (CS), computer supported cooperative work (CSCW), and information systems (IS). While citizens use personal ICT to respond to a disaster to cope with uncertainty, emergency services such as fire and police departments started using available online data to increase situational awareness and improve decision making for a better crisis response (Palen & Anderson, 2016).

When looking at even larger crises, such as the ongoing COVID-19 pandemic, it becomes apparent that the challenges of crisis informatics are amplified (Xie et al., 2020). Notably, information is often not available in perfect shape to assist crisis response: the dissemination of high-volume, heterogeneous and highly semantic data by citizens, often referred to as *big social data* (Olshannikova et al., 2017), poses challenges for emergency services in terms of access, quality, and quantity of information. In order to achieve situational awareness or even actionable information, meaning the right information for the right person at the right time (Zade et al., 2018), information must be refined according to event-based factors, organizational requirements, societal boundary conditions, and technical feasibility. In order to research the topic of *information refinement*, this dissertation combines the methodological framework of *design case studies* (Wulf et al.,

2011) with principles of *design science research* (Hevner et al., 2004). These extended design case studies consist of four phases, each contributing to research with distinct results.

This thesis first *reviews* existing research on use, role, and perception patterns in crisis informatics, emphasizing the increasing potentials of public participation in crisis response using social media. Then, *empirical studies* conducted with the German population reveal positive attitudes and increasing use of mobile and social technologies during crises, but also highlight barriers of use and expectations towards emergency services to monitor and interact in media. The findings led to the *design* of innovative ICT artefacts, including visual guidelines for citizens' use of social media in emergencies (SMG), an emergency service web interface for aggregating mobile and social data (ESI), an efficient algorithm for detecting relevant information in social media (SMO), and a mobile app for bidirectional communication between emergency services and citizens (112.social). The *evaluation* of artefacts involved the participation of end-users in the application field of crisis management, pointing out potentials for future improvements and research potentials. The thesis concludes with a *framework* on information refinement for crisis informatics, integrating event-based, organizational, societal, and technological perspectives.

Zusammenfassung

In den letzten 20 Jahren haben sich mobile Technologien und soziale Medien nicht nur im Alltag, sondern auch in Krisensituationen etabliert. Insbesondere großflächige Ereignisse wie der Hurrikan Sandy (2012) oder das mitteleuropäische Hochwasser (2013) haben gezeigt, dass sich die Bevölkerung aktiv mit Informations- und Kommunikationstechnologie (IKT) an der Schadensbewältigung beteiligt (Reuter, Hughes, et al., 2018). Daraus ist das Forschungsfeld der Kriseninformatik entstanden, welches Wissen der Informatik und Gesellschaftswissenschaften kombiniert und zudem in Disziplinen wie der Mensch-Maschine-Interaktion (HCI), Computerunterstützten Gruppenarbeit (CSCW) und Wirtschaftsinformatik (WI) verankert ist. Während die Bevölkerung IKT einsetzt, um die Unsicherheiten in Krisen zu bewältigen, nutzen Behörden und Organisationen mit Sicherheitsaufgaben (BOS), etwa Feuerwehr und Polizei, öffentliche Daten, um das Situationsbewusstsein und die Entscheidungsfindung für eine bessere Schadensbewältigung zu verbessern (Palen & Anderson, 2016).

Noch größere Katastrophen wie die aktuelle COVID-19-Pandemie verstärken dabei die Herausforderungen der Kriseninformatik (Xie et al., 2020). Für BOS stellt die umfangreiche Menge heterogener und semantisch verknüpfter Daten, auch *Social Big Data* genannt (Olshannikova et al., 2017), eine große Herausforderung im Hinblick auf die Qualität, Quantität und den Zugriff auf relevante Informationen dar. Um ein Situationsbewusstsein und nutzbare Informationen, d. h. die richtigen Informationen zur richtigen Zeit bei der richtigen Person, zu erhalten (Zade et al., 2018), müssen Informationen auf die Bedingungen des Ereignisses, organisationale Anforderungen, soziale Rahmenbedingungen und technische Möglichkeiten verfeinert werden. Diese Dissertation kombiniert das methodische Framework der *Designfallstudien* (Wulf et al., 2011) mit den Prinzipien der *Design-Science-Forschung* (Hevner et al., 2004), um das

Thema der *Informationsverfeinerung* (Information Refinement) in vier Phasen zu untersuchen, wovon jede unterschiedliche Forschungsbeiträge hervorbringt.

Die Arbeit *begutachtet* zunächst Nutzungs-, Rollen- und Wahrnehmungsmuster in der Kriseninformatik und stellt die Potenziale sozialer Medien zur öffentlichen Teilhabe an der Krisenbewältigung heraus. Die *empirische Studien* mit der deutschen Bevölkerung zeigen die positiven Einstellungen und die steigende Nutzung mobiler und sozialer Technologien in Krisen, stellen aber auch Barrieren heraus und zeigen die Erwartung, dass BOS in soziale Medien aktiv sind. Die Ergebnisse fundieren das *Design* innovativer IKT-Artefakte, darunter visuelle Bevölkerungsrichtlinien für soziale Medien in Krisen (SMG), ein Web-Interface für BOS zur Aggregation mobiler und sozialer Daten (ESI), ein Algorithmus zur Extraktion relevanter Informationen in sozialen Medien (SMO), und eine mobile App für die bidirektionale Kommunikation zwischen BOS und Bevölkerung (112.social). Die *Evaluation* der Artefakte involviert EndnutzerInnen aus dem Anwendungsfeld des Krisenmanagements, um potenziale für Verbesserungen und zukünftige Forschung zu identifizieren. Die Arbeit schließt mit einem Framework zur Informationsverfeinerung für die Kriseninformatik ab, welche die event-, gesellschafts-, organisation- und technologiebasierte Perspektive integriert.

My Contribution

The contributions as outlined in the last paragraph of the abstract and my cumulative dissertation as a whole is based on collaborative work with my colleagues in complex national and international research projects at the Technical University of Darmstadt (ATHENE, MAKI) and the University of Siegen (EmerGent, KontiKat). In the following, I want to highlight how the involved co-authors contributed to the research papers that are published in both findings' parts of my dissertation.

Chapter 4 was published as the research paper "Fifteen Years of Social Media in Emergencies: A Retrospective Review and Future Directions for Crisis Informatics" (Reuter & Kaufhold, 2018) in the "Journal of Contingencies and Crisis Management" [JIF 1.365]. It constitutes a joint work with Christian Reuter, which was funded within the research projects EmerGent and KontiKat (Table 0.1). As corresponding and leading author, Christian led the overall research design, management, and introduction (section 4.1) of the paper. The analysis of published cases (section 4.2) and usage patterns (section 4.3) of social media in emergencies were based on previous work by Christian and extended with state-of-the-art literature by both of us. While I focused on the analysis of role patterns (section 4.4) of involved actors, Christian concentrated on perception patterns (section 4.5) of authorities and citizens. The discussion and conclusion (section 4.6) were composed by contributions and reflections of Christian and me.

Chapter 5 was published as the research paper "Potentiale von IKT beim Ausfall kritischer Infrastrukturen: Erwartungen, Informationsgewinnung und Mediennutzung der Zivilbevölkerung in Deutschland" (Kaufhold, Grinko, et al., 2019) at the "International Conference on Wirtschaftsinformatik" [CORE-C, WKWI-A]. It constitutes a joint work with Margarita Grinko, Christian Reuter, Marén Schorch, Amanda Langer, Sascha Skudelny, and Matthias Hollick, which was

Table 0.1 Contribution summary of chapter 4: Paper title, conference, involved actors (surname) and their contributions, separated by categories and ordered decreasingly by contribution from left to right

Paper Title	Fifteen Years of Social Media in Emergencies: A Retrospective Review and Future Directions for Crisis Informatics
Journal	Journal of Contingencies and Crisis Management, 2018, 26(1)
Involved Authors	Reuter, Kaufhold
Scientific Lead and Management	Reuter
Paper Design and Introduction	Reuter, Kaufhold (ch. 4.1)
Literature Review	Reuter (ch. 4.2, 4.3, 4.5), Kaufhold (ch. 4.2, 4.3, 4.4)
Discussion and Implications	Reuter, Kaufhold (ch. 4.6)
Enabling Project Funding	EmerGent, KontiKat

funded within the research projects KontiKat, HyServ, and MAKI (Table 0.2). As corresponding author, I led the overall research design, management, and introduction (section 5.1) of the paper. All authors made contributions to the literature review (section 5.2), ensuring multidisciplinary quality. While Christian and I designed the representative survey (section 5.3) including method and questionnaire, the qualitative and quantitative statistical analysis (section 5.4) of collected data was strongly driven by Margarita as well as supported and supervised by me. The discussion and conclusion (section 5.5) were composed by contributions and reflections of all authors.

Chapter 6 was published as the research paper "Social Media in Emergencies: A Representative Study on Citizens' Perception in Germany" (Reuter, Kaufhold, Spielhofer, et al., 2017) at the "ACM Conference on Computer Supported Cooperative Work and Social Computing" [CORE-A]. It constitutes a joint work with Christian Reuter, Thomas Spielhofer, and Anna Sophie Hahne, which was funded within the research projects EmerGent and KontiKat (Table 0.3). As corresponding and leading author, Christian led the overall research design, management, and introduction (section 6.1) of the paper. I conducted the literature review (section 6.2) together with Christian and the representative survey (section 6.3), including method and questionnaire, was mutually designed by all of us. While I focused on the qualitative analysis, Thomas and Anna Sophie led the quantitative statistical analysis of collected data (section 6.4). The discussion, conclusion (section 6.5) and limitations (section 6.6) were driven by Christian and me.

Table 0.2 Contribution summary of chapter 5: Paper title, conference, involved actors (surname) and their contributions, separated by categories and ordered decreasingly by contribution from left to right

Paper Title	Potentiale von IKT beim Ausfall kritischer Infrastrukturen: Erwartungen, Informationsgewinnung und Mediennutzung der Zivilbevölkerung in Deutschland
Conference Proceedings	International Conference on Wirtschaftsinformatik, 2019
Involved Authors	Kaufhold, Grinko, Reuter, Schorch, Langer, Skudelny, Hollick
Scientific Lead and Management	Kaufhold
Paper Design and Introduction	Kaufhold (ch. 5.1)
Literature Review	Grinko, Kaufhold, Reuter, Schorch, Langer, Skudelny, Hollick (ch. 5.2)
Empirical Design	Kaufhold, Reuter (ch. 5.3)
Empirical Enquiry	GapFish GmbH
Empirical Analysis	Grinko, Kaufhold (ch. 5.4)
Discussion and Implications	Kaufhold, Grinko, Reuter, Schorch, Langer, Skudelny, Hollick (Ch 5.5)
Enabling Project Funding	KontiKat, HyServ, MAKI

Chapter 7 was published as the research paper "Adoption, Use and Diffusion of Crisis Apps in Germany: A Representative Survey" (Grinko et al., 2019) at the conference "Mensch und Computer" [GI-CSCW-B]. It constitutes a joint work with Margarita Grinko and Christian Reuter, which was funded within the research projects KontiKat and MAKI (Table 0.4). Although I led the overall research design, management, and introduction (section 7.1) of the paper, Margarita's substantial contributions to the literature review (section 7.2) as well as qualitative and quantitative statistical analysis of collected data (section 7.4) qualified her as first author of the paper. The representative survey (section 7.3), including method and questionnaire, were designed by Christian and me. The discussion and conclusion (section 7.5) were composed by contributions and reflections of all involved authors.

Table 0.3 Contribution summary of chapter 6: Paper title, conference, involved actors (surname) and their contributions, separated by categories and ordered decreasingly by contribution from left to right

Paper Title	Social Media in Emergencies: A Representative Study on Citizens' Perception in Germany
Conference Proceedings	ACM Conference on Computer Supported Cooperative Work and Social Computing, 2018
Involved Authors	Reuter, Kaufhold, Spielhofer, Hahne
Scientific Lead and Management	Reuter
Paper Design and Introduction	Reuter, Kaufhold (ch. 6.1)
Literature Review	Kaufhold, Reuter (ch. 6.2)
Empirical Design	Reuter, Spielhofer, Kaufhold, Hahne (ch. 6.3)
Empirical Enquiry	GapFish GmbH
Empirical Analysis	Spielhofer, Hahne (ch. 6.4.1–6.4.4), Kaufhold, Reuter (ch. 6.4.5)
Discussion and Implications	Reuter, Kaufhold (ch. 6.5, 6.6)
Enabling Project Funding	EmerGent, KontiKat

Table 0.4 Contribution summary of chapter 7: Paper title, conference, involved actors (surname) and their contributions, separated by categories and ordered decreasingly by contribution from left to right

Paper Title	Adoption, Use and Diffusion of Crisis Apps in Germany: A Representative Survey
Conference Proceedings	Mensch und Computer, 2019
Involved Authors	Grinko, Kaufhold, Reuter
Scientific Lead and Management	Kaufhold
Paper Design and Introduction	Kaufhold, Grinko (ch. 7.1)
Literature Review	Grinko, Kaufhold (ch. 7.2)
Empirical Design	Reuter, Kaufhold (ch. 7.3)
Empirical Enquiry	GapFish GmbH
Empirical Analysis	Grinko, Kaufhold (ch. 7.4)
Discussion and Implications	Kaufhold, Grinko, Reuter (ch. 7.5)
Enabling Project Funding	KontiKat, MAKI

Chapter 8 was published as the research paper "Avoiding Chaotic Use of Social Media before, during, and after Emergencies: Design and Evaluation of Citizens' Guidelines" (Kaufhold, Gizikis, et al., 2019) in the "Journal of Contingencies and Crisis Management" [JIF 1.365]. It constitutes joint work with Alexis Gizikis, Christian Reuter, Matthias Habdank, and Margarita Grinko, which was funded within the research projects EmerGent and KontiKat (Table 0.5). As corresponding and leading author, I led the overall research design, management, and introduction (section 8.1) of the paper. While the literature review was a joint effort of all authors (section 8.2), Alexis was in charge of the design of citizen guidelines (section 8.3). The representative survey (section 8.3), including method and questionnaire, were designed by Christian and me. I and Margarita focused on the qualitative and quantitative statistical analysis of collected data (section 8.4). The discussion and conclusion (section 8.5) were composed by contributions and reflections of all involved authors.

Table 0.5 Contribution summary of chapter 8: Paper title, conference, involved actors (surname) and their contributions, separated by categories and ordered decreasingly by contribution from left to right

Paper Title	Avoiding Chaotic Use of Social Media before, during, and after Emergencies: Design and Evaluation of Citizens' Guidelines
Journal	Journal of Contingencies and Crisis Management, 2019, 27(3)
Involved Authors	Kaufhold, Gizikis, Reuter, Habdank, Grinko
Scientific Lead and Management	Kaufhold
Paper Design and Introduction	Kaufhold (ch. 8.1)
Literature Review	Kaufhold, Gizikis, Habdank, Reuter (ch. 8.2)
Empirical Design	Gizikis (ch. 8.3.1)
Empirical Analysis	Gizikis (ext. D2.6, 2.7)
Evaluation Design	Kaufhold, Reuter (ch. 8.3.1)
Evaluation Enquiry	GapFish GmbH
Evaluation Analysis	Kaufhold, Grinko (ch. 8.4)
Discussion and Implications	Kaufhold, Grinko, Reuter (ch. 8.5)
Enabling Project Funding	EmerGent, KontiKat

Chapter 9 was published as the research paper "Mitigating Information Overload in Social Media during Conflicts and Crises: Design and Evaluation of a

Cross-Platform Alerting System" (Kaufhold, Rupp, et al., 2020) in the journal "Behaviour & Information Technology" [CORE-A, JIF 1.429]. It constitutes a joint work with Nicola Rupp, Christian Reuter, and Matthias Habdank, which was funded within the research projects EmerGent, KontiKat, and MAKI (Table 0.6). As corresponding and leading author, I led the overall research design, management, and introduction (section 9.1) of the paper. Christian supported me in refining the literature review (section 9.2). The overall requirement analysis (section 9.3) as well as architecture and development of the system (section 9.4) was led by Matthias and complemented by me. All authors were involved in the refinement of the interview guideline, while Nicola and I conducted the stakeholder interviews (section 9.5). Furthermore, Nicola was involved in the evaluation of interview results (section 9.6) and Christian assisted in the discussion (section 9.7) and conclusion (section 9.8) of the paper.

Table 0.6 Contribution summary of chapter 9: Paper title, conference, involved actors (surname) and their contributions, separated by categories and ordered decreasingly by contribution from left to right

Paper Title	Mitigating Information Overload in Social Media during Conflicts and Crises: Design and Evaluation of a Cross-Platform Alerting System
Journal	Behaviour & Information Technology, 2020, 39(3)
Involved Authors	Kaufhold, Rupp, Reuter, Habdank
Scientific Lead and Management	Kaufhold
Paper Design and Introduction	Kaufhold (ch. 9.1)
Literature Review	Kaufhold, Reuter (ch. 9.2)
Technology Design	Habdank, Kaufhold (ch. 9.3)
Technology Implementation	Habdank, Kaufhold (ch. 9.4)
Evaluation Design	Habdank, Reuter, Rupp, Kaufhold (ch. 9.5)
Evaluation Enquiry	Rupp, Kaufhold
Evaluation Analysis	Kaufhold, Rupp (ch. 9.6)
Discussion and Implications	Kaufhold, Reuter (ch. 9.7, 9.8)
Enabling Project Funding	EmerGent, KontiKat, MAKI

Chapter 10 was published as the research paper "Rapid relevance classification of social media posts in disasters and emergencies: A system and evaluation featuring active, incremental and online learning" (Kaufhold, Bayer, et al., 2020) in the journal "Information Processing & Management" [CORE-A, JIF 3.892].

It constitutes a joint work with Markus Bayer and Christian Reuter, which was funded within the research projects KontiKat, CRISP, and MAKI (Table 0.7). As corresponding and leading author, I led the overall research design, management, and introduction (section 10.1) of the paper. The literature review (section 10.2) was a joint work of me, Markus and Christian, while I focused on the technological foundation (section 10.3) and Markus on the evaluation of designed algorithms (section 10.4, 10.5). The discussion and conclusion (section 10.6) were composed by contributions and reflections of all involved authors.

Table 0.7 Contribution summary of chapter 10: Paper title, conference, involved actors (surname) and their contributions, separated by categories and ordered decreasingly by contribution from left to right

Paper Title	Rapid relevance classification of social media posts in disasters and emergencies: A system and evaluation featuring active, incremental and online learning
Journal	Information Processing & Management, 2020, 57(1)
Involved Authors	Kaufhold, Bayer, Reuter
Scientific Lead and Management	Kaufhold
Paper Design and Introduction	Kaufhold (ch. 10.1)
Literature Review	Kaufhold, Bayer, Reuter (ch. 10.2)
Technology Design	Kaufhold, Bayer (ch. 10.3)
Technology Implementation	Kaufhold, Bayer (ch. 10.3)
Evaluation Design	Bayer, Kaufhold (ch. 10.4, 10.5)
Evaluation Enquiry	Bayer
Evaluation Analysis	Bayer, Kaufhold (ch. 10.4, 10.5)
Discussion and Implications	Kaufhold, Bayer, Reuter (ch. 10.6)
Enabling Project Funding	KontiKat, CRISP, MAKI

Chapter 11 was published as the research paper "112.social: Design and Evaluation of a Mobile Crisis App for Bidirectional Communication between Emergency Services and Citizen" (Kaufhold et al., 2018) at the "European Conference on Information Systems" [CORE-A]. It constitutes a joint work with Nicola Rupp, Christian Reuter, Christoph Amelunxen, and Massimo Cristaldi, which was funded within the research projects EmerGent and KontiKat (Table 0.8). As corresponding and leading author, I led the overall research design, management, and introduction (section 11.1) of the paper. Nicola and I were involved in the refinement of the interview guideline and conducted all stakeholder interviews (section 11.3). Furthermore, Nicola was involved in the evaluation of interview results and led the design and evaluation of conducted surveys together with Christoph (section 11.4). Christian supported me especially in refining the literature review (section 11.2), discussion and conclusion (section 11.5) of the paper, while Massimo was in charge of the design and implementation of the 112.social app (section 11.3).

Table 0.8 Contribution summary of chapter 11: Paper title, conference, involved actors (surname) and their contributions, separated by categories and ordered decreasingly by contribution from left to right.

Paper Title	112.social: Design and Evaluation of a Mobile Crisis App for Bidirectional Communication between Emergency Services and Citizen
Conference Proceedings	European Conference on Information Systems, 2018
Involved Authors	Kaufhold, Rupp, Reuter, Amelunxen, Cristaldi
Scientific Lead and Management	Kaufhold
Paper Design and Introduction	Kaufhold (ch. 11.1)
Literature Review	Kaufhold, Reuter (ch. 11.2)
Technology Design	Cristaldi (ch. 11.3.1)
Technology Implementation	Cristaldi (ch. 11.3.2)
Evaluation Design	Amelunxen, Reuter, Rupp, Kaufhold (ch. 11.4.1)
Evaluation Enquiry	Rupp, Kaufhold, Amelunxen
Evaluation Analysis	Kaufhold, Rupp (ch. 11.4.1)
Discussion and Implications	Kaufhold, Reuter (ch. 11.5)
Enabling Project Funding	EmerGent, KontiKat

As a closing note, in order to ensure a consistent appearance of all research papers across the document, they were reformatted using the dissertation's stylesheet. If it served the reading flow, their inherent figures, footnotes, and tables were shifted to more fitting positions. Furthermore, chapter 5 was translated from German to English and detected misspellings, even if they made it into the original published articles, were corrected in a last round of proofreading. Existing chapter references were adapted to fit their position in this thesis. With the exception of established phrases and terms, such as paper titles and research projects, the document was adapted to British English.

Publications of the Author

In sum, 63 publications emerged in the context of my work. The publications are rated according to the CORE (Computing Research and Education Association of Australasia, 2018), GI-CSCW (Leitungsgremium der GI-Fachgruppe CSCW, 2009), WKWI (Wissenschtliche Kommission Wirtschaftsinformatik, 2008) rankings, and current JIF journal impact factor (Clarivate Analytics, 1994). The following eight publications are included as chapters in this thesis, whereof four are ranked CORE-A:

Reuter, C. & **Kaufhold, M.-A.** (2018). Fifteen Years of Social Media in Emergencies: A Retrospective Review and Future Directions for Crisis Informatics. *Journal of Contingencies and Crisis Management (JCCM)*, 26 (1), 41–57. **[ch. 4, JIF 1.365]**

Kaufhold, M.-A., Grinko, M., Reuter, C., Schorch, M., Langer, A., Skudelny, S. & Hollick, M. (2019). Potentiale von IKT beim Ausfall kritischer Infrastrukturen: Erwartungen, Informationsgewinnung und Mediennutzung der Zivilbevölkerung in Deutschland. *International Conference on Wirtschaftsinformatik (WI)*. Siegen, Germany: AIS. **[ch. 5, CORE-C, WKWI-A]**

Reuter, C., **Kaufhold, M.-A.**, Spielhofer, T. & Hahne, A. S. (2017). Social Media in Emergencies: A Representative Study on Citizens' Perception in Germany. *Proceedings of the ACM on Human Computer Interaction (PACM)*. New York: ACM, (pp. 1–19). **[ch. 6, CORE-A]**

Grinko, M., **Kaufhold, M.-A.**, & Reuter, C. (2019). Adoption, Use and Diffusion of Crisis Apps in Germany: A Representative Survey. In F. Alt, A. Bulling, & T. Döring (Eds.), *Mensch und Computer 2019 (MuC)* (pp. 263–274). Hamburg, Germany: ACM. **[ch. 7, GI-CSCW-C]**

Kaufhold, M.-A., Gizikis, A., Reuter, C., Habdank, M. & Grinko, M. (2018). Avoiding chaotic use of social media before, during, and after emergencies:

Design and evaluation of citizens' guidelines. *Journal of Contingencies and Crisis Management (JCCM)*, 198–213. **[ch. 8, JIF 1.365]**

Kaufhold, M.-A., Rupp, N., Reuter, C. & Habdank, M. (2020). Mitigating Information Overload in Social Media during Conflicts and Crises: Design and Evaluation of a Cross-Platform Alerting System. *Behaviour & Information Technology (BIT)*, 39 (3), 319–342. **[ch. 9, CORE-A, JIF 1.429]**

Kaufhold, M.-A., Bayer, M. & Reuter, C. (2020). Rapid relevance classification of social media posts in disasters and emergencies: A system and evaluation featuring active, incremental and online learning. *Information Processing & Management (IPM)*, 57 (1), 102–132. **[ch. 10, CORE-A, JIF 3.892]**

Kaufhold, M.-A., Rupp, N., Reuter, C. Amelunxen, C. & Cristaldi, M. (2018). 112.social: Design and Evaluation of a Mobile Crisis App for Bidirectional Communication between Emergency Services and Citizens. *European Conference on Information Systems (ECIS)*. Portsmouth, UK: AIS. **[ch. 11, CORE-A]**

The following publications contain related findings, but are not explicitly included as chapters:

Bernardi, A., Reuter, C., Schneider, W., Linsner, S., & **Kaufhold, M.-A.** (2019). Hybride Dienstleistungen in digitalisierten Kooperationen in der Landwirtschaft. In A. Meyer-Aurich (Ed.), *38. GIL-Jahrestagung: Digitalisierung in kleinstrukturierten Regionen, Lecture Notes in Informatics (LNI)* (pp. 25–30). **[CORE-C]**

Haunschild, J., **Kaufhold, M.-A.** & Reuter, C. (2020). Cultural Violence and Peace in Social Media: Interventions by Humans and Social Bots. In Dunn, M. & Wenger, A. (Ed.), *Cyber Security Meets Security Policy: Socio-Technological Dynamics, Political Responses*.

Haunschild, J., **Kaufhold, M.-A.** & Reuter, C. (2020). Sticking with Landlines? Citizens' and Police Social Media Use and Expectation During Emergencies, *Proceedings of the International Conference on Wirtschaftsinformatik (WI)*, Potsdam, Germany: AIS Electronic Library (AISel) **[CORE-C, WKWI-A]**

Kalle, T., **Kaufhold, M.-A.**, Kuntke, F., Reuter, C., Rizk, A., & Steinmetz, R. (2019). Resilience in Security and Crises through Adaptions and Transitions. In C. Draude, M. Lange, & B. Sick (Eds.), *INFORMATIK 2019: 50 Jahre Gesellschaft für Informatik – Informatik für Gesellschaft (Workshop-Beiträge), Lecture Notes in Informatics (LNI)* (pp. 571–584). Gesellschaft für Informatik e. V. **[CORE-C]**

Kaufhold, M.-A., & Reuter, C. (2014). Vernetzte Selbsthilfe in Sozialen Medien am Beispiel des Hochwassers 2013 / Linked Self-Help in Social Media using the example of the Floods 2013 in Germany. *I-Com – Zeitschrift für interaktive und kooperative Medien*, *13*(1), 20–28. **[GI-CSCW-B]**

Kaufhold, M.-A., & Reuter, C. (2015). Konzept und Evaluation einer Facebook-Applikation zur crossmedialen Selbstorganisation freiwilliger Helfer. In O. Thomas & F. Teuteberg (Eds.), *Proceedings of the International Conference on Wirtschaftsinformatik (WI)* (pp. 1844–1858). AIS. **[CORE-C, GI-CSCW-A]**

Kaufhold, M.-A., & Reuter, C. (2016). The Self-Organization of Digital Volunteers across Social Media: The Case of the 2013 European Floods in Germany. *Journal of Homeland Security and Emergency Management (JHSEM)*, *13*(1), 137–166.

Kaufhold, M.-A., & Reuter, C. (2017). Integration von Flow in die Mensch- Computer-Interaktion? Potenziale für die Gestaltung interaktiver Systeme. *Mittelstand-Digital "Wissenschaft trifft Praxis,"* *7*(1), 78–88.

Kaufhold, M.-A., & Reuter, C. (2017). The Impact of Social Media in Emergencies: A Case Study with the Fire Department of Frankfurt. In T. Comes, F. Bénaben, C. Hanachi, & M. Lauras (Eds.), *Proceedings of the Information Systems for Crisis Response and Management (ISCRAM)* (pp. 603–612).

Kaufhold, M.-A., & Reuter, C. (2019). Cultural Violence and Peace in Social Media. In C. Reuter (Ed.), *Information Technology for Peace and Security—IT-Applications and Infrastructures in Conflicts, Crises, War, and Peace* (pp. 361–381). Springer Vieweg.

Kaufhold, M.-A., & Reuter, C. (2019). Social Media Misuse: Cultural Violence, Peace and Security in Digital Networks. In C. Reuter (Ed.), *SCIENCE PEACE SECURITY '19—Proceedings of the Interdisciplinary Conference on Technical Peace and Security Challenges*. Darmstadt, Germany: TUprints.

Kaufhold, M.-A., Haunschild, J. & Reuter, C. (2020). Warning the Public: A Survey on Attitudes, Expectations and Use of Mobile Crisis Apps in Germany, *Proceedings of the European Conference on Information Systems (ECIS) [CORE-A]*

Kaufhold, M.-A., Reuter, C., & Ermert, T. (2018). Interaktionsdesign eines Risiko-Bewertungskonzepts für KMU. In R. Dachselt & G. Weber (Eds.), *Mensch und Computer 2018: Tagungsband* (pp. 309–312). Dresden, Germany: Gesellschaft für Informatik e.V. **[GI-CSCW-B]**

Kaufhold, M.-A., Reuter, C., & Ludwig, T. (2019). Cross-Media Usage of Social Big Data for Emergency Services and Volunteer Communities: Approaches, Development and Challenges of Multi-Platform Social Media Services. *ArXiv:1907.07725 [Cs.SI]*, 1–11.

Kaufhold, M.-A., Reuter, C., & Ludwig, T. (2019). Flow Experience in Software Engineering: Development and Evaluation of Design Options for Eclipse. *Proceedings of the European Conference on Information Systems (ECIS).* **[CORE-A]**

Kaufhold, M.-A., Reuter, C., & Stefan, M. (2017). Gesellschaftliche Herausforderungen des Missbrauchs von Bots und sozialen Medien. In M. Burghardt, R. Wimmer, C. Wolff, & C. Womser-Hacker (Eds.), *Mensch & Computer: Workshopband* (pp. 51–58). **[GI-CSCW-B]**

Kaufhold, M.-A., Reuter, C., Ludwig, T., & Scholl, S. (2017). Social Media Analytics: Eine Marktstudie im Krisenmanagement. In M. Eibl & M. Gaedke (Eds.), *INFORMATIK 2017, Lecture Notes in Informatics (LNI), Gesellschaft für Informatik* (pp. 1325–1338). **[CORE-C]**

Kaufhold, M.-A., Reuter, C., Riebe, T., & Radziewski, E. von. (2018). Design eines BCM-Dashboards für kleine und mittlere Unternehmen. In R. Dachselt & G. Weber (Eds.), *Mensch und Computer 2018: Workshopband* (pp. 579–586). **[GI-CSCW-B]**

Kaufhold, M.-A., Riebe, T., Reuter, C., Hester, J., Jeske, D., Knüver, L., & Richert, V. (2018). Business Continuity Management in Micro Enterprises: Perception, Strategies and Use of ICT. *International Journal of Information Systems for Crisis Response and Management (IJISCRAM), 10*(1), 1–19.

Kaufhold, M.-A., Schmidt, A., Seifert, F., Riebe, T., & Reuter, C. (2019). SentiNet: Twitter-basierter Ansatz zur kombinierten Netzwerk- und Stimmungsanalyse in Katastrophenlagen. *Mensch und Computer 2019 – Workshopband.* Hamburg, Germany: ACM. **[GI-CSCW-B]**

Langer, A., **Kaufhold, M.-A.,** Runft, E., Reuter, C., Grinko, M., & Pipek, V. (2019). Counter Narratives in Social Media: An Empirical Study on Combat and Prevention of Terrorism. In Z. Franco, J. J. González, & J. H. Canós (Eds.), *Proceedings of the Information Systems for Crisis Response and Management (ISCRAM).*

Mentler, T., Reuter, C., Nestler, S., **Kaufhold, M.-A.,** Herczeg, M. & Pottebaum, J. (2020). 7. Workshop Mensch-Maschine-Interaktion in sicherheitskritischen Systemen, *Mensch & Computer 2020*, Magdeburg, Germany: Gesellschaft für Informatik e. V. **[GI-CSCW-B]**

Reuter, C., & **Kaufhold, M.-A.** (2016). Warum Katastrophenschutzbehörden soziale Medien nicht nutzen wollen. In W. Prinz, J. Borchers, & M. Jarke (Eds.), *Mensch & Computer: Tagungsband.* **[GI-CSCW-B]**

Reuter, C., & **Kaufhold, M.-A.** (2018). Informatik für Frieden und Sicherheit. In C. Reuter (Ed.), *Sicherheitskritische Mensch-Computer-Interaktion: Interaktive Technologien und Soziale Medien im Krisen- und Sicherheitsmanagement* (pp. 573–595).

Reuter, C., & **Kaufhold, M.-A.** (2018). Soziale Medien in Notfällen, Krisen und Katastrophen. In C. Reuter (Ed.), *Sicherheitskritische Mensch-Computer-Interaktion: Interaktive Technologien und Soziale Medien im Krisen- und Sicherheitsmanagement* (pp. 379–402).

Reuter, C., & **Kaufhold, M.-A.** (2018). Usable Safety Engineering sicherheitskritischer interaktiver Systeme. In C. Reuter (Ed.), *Sicherheitskritische Mensch-Computer-Interaktion: Interaktive Technologien und Soziale Medien im Krisen- und Sicherheitsmanagement* (pp. 17–40). Springer Vieweg.

Reuter, C., Aal, K., Aldehoff, L., Altmann, J., Buchmann, J., Bernhardt, U., Denker, K., Herrmann, D., Hollick, M., Katzenbeisser, S., **Kaufhold, M.-A.**, Nordmann, A., Reinhold, T., Riebe, T., Ripper, A., Ruhmann, I., Saalbach, K.-P., Schörnig, N., Sunyaev, A., & Wulf, V. (2019). The Future of IT in Peace and Security. In C. Reuter (Ed.), *Information Technology for Peace and Security—IT-Applications and Infrastructures in Conflicts, Crises, War, and Peace* (pp. 405–413). Springer Vieweg.

Reuter, C., Aal, K., Beham, F., Boden, A., Brauner, F., Ludwig, T., Lukosch, S., Fiedrich, F., Fuchs-Kittowski, F., Geisler, S., Gennen, K., Herrmann, D., **Kaufhold, M.-A.**, Klafft, M., Lipprandt, M., Lo Iacono, L., Pipek, V., Pottebaum, J., Mentler, T., Nestler, S., Stieglitz, S., Sturm, C., Rusch, G., Sackmann, S., Volkamer, M., & Wulf, V. (2018). Die Zukunft sicherheitskritischer Mensch-Computer-Interaktion. In C. Reuter (Ed.), *Sicherheitskritische Mensch-Computer-Interaktion: Interaktive Technologien und Soziale Medien im Krisen- und Sicherheitsmanagement* (pp. 621–630). Springer Vieweg.

Reuter, C., Aldehoff, L., Riebe, T., & **Kaufhold, M.-A.** (2019). IT in Peace, Conflict, and Security Research. In C. Reuter (Ed.), *Information Technology for Peace and Security—IT-Applications and Infrastructures in Conflicts, Crises, War, and Peace* (pp. 11–37).

Reuter, C., Backfried, G., **Kaufhold, M.-A.**, & Spahr, F. (2018). ISCRAM turns 15: A Trend Analysis of Social Media Papers 2004–2017. In K. Boersma & B. Tomaszewski (Eds.), *Proceedings of the Information Systems for Crisis Response and Management (ISCRAM).*

Reuter, C., Hughes, A. L., & **Kaufhold, M.-A.** (2018). Social Media in Crisis Management: An Evaluation and Analysis of Crisis Informatics Research. *International Journal on Human-Computer Interaction (IJHCI), 34*(4), 280–294. **[CORE-B, GI-CSCW-B, WKWI-B, Five Year JIF 1.905]**

Reuter, C., **Kaufhold, M.-A.** & Schmid, S. (2020). Risikokulturen bei der Nutzung Sozialer Medien in Katastrophenlagen, *BBK Bevölkerungsschutz*

Reuter, C., **Kaufhold, M.-A.**, & Klös, J. (2017). Benutzbare Sicherheit: Usability, Safety und Security bei Passwörtern. In M. Burghardt, R. Wimmer, C. Wolff,

& C. Womser-Hacker (Eds.), *Mensch & Computer: Workshopband* (pp. 33–41). **[GI-CSCW-B]**

Reuter, C., **Kaufhold, M.-A.**, & Ludwig, T. (2017). End-User Development and Social Big Data—Towards Tailorable Situation Assessment with Social Media. In F. Paternò & V. Wulf (Eds.), *New Perspectives in End-User Development* (pp. 307–332).

Reuter, C., **Kaufhold, M.-A.**, & Steinfort, R. (2017). Rumors, Fake News and Social Bots in Conflicts and Emergencies: Towards a Model for Believability in Social Media. In T. Comes, F. Bénaben, C. Hanachi, & M. Lauras (Eds.), *Proceedings of the Information Systems for Crisis Response and Management (ISCRAM)* (pp. 583–591).

Reuter, C., **Kaufhold, M.-A.**, Comes, T., Knodt, M. & Mühlhäuser, M. (2020) Designing Mobile Interactive Systems for Societal and Technical Resilience, *Proceedings of the International Conference on Human-Computer Interaction with Mobile Devices and Services (MobileHCI)*[**CORE-B, GI-CSCW-A**]

Reuter, C., **Kaufhold, M.-A.**, Leopold, I., & Knipp, H. (2017). Informing the Population: Mobile Warning Apps. In M. Klafft (Ed.), *Risk and Crisis Communication in Disaster Prevention and Management* (pp. 31–41).

Reuter, C., **Kaufhold, M.-A.**, Leopold, I., & Knipp, H. (2017). Katwarn, NINA or FEMA? Multi-Method Study on Distribution, Use and Public Views on Crisis Apps. *European Conference on Information Systems (ECIS)*, 2187–2201. **[CORE-A]**

Reuter, C., **Kaufhold, M.-A.**, Schmid, S., Hahne, A. S., & Spielhofer, T. (2019). The Impact of Risk Cultures: Citizens' Perception of Social Media Use in Emergencies across Europe. *Technological Forecasting and Social Change.* **[JIF 3.815]**

Reuter, C., **Kaufhold, M.-A.**, Schmid, S., Hahne, A. S., & Spielhofer, T. (2019). The Impact of Risk Cultures: Citizens' Perception of Social Media Use in Emergencies across Europe. *Technological Forecasting and Social Change (TFSC),148* (119724), 168–180. **[JIF 3.815]**

Reuter, C., **Kaufhold, M.-A.**, Schorch, M., Gerwinski, J., Soost, C., Hassan, S. S., ... Wulf, V. (2017). Digitalisierung und Zivile Sicherheit: Zivilgesellschaftliche und betriebliche Kontinuität in Katastrophenlagen (KontiKat). In G. Hoch, H. Schröteler von Brandt, V. Stein, & A. Schwarz (Eds.), *Sicherheit (DIAGONAL Jahrgang 38)* (pp. 207–224).

Reuter, C., **Kaufhold, M.-A.**, Spahr, F., Spielhofer, T. & Hahne, A. S. (2020). Emergency Service Staff and Social Media—A Comparative Empirical Study of the Perception by Emergency Services Members in Europe in 2014 and 2017, *International Journal of Disaster Risk Reduction (IJDRR), 46* (101516)[**JIF 2.568**]

Reuter, C., **Kaufhold, M.-A.**, Spielhofer, T., & Hahne, A. S. (2018). Soziale Medien und Apps in Notsituationen: Eine repräsentative Studie über die Wahrnehmung in Deutschland. *BBK Bevölkerungsschutz, 2,* 22–24.

Reuter, C., Ludwig, T., **Kaufhold, M.-A.**, & Hupertz, J. (2017). Social Media Resilience during Infrastructure Breakdowns using Mobile Ad-Hoc Networks. In V. Wohlgemuth, F. Fuchs-Kittowski, & J. Wittmann (Eds.), *Advances and New Trends in Environmental Informatics—Proceedings of the 30th EnviroInfo Conference* (pp. 75–88).

Reuter, C., Ludwig, T., **Kaufhold, M.-A.**, & Pipek, V. (2015). XHELP: Design of a Cross-Platform Social-Media Application to Support Volunteer Moderators in Disasters. *Proceedings of the Conference on Human Factors in Computing Systems (CHI)*, 4093–4102. [CORE-A*, GI-CSCW-A]

Reuter, C., Ludwig, T., **Kaufhold, M.-A.**, & Spielhofer, T. (2016). Emergency Services Attitudes towards Social Media: A Quantitative and Qualitative Survey across Europe. *International Journal on Human-Computer Studies (IJHCS), 95,* 96–111. [CORE-A, WKWI-A, GI-CSCW-B, JIF 2.006]

Reuter, C., Ludwig, T., **Kaufhold, M.-A.**, & Spielhofer, T. (2018). Studie: Wie sehen Mitarbeiter von Feuerwehr und THW den Einsatz sozialer Medien in Gefahrenlagen? *Crisis Prevention – Fachmagazin für Innere Sicherheit, Bevölkerungsschutz und Katastrophenhilfe, 1,* 64–66.

Reuter, C., Ludwig, T., Kotthaus, C., **Kaufhold, M.-A.**, Radziewski, E. von, & Pipek, V. (2016). Big Data in a Crisis? Creating Social Media Datasets for Emergency Management Research. *I-Com: Journal of Interactive Media, 15*(3), 249–264.

Reuter, C., Mentler, T., Nestler, S., Herczeg, M., Ludwig, T., Pottebaum, J., & **Kaufhold, M.-A.** (2019). 6. Workshop Mensch-Maschine-Interaktion in sicherheitskritischen Systemen – Neue digitale Realitäten. *Mensch Und Computer 2019 – Workshopband.* Hamburg, Germany: ACM. [GI-CSCW-B]

Reuter, C., Riebe, T., Aldehoff, L., **Kaufhold, M.-A.**, & Reinhold, T. (2019). Cyberwar – die Digitalisierung der Kriegsführung. In I.-J. Werkner & N. Schörnig (Eds.), Cyberwar – die Digitalisierung der Kriegsführung? (pp. 15–38). Springer VS.

Riebe, T., **Kaufhold, M.-A.**, Kumar, T., & Reuter, C. (2019). Threat Intelligence Application for Cyber Attribution. In C. Reuter (Ed.), *SCIENCE PEACE SECURITY '19—Proceedings of the Interdisciplinary Conference on Technical Peace and Security Challenges.* Darmstadt, Germany: TUprints.

Riebe, T., Langer, A., **Kaufhold, M.-A.**, Kretschmer, N. K., & Reuter, C. (2019). Werte und Wertekonflikte in sozialen Medien für die Vernetzung ungebundener Helfer in Krisensituationen – Ein Value-Sensitive Design Ansatz. *Mensch*

Und Computer 2019 – Workshopband, 308–318. Hamburg, Germany: Gesellschaft für Informatik e.V. **[GI-CSCW-B]**

Riebe, T., Pätsch, K., **Kaufhold, M.-A.**, & Reuter, C. (2018). From Conspiracies to Insults: A Case Study of Radicalisation in Social Media Discourse. In R. Dachselt & G. Weber (Eds.), *Mensch und Computer 2018: Workshopband* (pp. 595–603). **[GI-CSCW-B]**

Scholl, S., Reuter, C., Ludwig, T., & **Kaufhold, M.-A.** (2018). SocialML: EUD im Maschine Learning zur Analyse sozialer Medien. *Mensch Und Computer 2018: Tagungsband*, 443–446. **[GI-CSCW-B]**

Spielhofer, T., Hahne, A. S., Reuter, C., **Kaufhold, M.-A.**, & Schmid, S. (2019). Social Media Use in Emergencies of Citizens in the United Kingdom. In Z. Franco, J. J. González, & J. H. Canós (Eds.), *Proceedings of the Information Systems for Crisis Response and Management (ISCRAM)*.

Stute, M., Maass, M., Schons, T., **Kaufhold, M.-A.**, Reuter, C. & Hollick, M. (2020). Empirical Insights for Designing Information and Communication Technology for International Disaster Response, *International Journal of Disaster Risk Reduction (IJDRR)*, *47*(101598)**[JIF 2.568]**

Contents

Part I   Outline

1   Introduction 3
  1.1   Motivation 3
  1.2   Aims and Objectives 5
  1.3   Structure of the Work 6

2   Related Work 11
  2.1   Foundations and Terms of Crisis Management 11
  2.2   Research Domain and Technologies of Crisis Informatics 15
  2.3   Adoption of Social Media Analytics in Crisis Informatics 18
  2.4   Towards Information Refinement in Crisis Informatics 21
  2.5   Research Gaps and Potentials 23

3   Research Design 27
  3.1   Research Field and Foundations 27
  3.2   Research Approach 30
  3.3   Research Context 33
  3.4   Methods 35

Part II   Theoretical and Empirical Findings

**4   Retrospective Review and Future Directions for Crisis
    Informatics** 47
  4.1   Introduction and Brief History 47
  4.2   Published Cases of Social Media in Emergencies 49
  4.3   Usage Patterns—Types of Interaction in Social Media 54
  4.4   Role Patterns—Types of Users in Social Media 61

4.5   Perception Patterns—Views on Social Media   66
4.6   The Past and the Future: Discussion and Conclusion   70

5   Survey on Media Perception and Use During Infrastructure
Failure   75
5.1   Introduction   76
5.2   State of the Art   77
5.3   Methodology   79
5.4   Results   81
5.5   Discussion and Conclusion   87
5.6   Appendix: Survey Questions   90

6   Survey on Citizens' Perception and Use of Social Media   93
6.1   Introduction   93
6.2   Perception of the Use of Social Media in Emergencies   95
6.3   Methodology   98
6.4   Empirical Results   102
6.5   Discussion and Conclusion   111
6.6   Limitations   117
6.7   Appendix: Survey Questions   117

7   Survey on the Adoption, Use and Diffusion of Crisis Apps   119
7.1   Introduction   119
7.2   Background and Related Work   121
7.3   Representative Study: Methodology   128
7.4   Results   130
7.5   Discussion and Conclusion   138

Part III   Design and Evaluation Findings

8   Design and Evaluation of Social Media Guidelines for Citizens ...   147
8.1   Introduction   147
8.2   Related Work: Challenges of Social Media
in Emergencies and Existing Guidelines   149
8.3   Methodology   154
8.4   Results   158
8.5   Discussion and Conclusion   168
8.6   Appendix: Visualization of the Guidelines   172
8.7   Appendix: Correlations between Responses for all
Guidelines   173

9 Design and Evaluation of a Cross Social Media Alerting System 175
   9.1 Introduction 175
   9.2 Conceptual Framing and Related Work 177
   9.3 Requirements Analysis: Methodology, Pre-Studies and Workshops 189
   9.4 Development and Architecture of a Cross-Platform Social Media Based Alerting System 191
   9.5 Methodology of the Systems' Evaluation 197
   9.6 Results of the Systems' Evaluation 202
   9.7 Discussion 208
   9.8 Conclusion 212

10 Design and Evaluation of an Active Relevance Classification System 215
   10.1 Introduction 216
   10.2 Literature Review 218
   10.3 Technological Basis: Social Data Management and Analysis 230
   10.4 Evaluation I: Relevance Classification via Batch Learning 240
   10.5 Evaluation II: Relevance Classification via Active and Online Learning 257
   10.6 Discussion and Conclusion 265

11 Design and Evaluation of a Mobile Crisis App 273
   11.1 Introduction 274
   11.2 Related Work and Comparison of Crisis Apps 275
   11.3 Development and Architecture of the Emergency Mobile App 279
   11.4 Evaluation of the Emergency Mobile App 284
   11.5 Discussion and Conclusion 293

Part IV Conclusion and Outlook

12 Information Refinement Technologies for Crisis Response 299
   12.1 Overview of the Information Refinement Framework 299
   12.2 Event Perspective 301
   12.3 Organisational Perspective 305
   12.4 Societal Perspective 308
   12.5 Technological Perspective 312

13   Conclusion and Future Work 333
       13.1   Main Findings 333
       13.2   Empirical and Theoretical Contributions 339
       13.3   Design and Practical Contributions 341
       13.4   Limitations and Future Work 343

References 347

Abbreviations

112.social	Emergency Mobile App (ICT artefact)
A2A	Authorities to Authorities: Inter-organizational Crisis Management
A2C	Authorities to Citizens: Crisis Communication
API	Application Programming Interface
Apps	Mobile smartphone applications
BIWAPP	Citizen Information and Warning App (German: Bürger Info- & Warn-App; mobile crisis app)
BOS	Public Authorities and Organizations with Security Responsibilities (German: Behörden und Organisationen mit Sicherheitsaufgaben)
C2A	Citizens to Authorities: Integration of Citizen Generated Content
C2C	Citizens to Citizens: Self-Help Communities
CAP	Common Alerting Protocol
CS	Computer Science
CSCW	Computer-Supported Cooperative Work
EMA	Emergency Management Agency
EMC	Emergency Management Cycle
ES	Emergency Service
ESI	Emergency Service Interface (ICT artefact)
FD	Fire Department
HCI	Human-Computer Interaction
ICT	Information and Communications Technology
IS	Information Systems
JSON	JavaScript Object Notation
Katwarn	Disaster Warning (German: Katastrophen-Warnung; mobile crisis app)
ML	Machine Learning

NINA	Emergency Information and News App (German: Notfall-Informations- und Nachrichten-App; mobile crisis app)
REST	Representational State Transfer
SMA	Social Media API (ICT artefact)
SMG	Social Media Guidelines (ICT artefact)
SMO	Social Media Observatory (ICT artefact)
THW	Federal Agency for Technical Relief (German: Technisches Hilfswerk)
UX	User Experience
V&TC	Virtual & Technical Communities
VOST	Virtual Operations Support Team
XML	Extensible Markup Language

List of Figures

Figure 3.1 Schematic display of the structure of a design case study, illustration reproduced from Wulf et al. (2015) 31

Figure 3.2 Three cycle view of design science research, informed by Hevner (2007) 32

Figure 3.3 Extended and modified research design based on design case studies 33

Figure 3.4 Overview of artefacts: graphical version of the social media guidelines (SMG, top-left), the emergency service interface (ESI, top-right), the mobile crisis app (112.social, bottom-left), and the social media observatory (SMO, bottom-right) 41

Figure 4.1 Crisis Communication Matrix (Reuter et al., 2012), adapted concerning the terminology 55

Figure 4.2 Role Typology Matrix 65

Figure 6.1 Please indicate how often, on average, you do the following things (Q1) 100

Figure 6.2 Have you ever used social media (such as Facebook, Twitter, Instagram) to find out or share information in an emergency such as an accident, power cut, severe weather, flood, or earthquake close to you? (Q2) 103

Figure 6.3 What types of information did you share? (Select as many as apply) (Q3) 103

Figure 6.4　　　Imagine that you posted an urgent request for help
or information on a social media site of a local
emergency service, such as your local police,
coastguard, fire, or medical emergency service.
To what extent do you agree with the following
statements (Q4) 105

Figure 6.5　　　What might put you off using social media
during an emergency? (Q5) 105

Figure 6.6　　　Have you ever downloaded a smartphone app
that could help in a disaster or emergency? (Q6) 106

Figure 6.7　　　What type of app did you download? (Q7) 107

Figure 6.8　　　Please indicate how likely you are in future to use
a smartphone app for each of the following purposes
as a result of an emergency? (Q8) 107

Figure 6.9　　　Infographic on Citizens' Perception of Social Media
in Emergencies in Germany (2017) 112

Figure 8.1　　　How do you rate the following recommendations
for usage of social media before an emergency? (Q22) .. 159

Figure 8.2　　　How do you rate the following recommendations
for usage of social media during an emergency?
(Q23) 160

Figure 8.3　　　How do you rate the following recommendations
for usage of social media after an emergency? (Q24) 161

Figure 8.4　　　How do you rate the following, general aspects
for using social media in an emergency? (Q25) 162

Figure 8.5　　　Visualisation of the Guidelines 172

Figure 9.1　　　The backend with C2A (blue) and A2C (red)
information flows 192

Figure 9.2　　　The Emergency Service Interface (ESI): dashboard
view 197

Figure 9.3　　　ESI: details of an alert 198

Figure 9.4　　　ESI in a simulated C&C room during the workshop 201

Figure 10.1　　Supervised machine learning steps 228

Figure 10.2　　Social Media Architecture comprising Social Media
Observatory and Social Media Service 232

Figure 10.3　　Dashboard overview for administrators 237

Figure 10.4　　Overview of crawljobs 237

Figure 10.5　　Creation of a crawljob 238

Figure 10.6　　Relevance labelling for a crawljob 239

Figure 10.7   Create a relevance classifier based on labelled crawljob posts 239

Figure 10.8   Filtered List of crawljob posts 240

Figure 10.9   Incremental active learning in the labelling process of SMO 259

Figure 10.10   Real-time evaluation during the labelling process in SMO 261

Figure 10.11   The classification process comprising active learning, real-time evaluation and feedback classification 262

Figure 10.12   Comparison of the incremental learning methods iNB, HT, and IBk using AUC values from the real-time evaluation in SMO 264

Figure 10.13   Comparison of AUC values of RF classifiers on the flooding dataset with (solid) or without (dashed) active learning 264

Figure 10.14   Comparison of AUC values of RF classifiers on the BASF SE dataset with (solid) or without (dashed) active learning 265

Figure 11.1   112.social-1: (I) start screen, (II) multimedia dialogue, (III) alert sent confirmation 282

Figure 11.2   112.social-2: (IV) start screen, (V) communication threads and (VI) details 283

Figure 12.1   Social media guidelines for emergency services. From Gizikis, O'Brien, et al., (2017) 305

Figure 12.2   The 112.social mobile crisis app for reporting (left), overview (centre) and bidirectional communication (right) of incidents 316

Figure 12.3   The relevance classification process of the SMO comprising active learning, real-time evaluation, and feedback classification 323

Figure 12.4   Dashboard of the SMO without (background) and with (foreground) enabled interactive chart filtering, showing only information of negative sentiment and in a specific timeframe 325

Figure 12.5   Dashboard of the ESI featuring mobile app and grouped social media alert in a map and list view; upon the click on a social media alert, the individual messages are displayed (right) 326

Figure 12.6   Relevance labelling of the SMO, which predicts
               the accuracy, precision, and recall of a model
               by the increasing number of labelled posts using
               incremental learning   329
Figure 12.7   The browser plugin TrustyTweet detects
               and visualises indicators of fake news in Twitter
               posts. Illustration from Hartwig and Reuter (2019)   330

List of Tables

Table 2.1   Basic terms and definitions for crisis management,
             increasing by scale and severity 13
Table 2.2   Crisis communication matrix, from Reuter, Marx,
             and Pipek (2012) 14
Table 2.3   Social media classification adapted from Kaplan
             and Haenlein (2010) 15
Table 2.4   Social media analytics frameworks of Fan and Gordan
             (2014) and Stieglitz et al. (2018c) 19
Table 2.5   The information challenges of a crisis, reproduced
             from Hagar (2010, p. 10) 23
Table 3.1   Epistemological framework, from Becker
             and Niehaves (2007); the epistemological foundations
             of this thesis are marked in bold 29
Table 3.2   Applied methods, target audiences (A for authorities,
             C for citizens), content focuses, years of application
             and count (of literature or requirement documents,
             empiricism or evaluation participants, or designed
             or redesigned systems) with different research
             outputs (D for project deliverables in EmerGent, P
             for scientific publications in conferences or journals)
             across three development cycles and three research
             projects, whereof the research contributions of this
             dissertation are marked in italics 36
Table 3.3   Representative survey on social media and crisis apps 38
Table 3.4   Representative survey on infrastructure, crisis apps,
             and social media guidelines 39

Table 3.5   Non-technical (SMG) and technical artefacts
               (112.social, ESI, SMO) within this thesis   40
Table 3.6   112.Social and ESI evaluation: personal details
               of participants   42
Table 3.7   112.Social evaluation: OSCE survey participants
               by occupation, experience (multiple selection)
               and device   43
Table 3.8   112.Social evaluation: treasure hunt survey
               participants with occupation and device information   44
Table 3.9   SMO dataset evaluation: comparison of characteristics   44
Table 4.1   Overview of cases in literature   51
Table 4.2   Public perspective on social media roles   62
Table 4.3   Organizational perspective on social media roles   63
Table 5.1   Results for Q1: Which of the following options have
               you used in an acute crisis situation (when you
               were concerned or a helper) to obtain information
               about the events, and which was especially useful
               for you?   81
Table 5.2   Significant results of chi-squared tests on media usage
               in crisis situations (*p < .05, **p < .01, N = 1,024)   82
Table 5.3   Results for Q2: What do you expect on part
               of the operator (e.g. Telekom, energy suppliers)
               in case of infrastructure failures (e.g. electricity, gas
               or telecommunications)?   84
Table 5.4   Significant chi-squared results on communication
               expectations in crises (*p < .05, **p < .01, N = 1,024),
               see table 5.2 for explanations of abbreviations   85
Table 5.5   Results for Q3: In the case of a 112 emergency
               call failure, various measures are taken to enable
               the population to report an emergency. To what extent
               do you know the following measures or locations?
               (THW = "Technisches Hilfswerk", Federal Agency
               for Technical Relief; BOS = "Behörden und
               Organisationen mit Sicherheitsaufgaben", authorities
               and organisations with safety tasks)   86
Table 7.1   Comparison of free crisis apps in Germany   125
Table 7.2   Main functions of the crisis app prototype   130

Table 7.3   Have you ever used one or more of the following crisis apps, are you currently using them or planning to use them in the future? (Q4) 132

Table 7.4   What do you think of crisis apps that you can install on your smartphone? (Q6) 134

Table 7.5   Chi-squared results for crisis app expectations. Values which were not significant are indicated in grey. * p<.05, **p<.01 135

Table 7.6   Overview of the code categories in open responses (Q5). Only the five most frequent codes are displayed with the according number of responses in parentheses ... 138

Table 8.1   List of guidelines for the use of social media in general .. 152

Table 8.2   List of guidelines for the use of social media in emergencies (ES = emergency service guidelines, C = guidelines for citizens, EMC = emergency management cycle) 153

Table 8.3   Social media guidelines 155

Table 8.4   Correlations between responses for the individual questions 163

Table 8.5   Results of chi-squared tables for all questions, with correlations where appropriate. Non-significant results are displayed on grey background 165

Table 8.6   Correlation coefficients (Spearman's Rho) between responses for all guidelines (p < .001) 173

Table 9.1   Overview of Intelligence, Management and Special Systems, adapted from Kaufhold et al. (2017), *de facto non-operating systems, **architecture but no implemented system 186

Table 9.2   Empirical pre-studies and workshops 190

Table 9.3   Abstraction of the system requirements 191

Table 9.4   Comparison of two exemplary pipelines and their characteristics based on simplified fire and flood scenarios. The affected component, if any, is indicated in parentheses within the first column 193

Table 9.5   The information quality framework with criteria and indicators 194

Table 9.6   The contextual filters influencing the generation of alerts. Some factors are configurable in the backend (BE) and some in the frontend (FE) 196

Table 9.7   Second evaluation: personal details of participants   199
Table 9.8   Indicated benefits of social media alerts (Q3)
              and information quality (Q4)   204
Table 9.9   Indicated importance of functionality (very important
              to not important at all)   206
Table 9.10   Outline of requested features in terms of display,
              alerts, filters, and map   209
Table 10.1   Abstract and interpretative relevance criteria   223
Table 10.2   Precise and factual relevance criteria   225
Table 10.3   Overview of works examining textual classification
              problems in disasters or emergencies. If a comparison
              was conducted, the best performing method is
              marked bold; abbreviations: Artificial/Convolutional
              Neuronal Networks (ANN/CNN), Decision Trees
              (DT), Jaccard Similarity (JS), K-Nearest Neighbours
              (KNN), Logistic Regression (LR), Maximum Entropy
              (ME), Naïve Bayes (NB), Neuronal Networks
              (NN), Relevance-Based Language Models (RBLM),
              Random Forest (RF), Support Vector Machines (SVM) . . .   227
Table 10.4   Required and optional query parameters   233
Table 10.5   Activity Streams with EnrichedData object   235
Table 10.6   Features, description and internal representation
              in the classifier (*means optional, [] means that it is
              no parameter feature)   246
Table 10.7   Classification quality of TF and TF.IDF vectorizations
              based on a textual Random Forest   250
Table 10.8   Classification quality and time of the word stem
              creations stemming and lemmatization on the basis
              of a textual Random Forest   251
Table 10.9   Classification quality and time
              with and without the NER classification
              feature based on a textual Random Forest   251
Table 10.10   Classifications with different feature set combinations.
              A Random Forest with the respective classification
              features is used   252
Table 10.11   Modification of the threshold value (threshold-moving)
              of the Random Forest with the classification features
              words, geographical distance, temporal distance
              and length   253

Table 10.12 Classification quality and time of the Naïve
Bayes and Random Forest classifier based
on the classification features words, geographical
distance, temporal distance and length 254

Table 10.13 Classification quality and time
with and without a feature set elimination
procedure based on the classification features words,
geographical distance, temporal distance and length 255

Table 10.14 Classification quality and time of two different
additional features sets and a feature reduction
on the BASF incident dataset 256

Table 10.15 Time for the retraining of the classifier with Random
Forest Naïve Bayes, Hoeffding Tree, and k-Nearest
Neighbour (IBk) 258

Table 10.16 Comparison of time for the recreation of the RF
classifier with and without Feature Subset Selection
(FSS) using the 2013 European floods dataset 261

Table 11.1   Empirical pre-studies and workshops 280

Table 11.2   Abstraction of requirements comprising 112.social,
ESI and underlying architecture 281

Table 11.3   OSCE survey participants by occupation, experience
(multiple selection) and device 284

Table 11.4   Demonstration, field trials and workshop participants
(I1-I22) 286

Table 11.5   Treasure hunt survey participants with occupation
and device information 286

Table 11.6   Refined categorisation of emergencies as implemented
in version two of 112.social 288

Table 11.7   Indicated importance of information flows and benefit
of functionality 289

Table 11.8   Indicated quality of functionalities 292

Table 11.9   Indicated importance of functionalities 292

Table 11.10 Summary of features, requirements (C2)
and contextual factors (C3) 294

Table 12.1   Boundary conditions influencing the information
refinement process 300

Table 12.2   Event-based categories, features, and conditions
for information refinement 301

Table 12.3   Hazard categories and sub-categories (left).
               Reproduced from Olteanu et al. (2015),
               and categorisation used in 112.social (chapter 11)   302
Table 12.4   Compilation of some social media guidelines, reduced
               version of Table 8.3   304
Table 12.5   Organisational categories, features, and conditions
               for information refinement   306
Table 12.6   Societal categories, features, and conditions
               for information refinement   309
Table 12.7   Technological steps, categories, features,
               and conditions for information refinement   314
Table 12.8   Challenges of using various social media platform
               APIs   317
Table 12.9   Dimensions of information. Own illustration based
               on Imran et al. (2015), Olteanu et al. (2015)
               and Starbird et al. (2010)   319
Table 12.10   Exemplary application of machine learning
               techniques in crisis informatics. Own overview based
               on the referenced literature   322
Table 12.11   Social media information quality framework
               with criteria (left) and indicators (right). Reproduced
               from Moi et al. (2017)   327

Part I
Outline

Introduction

1

1.1 Motivation

The investigation of crisis communication, management and technology has become more common in domains such as Computer Supported Cooperative Work (CSCW), Human Computer Interaction (HCI), Computer Science (CS) and Information Systems (IS) as the work of citizens, emergency services, and other organizations is increasingly mediated by computer technology, notably by the use of mobile technologies and social media (Reuter, Kaufhold, Spielhofer, et al., 2017; Tan et al., 2017). In academics, social media is often defined as a "group of internet-based applications that build on the ideological and technological foundations of Web 2.0, and that allow the creation and exchange of user-generated content" (Kaplan & Haenlein, 2010, p. 61). Whereas Web 2.0 was conceptualized as an architecture for participation with new possibilities for social interaction (O'Reilly, 2007), user-generated content refers to "the various forms of media content that are publicly available and created by end-users" (Kaplan & Haenlein, 2010, p. 61).

Social media is not only part of everyday live but also used in crises, disasters, or emergencies (Reuter, Hughes, et al., 2018): Already after the 2001 September 11 attacks, emergency services used web-based technologies to inform the public and to provide status reports internally and externally (Harrald et al., 2002), but also citizens created wikis to collect information about missing people (Palen & Liu, 2007). In large-scale events, such as the 2012 Hurricane Sandy or the 2013 European Floods, citizens were active participants in both the real world, organized in emergent groups (Stallings & Quarantelli, 1985), and virtual realm, often called digital volunteers (Starbird & Palen, 2011), utilizing mobile and social

© The Author(s), under exclusive license to Springer Fachmedien Wiesbaden GmbH, part of Springer Nature 2021
M.-A. Kaufhold, *Information Refinement Technologies for Crisis Informatics*,
https://doi.org/10.1007/978-3-658-33341-6_1

information and communication technologies (ICT) for crisis response (Starbird & Palen, 2011). From the professional perspective, emergency services such as fire and police departments started to analyse social media streams to enhance crisis communication, increase situational awareness and inform decision making (Eismann et al., 2018; Reuter, Ludwig, Kaufhold, et al., 2016). Furthermore, public authorities started to deploy mobile crisis and warning apps specifically designed for "the distribution of disaster-related information and communication between authorities, organizations and citizens across different phases of the emergency management cycle" (Kaufhold, Haunschild, et al., 2020, p. 2).

The resulting research area of *crisis informatics* "views emergency response as an expanded social system where information is disseminated within and between official and public channels and entities" (Palen et al., 2009, p. 469). It now is established as a "multidisciplinary field combining computing and social science knowledge of disasters; its central tenet is that people use personal information and communication technology to respond to disaster in creative ways to cope with uncertainty" (Palen & Anderson, 2016, p. 224). Today, the research area comprises a wealth of studies on mobile technologies and social media in conflicts and crises (Reuter, Hughes, et al., 2018), supported by international journals with special issues (Hiltz, Diaz, et al., 2011; Pipek et al., 2014; Reuter, Mentler, et al., 2015; Reuter, Stieglitz, et al., 2020) and tracks at various conferences, such as CSCW or ISCRAM (Reuter, Backfried, et al., 2018).

One important research stream of crisis informatics is the *integration of citizen-generated content* into professional emergency response (Reuter et al., 2012), which utilizes methods and techniques of machine learning (M. Imran et al., 2015, 2018) and social media analytics (Stieglitz et al., 2014; Stieglitz, Mirbabaie, Fromm, et al., 2018) to extract relevant information from social media, including situational updates, multimedia files and public mood, amongst others (Meurisch et al., 2019; Reuter, Ludwig, Kaufhold, et al., 2016). However, the analysis of high-volume, heterogeneous, and highly semantic data, or big social data, poses challenges to emergency services: relevant expertise and personnel resources are limited, and information is often not available in perfect shape to assist crisis response. This includes restricted access to social data (Olshannikova et al., 2017), the multitude of available information channels (Kaufhold, Reuter, et al., 2019), diverse characteristics of information (Olteanu et al., 2015), resources and strategies for information for information processing, the handling of information overload (Hiltz et al., 2020), and the quality of information, including the issues of fake news or online rumours (Krafft et al., 2017). Due to these challenges, without a proper curation of incoming social big data, emergency services might fail to identify safety-critical and relevant information for effective crisis response with the ultimate goal of rescuing lives.

Furthermore, the analysis of social data is not only important for establishing situational awareness but also to inform *crisis communication*, including the dissemination of behavioural or preparatory crisis information (Reuter, Kaufhold, et al., 2020) or the correction of fake information (Reuter, Hartwig, et al., 2019). In order to communicate the "right information for the right person at the right time" (Zade et al., 2018, p. 1) on intra- and interorganizational level as well as for external stakeholders, obtained information must be refined according to human and organizational requirements as well as technological feasibility. Therefore, for the domain of crisis informatics, we understand information refinement, which is further elaborated in section 2.4, as a *process of applied practices and used technologies to refine obtained information according to contextual, i.e. event-based, human, organizational, societal, and technological, boundary conditions in order to improve crisis management, response and, as a desired consequence, social wellbeing.*

1.2   Aims and Objectives

This dissertation seeks to examine *information refinement practices and technologies* for crisis informatics. Based on theoretical considerations associated with the introduced issues and related work, the thesis presents empirical studies on the perception and use of media in crises, especially social media and crisis apps, the design and implementation of innovative ICT artefacts, and evaluations of their appropriation in the application field of crisis management (Reuter, 2014b). The research was accomplished primarily in the projects *EmerGent*, *KontiKat* and *MAKI*, which all included the examination of social media and crisis apps for crisis management and response. The purpose of this dissertation is to contribute to practices and technologies for information refinement by addressing two research questions in the field of crisis informatics, which are motivated by research gaps and potentials elaborated in section 2.5. In short, the first research question intends to tackle two important research gaps, including the lack of systematisation of the usage, roles and perceptions of social media and a lack of quantitative representative research on attitudes, expectations and use of social media and mobile crisis apps:

- *What are citizens' and emergency services' expectations, perceptions and use patterns with regard to media, especially social media and mobile apps in crises (RQ1)?*

The second research question is based upon the need for (semi-)automatic approaches to tackle the issues of information overload and information quality in crises and the lack of mobile technologies for bidirectional communication between emergency services and citizens:

- *What are implications for the design of social media technologies and mobile crisis apps to support information refinement for crisis response (RQ2)?*

In the synthesis of both research questions, the thesis aims to address the need for an information refinement framework which is not only capable of capturing nuances of information processing in crisis informatics but also the event-based, organisational, and societal boundary conditions of crisis management. The inclusion of human and organizational aspects calls for the integration of human-centred or participatory approaches into technology design. In order to research this topic, this dissertation combines *design case studies* (Stevens et al., 2018; Wulf et al., 2011), which are inspired by action research (Lewin, 1958), with principles of *design science research* (Hevner et al., 2004) as a methodological approach. This approach consist of four phases: the review of theoretical foundations (Hevner, 2007), an empirical analysis of the given social practices, the innovative design of ICT artefacts related to found empirical findings, and an evaluation of the appropriation of the artefacts involving participation of end-users (Stevens & Pipek, 2018).

1.3   Structure of the Work

This dissertation is organized into the four main sections of (I) outline, (II) theoretical and empirical findings, (III) design and evaluation findings, as well as (IV) conclusion and outlook.

Part I: Outline
The first part presents the conceptual foundations of this thesis by giving a short introduction into the topic, discussing related work, and presenting the research design of the dissertation.

Chapter 1 (Introduction) concisely motivates the research area of social media, mobile technologies and crisis informatics, introduces the notion of information refinement, outlines aims and objectives of the dissertation, and summarizes the structure of the document.

Chapter 2 (Related Work) defines relevant terms and discusses related work on crisis management and response, mobile technologies and social media, and the application of machine learning and social media analytics in the domain of crisis informatics. It further elaborates the concept of information refinement and outlines open research gaps and potentials.

Chapter 3 (Research Design) outlines the research design, including the research field and epistemological foundations, the overall approach used within this work, the research context, and gives an overview of the applied methods in terms of theoretical review, empirical study, design of ICT artefacts, and evaluation.

Part II: Theoretical and Empirical Findings
The second part presents theoretical and empirical findings related to the outlined research questions. The chapters of this part have been published and are based on the accepted versions of peer-reviewed journal articles or conference papers.

Chapter 4 (Retrospective Review and Future Directions for Crisis Informatics) analyses social media use in past emergencies and prevalent usage, role and perception patterns of both authorities and citizens, outlining future research directions. It has been published in the *Journal of Contingencies and Crisis Management (JCCM)* (Reuter & Kaufhold, 2018).

Chapter 5 (Survey on Media Perception and Use during Infrastructure Failure) analyses the results of a representative survey on German media use in crises, expectations during critical infrastructure failure, knowledge on emergency measures, and anticipated behaviour in a crisis. It has been published in the *Proceedings of the International Conference on Wirtschaftsinformatik (WI)* (Kaufhold, Grinko, et al., 2019).

Chapter 6 (Survey on Citizens' Perception and Use of Social Media) presents the results of a representative survey, which was conducted in Germany, on social media use, published types of information, expectations towards emergency services, as well as perceived benefits and drawbacks of social media during emergencies. It has been published in the *Proceedings of the ACM on Human Computer Interaction (PACM)* (Reuter, Kaufhold, Spielhofer, et al., 2017).

Chapter 7 (Survey on the Adoption, Use and Diffusion of Crisis Apps) examines results on the distribution of crisis app usage, attitudes towards crisis apps, and reasons for (non-)usage of crisis apps based on a representative survey in

Germany. It has been published in *Mensch und Computer 2019* (Grinko et al., 2019).

Part III: Design and Evaluation Findings
The third part presents design and evaluation findings related to the outlined research questions. The chapters of this part have been published and are based on the accepted versions of peer-reviewed journal articles or conference papers.

Chapter 8 (Design and Evaluation of Social Media Guidelines for Citizens) presents the design and survey-based evaluation of social media guidelines for citizens to reduce the chaotic use of social media and improve information quality before, during and after emergencies. It has been published in the *Journal of Contingencies and Crisis Management (JCCM)* (Kaufhold, Gizikis, et al., 2019).

Chapter 9 (Design and Evaluation of a Cross Social Media Alerting System) presents the design and qualitative interview-based evaluation of the Emergency Service Interface (ESI), a social media alerting system for emergencies managers, comprising filtering techniques, relevance and information quality classification, information grouping, and a usable interface to overcome information overload. It has been published in *Behaviour & Information Technology (BIT)* (Kaufhold, Rupp, et al., 2020).

Chapter 10 (Design and Evaluation of an Active Relevance Classification System) presents a system for social media monitoring (Social Media Observatory, short: SMO), which includes approaches and preliminary evaluations of rapid relevance classification algorithms to reduce the amount of required labelled data in emergencies and to correct misclassifications of the algorithm. It has been published in *Information Processing and Management (IPM)* (Kaufhold, Bayer, et al., 2020).

Chapter 11 (Design and Evaluation of a Mobile Crisis App) presents the design and qualitative evaluation of 112.social, a mobile crisis apps for bidirectional communication between authorities and citizens, based on interviews and surveys. It has been published in the *Proceedings of the European Conference on Information Systems (ECIS)* (Kaufhold et al., 2018).

Part IV: Outlook
The fourth part combines the different perspectives and already published papers to outline theoretical and practical implications of my dissertation using the frame of practices and technologies for information refinement.

Chapter 12 (Information Refinement Technologies for Crisis Response) integrates the results of the previous chapters to propose an information refinement framework which elaborates the channels, access, content, analysis, filtering, and evaluation of information, also considering event-based, organisational, and societal boundary conditions.

Chapter 13 (Conclusion and Future Work) summarizes the main findings by answering the two research questions, outlines empirical and theoretical contributions as well as design and practical contributions. It concludes with the discussion of limitations and future work.

Related Work

2

This chapter introduces the application field and research domain of this thesis. First, it will define and discuss different foundations and terms, such as incidents, emergencies, crises, disasters, and catastrophes, with regard to crisis management (Section 2.1). Thereafter, it will introduce crisis informatics, focusing on social media technologies and mobile crisis apps (Section 2.2). A further emphasis is set on social media analytics, which be discussed from the perspectives of algorithms and data, interfaces and systems, and human interaction (Section 2.3). Finally, the chapter will establish a basic definition and understanding for information refinement in crisis informatics (Section 2.4) and elaborate open research gaps and potentials that are addressed in this thesis (Section 2.5).

2.1   Foundations and Terms of Crisis Management

The term *crisis* originates from the Greek krisis (decision, judgment) and is defined as a "crucial point of [a] situation in the course of anything; a turning point; an unstable condition in which an abrupt or decisive change is imminent" (Institute for Crisis Disaster and Risk Management, 2009). Objectively, it is not a positive or negative term but "often connoted rather negatively in everyday use and expected to lead to an unstable situation affecting an individual, group, community, or whole society" (Ludwig, 2017, p. 22). For instance, the Federal Office for Information Security (German: Bundesamt für Sicherheit in der Informationstechnik, abbreviated as BSI) distinguishes between simple incidents, emergencies, crises, catastrophes by increasingly negative effects, whereof crisis is defined as "a limited and exacerbated emergency that threatens the existence of [an] institution or impairs the health or life of individuals" (Bundesamt für Sicherheit in der

M.-A. Kaufhold, *Information Refinement Technologies for Crisis Informatics*, https://doi.org/10.1007/978-3-658-33341-6_2

Informationstechnik, 2008). In accordance with the latter definition, this dissertation focuses on the potential negative impact of crises, including emergencies, disasters, and catastrophes.

According to the Institute for Crisis Disaster and Risk Management (2009), an *emergency* can be defined as a "hazard impact causing adverse physical, social, psychological, economic or political effects that challenges the ability to rapidly and effectively respond". Thus, in contrast to simple incidents, the effective response to emergencies requires a change from routine management methods to an incident command process (Ludwig, 2017), involving public authorities with safety and security responsibilities (German: Behörden und Organisationen mit Sicherheitsaufgaben, abbreviated as BOS). If the outcome of a crisis is worse than expected, an emergency develops into a disaster that far exceeds response capabilities (Quarantelli, 1985). The internationally agreed glossary of basic terms related to disaster management defines *disaster* as a "serious disruption of the functioning of a society, causing widespread human, material, or environmental losses which exceed the ability of [the] affected society to cope using only its own resources" (United Nations Department of Humanitarian Affairs, 2000). Furthermore, a large-scale disaster can lead to a *catastrophe*, which is defined as "any natural or manmade incident, including terrorism, that results in extraordinary levels of mass casualties, damage, or disruption severely affecting the population, infrastructure, environment, economy, national morale, and/or government functions" (Federal Emergency Management Agency, 2006). According to Quarantelli (1987), catastrophes are events *"of such impact upon a community that new organizations must be created in order to deal with the situation"*. The selected definitions of incident, emergency, crisis, disaster, and catastrophe are collected in Table 2.1.

Despite significant differences in scale and severity of incidents, emergencies, disasters, and catastrophes, which require different management processes, Hiltz et al. (2011b) outline that "disaster, crisis, catastrophe, and emergency management are sometimes used synonymously and sometimes with slight differences, by scholars and practitioners". This dissertation follows a general understanding of crisis management without a strong differentiation of managing emergencies, disasters, or catastrophes. Then, *crisis management* can be understood as "the coordination of efforts to control a crisis event consistent with strategic goals of an organization. Although generally associated with response, recovery and resumption operations during and following a crisis event, crisis management responsibilities extend to pre-event awareness, prevention and preparedness and post event restoration and transition" (Shaw & Harrald, 2004, p. 7). According to the Federal Ministry of the Interior, Building and Community (German: Bundesministerium des Innern, für Bau und Heimat, abbreviated BMI), a crisis comprises

Table 2.1 Basic terms and definitions for crisis management, increasing by scale and severity

Terms	Definitions
(Simple) Incident	Short-term outage of services or resources with minor damage (Bundesamt für Sicherheit in der Informationstechnik, 2008).
Emergency	Hazard impact causing adverse physical, social, psychological, economic or political effects that challenges the ability to rapidly and effectively respond (Institute for Crisis Disaster and Risk Management, 2009).
Crisis	Crucial point of situation in the course of anything; a turning point; an unstable condition in which an abrupt or decisive change is imminent (Institute for Crisis Disaster and Risk Management, 2009).
Disaster	Serious disruption of the functioning of a society, causing widespread human, material, or environmental losses which exceed the ability of an affected society to cope using only its own resources (United Nations Department of Humanitarian Affairs, 2000).
Catastrophe	Any natural or manmade incident, including terrorism, that results in extraordinary levels of mass casualties, damage, or disruption severely affecting the population, infrastructure, environment, economy, national morale, and/or government functions (Federal Emergency Management Agency, 2006).

the three basic dimensions of (1) the *real event with a concrete hazard or damage to protective goods*, such as persons, objects, material assets, or nature, (2) the *actions of the involved emergency services*, and (3) the *perception of the crisis by affected citizens and the general public* (BMI, 2014, p. 5), which affect the crisis management process.

First, considering the characteristics of the real event, existing research commonly differentiates between natural (e.g. earthquakes, floods, hurricanes, or wildfires and human-induced (e.g. building collapse, derailment, bombing, or shooting) *hazards*, varying in geographical spread and temporal development (Olteanu et al., 2015). Putting a stronger emphasis on the differentiation of human-induced events, the BMI distinguishes trinomial between natural hazards, (accidental) technical/human failure as well as (intentional) terrorism, crime and war (BMI, 2014, p. 4). Second, the actions of the involved emergency services demand an efficient and action-related communication between partners in different institutions, locations, and cultures across different phases of a disaster (Gründer-Fahrer et al., 2018). Besides formal emergency services, such as fire and police departments, collaborative technologies and social software strengthened the role of both affected citizens and volunteers during crises (Kaufhold & Reuter,

2016; Starbird & Palen, 2011). To acknowledge that development, Reuter, Marx, and Pipek (2012) developed a crisis communication matrix which distinguishes organizations, such as emergency services, and the public, including affected citizens and volunteers, which can both act as sender and receiver of information (Table 2.2). Besides crisis communication (A2C) and inter-organizational crisis management (A2A), citizens organize self-help communities (C2C) in response to crises and authorities are interested in integrating citizen-generated content (C2A) to enhance situation awareness and improve decision-making (Eismann et al., 2018; Kaufhold & Reuter, 2016; Reuter, 2014b).

Table 2.2   Crisis communication matrix, from Reuter, Marx, and Pipek (2012)

Receiver	**Public (C)**	Crisis Communication (A2C)	Self-Help Communities (C2C)
	Organizations (A)	Inter-organizational Crisis Management (A2A)	Integration of Citizen Generated Content (C2A)
		Organizations (A)	**Public (C)**
		Sender	

This leads to the third aspect of an event: the perception of the crisis by affected citizens and the general public, which is characterized by trust building, opinion formation, and emotion management processes among citizens and emergency services (BMI, 2014, p. 5). The importance of these processes is amplified by new media: Due to the emergence of social media, the widespread distribution of smartphones and mobile apps as well as their increasing use during emergencies, citizens are able to search for and publish their own information and opinions in near real-time (Reuter & Spielhofer, 2017). However, the chaotic use of social media may increase the task complexity for professional crisis management (Perng et al., 2012) and the publication of outdated information or even fake news requires appropriate crisis communication strategies (Kaufhold & Reuter, 2016; Reuter, Hartwig, et al., 2019). Furthermore, the sheer amount of information can lead to information overload during large-scale emergencies, hampering the emergency services' capability of extracting relevant information from social media (Plotnick & Hiltz, 2018).

2.2   Research Domain and Technologies of Crisis Informatics

The increasing relevance of ICT for crisis management, as indicated in Section 2.1, led to the establishment of the research domain of *crisis informatics*, which includes "the empirical study as well as socially and behaviourally conscious ICT development and deployment" (Palen et al., 2007, p. 9). In a later work, Palen and Anderson (2016, p. 224) define crisis informatics as "a multidisciplinary field combining social science knowledge of disasters; its central tenet is that people use personal information and communication technology to respond to disaster in create ways to cope with uncertainty". Despite the variety of different ICT, crisis informatics research put a strong emphasis on social media use before, during and after emergencies (Soden & Palen, 2018). However, social media was initially established for everyday life uses, such as self-promotion, relationship building, news posting or information searching (Robinson et al., 2017). Social media are interactive Web 2.0 applications, whereby Web 2.0 was defined as an architecture for participation with new possibilities for social interaction (O'Reilly, 2007). Over the years, this interaction was more and more summarized under the term social media (Kaplan and Haenlein, 2010, p. 61): "Social Media is a group of Internet-based applications that build on the ideological and technological foundations of Web 2.0, and that allow the creation and exchange of User Generated Content".

Table 2.3   Social media classification adapted from Kaplan and Haenlein (2010)

Social media		Social Presence/ Media richness		
		low	medium	high
Self-presentation/Self-disclosure	high	Blogs	Social network sites (e.g., Facebook)	Virtual social worlds (e.g., Second Life)
	low	Collaborative projects (e.g., Wikipedia)	Content communities (e.g., YouTube)	Virtual game worlds (e.g., World of Warcraft)

Due to the increasing numbers of social media, research started classifying functionality and types of social media. Kaplan and Haenlein (2010) ranged social media according to their social presence/media richness and degree of self-presentation/self-disclosure (Table 2.3). Furthermore, Kietzmann et al. (2011)

differentiate seven functional blocks of social media, i.e., identity, conversations, sharing, presence, relationships, reputation and groups, as well as their implications. However, not every functionality is present or pronounced in every social media. The widespread use of social media is facilitated by the distribution of smartphones worldwide, of which 3.5 billion devices are in service as of 2020 and estimated to reach 3.8 billion in 2021 (Statista, 2020b). In 2020, Facebook (2.498 billion), YouTube (2 billion), WhatsApp (2 billion), Facebook Messenger (1.3 billion), WeChat (1.165 billion), Instagram (1 billion), Reddit (430 million), and Twitter (386 million) are among the social media platforms with the most active users worldwide (Statista, 2020a).

Nowadays, social media is used in almost every major crisis worldwide (Reuter, Hughes, et al., 2018) and is approached from both the citizens' (C2C, C2A) and authorities' (A2A, A2C) perspective. The citizens' perspective was consolidated by the seminal work of Starbird and Palen (2011) on the phenomenon of *digital volunteers* or voluntweeters. Based on this, further research examined the collaboration and relationship between real and virtual volunteers, also identifying different social media roles in Twitter, such as helpers, reporters, retweeters, repeaters, and readers (Reuter et al., 2013). In order to further exploit the potential of *crowdsourcing* tasks to digital volunteers (Ludwig, Reuter, Siebigteroth, et al., 2015), attempts were undertaken to integrate digital volunteers into formal emergency response (St. Denis et al., 2012). These so-called trusted volunteers are nowadays often organized in Virtual Operations Support Teams (VOST) with the aim to leverage volunteers worldwide "in creating, managing and monitoring social media to maintain better engagement and communication with the public as well as rumour control during a disaster" (OECD, 2019). From a technical point of view, plenty of ICT was developed to facilitate citizens response in crises (Kaufhold, Reuter, et al., 2019; Pohl, 2013). For instance, *CrowdMonitor* is a mobile crowd sensing application for assessing physical and digital activities of citizens during emergencies (Ludwig, Reuter, Siebigteroth, et al., 2015) and *XHELP* allows digital volunteers to acquire, manage and distribute information across media, such as Facebook or Twitter (Reuter, Ludwig, Kaufhold, et al., 2015). However, since technologies for volunteer communities are not the focus of this work, this dissertation does not provide an in-depth examination of such ICT.

Besides the integration of trusted volunteers, research on the authorities' perspective comprise studies on the use and perception of social media by emergency managers to facilitate situational awareness. Apart from a facilitating organizational culture and necessary personal skills, emergency managers indicate that specific ICT is required as an enabler of social media use (Reuter, Ludwig,

Kaufhold, et al., 2016). Studies of Plotnick and Hiltz (2016; 2018) outline that information overload and trustworthiness of information are among the most critical challenges of social media that can be overcome by supportive ICT. Despite the availability of a multitude of social media analytics tools, most of them are not tailored to the requirements of emergency managers (Kaufhold et al., 2017). Furthermore, authorities or emergency services comprise different organizational roles and tasks (Bergstrand et al., 2013), such as incident managers or public relation officers, that require diverging information from social media and thus the tailorability of social media analytics tools in order to gather *actionable information* (Kaufhold, Rupp, et al., 2020; Zade et al., 2018). Within the frame of information refinement, this dissertation will examine the design and evaluation of ICT to mitigate information overload and improve information quality gathered from social media, amongst others.

While social media is characterized by its unstructured, less controllable nature and specific information challenges, plenty of *crisis apps*, which are "mobile apps providing specific functionality needed during crises, emergencies, or disasters" (Reuter, Kaufhold, Leopold, et al., 2017b, p. 2187), were established in practice and examined in research (Tan et al., 2017). In contrast to social media, they are tailored according to the needs of authorities or citizens and allow a more structured exchange of information (Kaufhold et al., 2018). In Germany, there is a focus on researching mobile warning apps due to the prevalence of Katastrophen-Warnung (Katwarn) and Notfall-Informations- und Nachrichten-App (NINA), which both provide area- and location-based warnings combined with a warning map view as their main features, supplemented by general disaster information and information sharing capabilities (Reuter, Kaufhold, Leopold, et al., 2017a). However, crisis apps are not limited to warning capabilities. For instance, Reuter and Ludwig (2013) compared 25 apps which are tailored for different *types of crises*, such as earthquakes, epidemics, floods, severe weather, tsunamis, or wildfires, and support different *functionalities*, such as the interactive display of crises on maps, sharing of information, collection of eyewitness reports, or live broadcasts by authorities or infrastructure providers.

In a study comprising both a representative sample in Germany and a snowball sample in Europe, it was found out that 16% of the German and European population already downloaded a crisis app and that 57% (60%) of the German (European) population are likely to use crisis apps in the future (Reuter, Kaufhold, Leopold, et al., 2017b). However, Kotthaus, Ludwig and Pipek (2016, p. 10) compared user comments from app stores on Katwarn and NINA, concluding that warning messages "lack in quality and timing", that "malfunctions [...] lead to a high amount of user complaints" and that "both apps [do not] aim at addressing users [individually]". With regard to contextual factors, Fischer et al. (2019, p. 639) conclude that "risk perception, trust, and subjective norms positively influence both use of a warning app and compliance intention, whereas concerns about data security have negative effects". Besides existing challenges and issues in functionality, perception, and usability, Groneberg et al. (2017) identified that there is a lack of scientific studies to elicit application requirements, documented experiences, scientific surveys, or evaluations of the actual use of mobile crisis apps.

2.3   Adoption of Social Media Analytics in Crisis Informatics

While the establishment of mobile crisis apps is considerably younger phenomenon, in the past two decades, social media was used in almost every major crisis and thus, researchers were able to collect large datasets, of up to multiple millions of social media posts per crisis, and conduct case studies across different types of crises (Olteanu et al., 2015; Reuter, Hughes, et al., 2018). Accordingly, crisis informatics research increasingly frames social media use from the perspective of big data, which is often denoted by high-*volume*, high-*velocity*, and high-*variety* data and is too large or complex to be dealt with by traditional data-processing approaches (McAfee & Brynjolfsson, 2012). Due to the unique characteristics of social media data, the notion of *big social data* was established and defined as "any high-volume, high-velocity, high-variety and/or highly semantic data that is generated from technology-mediated social interactions and actions in digital realm, and which can be collected and analysed to model social interactions and behaviour" (Olshannikova et al., 2017, p. 11). While there is a similar understanding of high-volume (large-scale) and high-velocity (high speed of data generation) regarding big data in general, an enhanced understanding of high-variety and highly semantic data in social media is required. In big social data, high-variety is characterized as "heterogeneous data with a high degree of complexity due to the underlying social relations" and *highly semantic data* as "manually created and highly symbolic content with various, often subjective meanings" (Olshannikova et al., 2017, p. 11). Furthermore, with regard to crisis management, Castillo (2016) introduced the notion of *big crisis data*, discussing further characteristics such as vagueness, virality, veracity, validity, visualisation, values and the contribution of digital volunteers.

In order to address these challenges and allow a meaningful analysis of big social data the research field of social media analytics emerged, which is *"an emerging interdisciplinary research field that aims on combining, extending, and adapting methods for analysis of social media data"* (Stieglitz et al., 2014). Therefore, multiple frameworks for the process of social media analytics were developed. For instance, Fan and Gordon (2014) differentiate the steps of capturing, understanding and presenting social media data (Table 2.4). Despite many similarities, Stieglitz, Mirbabaie, Ross, et al. (2018) add the phase of discovery and explicate the step of preparation, which is coded into the capture and understanding phases of Fan and Gordon's (2014) framework. To further elaborate big social data in the crisis informatics research field, the following paragraphs approach the topic from the perspectives of (1) *algorithms and data*, (2) *interfaces and systems*, and (3) *human interaction*.

Table 2.4 Social media analytics frameworks of Fan and Gordan (2014) and Stieglitz et al. (2018c)

Fan and Gordon (2014)	Stieglitz et al. (2018c)
(no equivalent)	**Discovery** • Research (crisis, brand or political communication) • Management (innovation, stakeholder, or reputation) • General Monitoring
Capture • Gather data from various sources • Preprocess data • Extract pertinent information from data	**Tracking** • Tracking approach (keyword-related, actor-related, URL-related) • Tracking method (APIs, RSS/HTML parsing)
(no equivalent, realized partially in both the capture and understand phases)	**Preparation** • Data preprocessing • Removal of low-quality data
Understand • Remove noisy data (optional) • Perform advanced analytics: opinion mining, sentiment analysis, topic modelling, social network analysis, trend analysis	**Analysis** • Structural attribute (statistical analysis, social network analysis) • Opinion-/Sentiment-related (sentiment analysis) • Topic-/Trend-related (content analysis, trend analysis)
Present • Summarize and evaluate findings from understand stage • Present findings	**[Evaluation]** • Method mixture • Report/Summary

First, from the algorithms and data perspective, the retrieval and processing of high volume datasets is largely dependent on the availability of stable application programming interfaces (APIs) that are maintained and regularly changed by social media providers (Reuter & Scholl, 2014). These REST or RESTful services provide different functionalities using different authorization methods (e.g., API key or OAuth authorization), data exchange formats (e.g., JSON, XML), metadata and structures (e.g., Activity Streams), which are also subject to various access and query limitations. Thus, successfully maintaining and managing data access is a critical challenge for (multi-platform) social media technologies (Reuter, Ludwig, Kaufhold, et al., 2015). Furthermore, crisis informatics is increasingly driven by the application of *artificial intelligence* solutions, often using supervised or unsupervised *machine learning* techniques (M. Imran et al., 2015). Although most research focuses on the creation of models for the automated analysis and prediction of textual content, such as clustering, event detection, relevance or sentiment classification, and topic modelling, a considerable amount of studies now contribute to research on imagery content as well, including duplicate detection and damage assessment (Alam et al., 2019). Based on past research on artificial intelligence in crisis informatics, Imran et al. (2018) outlined manifold research challenges and future directions for algorithmic optimization, such as domain adaptation and transfer learning to better utilize data from past events, online and active learning to reduce the training time of models, and applications of deep learning to increase the accuracy of models.

Second, a multitude of "systems, tools and algorithms performing social media analysis have been developed and implemented to automatize monitoring, classification or aggregation tasks" (Pohl, 2013). In her work, Pohl (2013) analysed 16 social media analytics tools that were developed and published in crisis informatics literature. These provided different configurations of single- or multi-platform support, specific functionality (crowdsourcing, organizational management, sentiment analysis, event detection, visual content), as well as filter and visualization options. A comparison by Trilateral Research (2015) examined 31 social media analytics tools with regard to their impact on crisis preparedness, which furthermore examined the range of functions, costs, and single- or multi-language support. Furthermore, a market study of Kaufhold et al. (2017) compared 35 social media analytics systems by *management* (cross-media, communication, monitoring, notifications, and collaboration functionality), *analysis* (influencer, sentiment, topic, and quality analysis) and *visualization* (diagrams, filters, maps) features. Despite the fact that some systems provided use cases for the public sector, most of them were commercially motivated, did not assist with regard to information overload and information quality problems, and were not tailored for use by emergency services.

Finally, configurability and tailorability are subject of the human perspective. In their seminal paper on social media processing techniques, Imran et al. (2015, p. 30) outline a lack of studies on the usefulness and usability of existing systems, emphasizing the need for human-centred or participatory design: "How should information be presented to users? How should users interact with it? The key to answering these question lies with the users themselves, who should be brought into the process of designing the systems, dashboards, and/or visualizations that they require to serve their needs". Based on a systematic literature review and ten expert interviews, Stieglitz et al. (2018a, p. 11) derived propositions from technological challenges supporting the need for user-centred design, such as that "technical solutions should support crisis managers in the sensemaking and information validation process for a high acceptance of social media analytics; [...] a customisation of filtering algorithms for the needs of emergency services might be needed for them to respond to their specific crisis situations; [...] and social media analytics tools should have good usability, in particular during stressful crisis situations". Furthermore, Zade et al. (2018) propose that a shift from situational awareness towards *actionable information* is required, meaning that information must be customized to the demands of specific individuals, organizations, and roles in different contexts, for instance, in terms of location and time.

2.4 Towards Information Refinement in Crisis Informatics

Existing research emphasised the chaotic and unorganised use of social media in emergencies (Kaewkitipong et al., 2012; Kaufhold & Reuter, 2016; Valecha et al., 2013), leading to rumours, uncertainty, and mistakes, and potentially increasing the task complexity of emergency services (Perng et al., 2012). Taking a different perspective on social media analytics applied in crisis informatics, one might understand the aim of these frameworks, methods and technologies is to reduce *entropy* or disorder of information induced in a crisis. Based on the Second Law of Thermodynamics, entropy can be considered as the amount of disorder in a system (Stephen et al., 2009). If entropy induces a critical instability to a system such as society, it is likely that new practices and structures *emerge* that are suited to disperse entropy. Whether affected citizens and digital volunteers self-organise using social media (Kaufhold & Reuter, 2016), featuring an improvised use of infrastructure or technology (Reuter, 2014b), or emergency services monitor social media for improving situational awareness, decision making and informed action (Zade et al., 2018): They aim on *ordering, organising,* or *structuring* the information in a way that is useful according to their objectives related to crisis response.

Furthermore, the domain of knowledge management establishes relationships between signs, data, information, knowledge, action, expertise and competitiveness (North, 2016). In this relation, signs with a syntax become *data*, and data with a meaning becomes *information*. Then, *knowledge* is the process of connecting information purposefully, i.e. by context, experiences, and expectations. The application of knowledge informs *action* while performing actions correctly and purposefully is considered as *expertise*. The increasing use of mobile and social ICT in crises leads to complex information environments, which are subject to diverse information challenges exacerbating the process of knowledge acquisition and, as a consequence, successful action (Hagar, 2010). As outlined in Table 2.5, this includes challenges such as the lack or overload of information, changing information needs at various stages of a crisis, issues related to the involved actors and channels of communication, challenges related to information quality or establishing the *fitness for use*, i.e. "getting the right information to the right person at the right time" (Hagar, 2010, p. 10). As already motivated in section 2.2, emergency managers indicated that supportive ICT, tailored to the domain of crisis management, is required as an enabler of social media use to solve issues related to the quantity and quality of available information (Kaufhold et al., 2017; Plotnick & Hiltz, 2016, 2018; Reuter, Ludwig, Kaufhold, et al., 2016). Thus, this thesis examines practices and technologies of information refinement to improve the quality, quantity, or *fitness for use* of information for emergency services (Table 2.5).

The term *refinement* can be defined as "a small change to something that improves it", "a thing that is an improvement on an earlier, similar thing; the quality of being improved in this way", or "the process of improving something or of making something pure", amongst others (Cambridge Dictionary, 2020; Oxford Learner's Dictionaries, 2020). The term of *information refinement* is used in different areas of computer science. For instance, information refinement techniques are used to improve the quality of distributed video coding (N. Imran et al., 2015; Taieb et al., 2013), to increase the precision of geolocation information in mobile systems (Bekele et al., 2010; Hermann et al., 2007) or to enhance situational awareness in emergencies (Botega et al., 2015, 2017; Pereira et al., 2015). In existing studies, information refinement techniques were related to the *context*, used for improving *quality* and *relevance* of information (Bekele et al., 2010), considered

Table 2.5   The information challenges of a crisis, reproduced from Hagar (2010, p. 10)

Information challenges of a crisis
Information overload or, conversely, lack of information
Changing information needs at various stages of a crisis: preparedness, warning, impact, response, recovery and reconstruction
The many diverse actors and agencies involved who increase the amount of information produced
Integration and coordination of information by these actors and agencies
The connection of informal and formal channels of information creation and dissemination
Information uncertainty
Trustworthy sources of information
Conflicting information
Getting the right information to the right person at the right time

as an (iterative) *process* (Tomé & Pereira, 2011) and comprised aspects of *human* processing (Botega et al., 2017; Pereira et al., 2015).

When coming back the information challenges outlined by Hagar (2010), it becomes apparent that these are subject to one or multiple contextual factors, including *event-based* (e.g. implications based on the phase and type of the incident), *human* (e.g. perceived information overload and trustworthiness), *organisational* (e.g. goals, objectives, roles, tasks as well as the integration and coordination of information internally and across involved actors and agencies), *societal* (e.g. diverse actors and agencies involved increasing the amount of information produced), and *technological* (e.g. connection of channels of information creation and dissemination) boundary conditions. In an effort to integrate those different characteristics, this dissertation understands information refinement as a *process of applied practices and used technologies to refine obtained information according to contextual, i.e. event-based, human, organizational, societal, and technological, boundary conditions in order to improve crisis management, response and, as a desired consequence, social wellbeing.*

2.5   Research Gaps and Potentials

Crisis informatics has been a steadily growing research area in the past two decades (Palen & Anderson, 2016). Despite the wealth of research studies in the domains of HCI, CSCW, CS, and IS, there are still major research gaps that need to be closed to leverage emergency services utilization of mobile technologies and social media (Reuter, Hughes, et al., 2018). This dissertation seeks to contribute to five distinct research gaps highlighted in the following.

First, there is a lack of systematisation of the usage, roles, and perceptions of social media in emergencies. Therefore, new research on social media use will be contextualized into the crisis communication matrix of Reuter et al. (2012). Furthermore, plenty of existing studies identified diverse roles of both amateur and professional social media users in emergencies (Bergstrand et al., 2013; Reuter et al., 2013), however lacking an integrated role typology integrating both types of actors in the real and virtual realm. Although the perception of the crisis was identified as basic dimension of a crisis, research is required to understand the attitudes and perceptions of emergency services and citizens towards social media to identify barriers and potentials for a successful implementation of such technologies in crises (Reuter, Ludwig, Kaufhold, et al., 2016). Accordingly, this dissertation contributes with a role typology matrix and perception patterns of social media use (chapter 4).

Second, there is a plenitude of qualitative case studies and only a small number of quantitative studies on the attitudes, expectations and use of infrastructures such as mobile technologies and social media (Reuter, Ludwig, Friberg, et al., 2015; Reuter & Spielhofer, 2017). Furthermore, existing quantitative study largely rely on opportunity-based samples, which does not guarantee that they cover the population in terms of demography, for instance gender, age, region, education, and income. To some extent, this limits the generalizability of statements in these studies and provides an opportunity to conduct representative surveys. To address this gap, this dissertation contributes with three representative surveys for the German population on infrastructure, social media and crisis apps use in emergencies (chapters 5, 6, and 7).

Third, there is a lack of (semi-)automatic approaches to tackle the issues of information overload and information quality in crises. While there are plenty of tools for social media analytics, most of them are not tailored to the domain and requirements of crisis management (Kaufhold et al., 2017). When social big data is disseminated in large-scale emergencies, the high volume of noisy data has to be transformed into a low volume of high-quality content that is useful to emergency services (Moi et al., 2015). Recent research emphasized the potentials of supervised machine learning to automatically filter out irrelevant information (Habdank et al., 2017). However, the creation of a machine learning classifier is a time-consuming task, which is a crucial issue under the time-critical constraints of emergencies (M. Imran et al., 2018). This dissertation contributes to research by proposing social media guidelines for citizens to reduce the creation of noise, i.e.

irrelevant information, in emergencies (chapter 8), a cross-media alerting system for emergency services to mitigate the issue of information overload (chapter 9) and an efficient relevance classification algorithm for social media posts using active, incremental and batch learning (chapter 10).

Fourth, research indicates that crisis apps and social media are mostly utilised for information dissemination by emergency services but less for situational awareness, including the integration of citizen-generated content, and bidirectional communication (Reuter, Ludwig, Kaufhold, et al., 2016). While social media potentially allow the application of monitoring and analysis as well as bidirectional communication strategies (Wukich, 2015), this is not the case for mobile crisis apps, such as the established Katwarn and NINA apps in Germany. They focus on crisis warning functionality, extended by recommendations for action, but do not provide crisis reporting or bidirectional communication feature (Groneberg et al., 2017; Reuter, Kaufhold, Leopold, et al., 2017b). In order to contribute to this gap, the dissertation presents the design and evaluation of a mobile crisis app for bidirectional communication between emergency services and citizens (chapter 11).

Finally, there are existing frameworks for the research field of social media analytics (Fan & Gordon, 2014; Stieglitz et al., 2014). These are general-purpose frameworks adaptable to diverse domains such as crisis, brand and political communication or innovation, stakeholder and reputation management (Stieglitz, Mirbabaie, Ross, et al., 2018). In this way, however, these frameworks are not designed to capture the event-based, organisational, and societal boundary conditions of crisis management as well as the nuances of information processing in crisis informatics. Based on the findings of published articles, the dissertation proposes an information refinement framework tailored to the domain of crisis management (chapter 12).

Research Design

3

This chapter presents the overall research design of the dissertation. Based on the research field and epistemological foundations (Section 3.1), I argue for the research approach of extended design case studies (Section 3.2), which were app-lied in three research projects (Section 3.3) using a multitude of methods for literature reviews, empirical inquiry as well as design and evaluation of artefacts (Section 3.4).

3.1   Research Field and Foundations

As outlined, characteristics and challenges of information refinement are sub-ject to human requirements and technical feasibility, amongst others. Thus, this dissertation seeks to contribute to the research field of human-computer interac-tion (HCI), which is "concerned with the design, evaluation and implementation of interactive computing systems for human use and with the study of major phenomena surrounding them" (Hewett et al., 1992, p. 5). In a review of exis-ting HCI literature, Wobbrock and Kientz (2016) outline that the major research contribution types covered by HCI are empirical, artefact, methodological, theo-retical, dataset, survey, and opinion contributions. So far, different types of crisis informatics research have been embedded into the HCI discourse, such as empi-rical investigation of social media use, collection and processing of social media data, system design, building and evaluation, as well as cumulative and longitu-dinal research (Reuter, Hughes, et al., 2018). Since machine learning techniques are used and continuously improved to analyse data and refine information in a more efficient manner (M. Imran et al., 2015), crisis informatics is related to the

M.-A. Kaufhold, *Information Refinement Technologies for Crisis Informatics*,
https://doi.org/10.1007/978-3-658-33341-6_3

domain of computer science (CS), which is "the study of computers and algorithmic processes, including their principles, their hardware and software designs, their applications, and their impact on society" (A. Tucker et al., 2006, p. 2). The process of information refinement is furthermore related to the research field of computer-supported cooperative work (CSCW). According to Bannon and Schmidt (1989, p. 360), the intent of CSCW is to "understand the nature and characteristics of cooperative work with the objective of designing adequate computer-based technologies". Due to the collective nature of social media involving communication, coordination, cooperation and collaboration among diverse actors and across different scenarios, research on crisis informatics and social media has become common in CSCW (Soden & Palen, 2018).

Furthermore, HCI has a strong intersection with the research field of information systems (IS), in whose context user-centred models such as the technology acceptance model (Davis, 1993) and the unified theory of acceptance and use of technology (Venkatesh et al., 2003) have been developed. Hevner outlines that the discipline is characterized by the two paradigms of behavioural science, which *"seeks to develop and verify theories that explain or predict human or organizational behaviour"*, and design science, which *"seeks to extend the boundaries of human and organizational capabilities by creating new and innovative artifacts"* (Hevner, 2007). Thus, IS includes empirical and ethnographic fieldwork, design of artefacts and exchange with the knowledge base (Hevner, 2007). With regard to crisis informatics, IS researches diverse phenomena such as the distribution and use of crisis and warning apps, collective behaviour in social media, and the adoption of social media analytics for crisis management (Eismann et al., 2016; Reuter, Kaufhold, Leopold, et al., 2017b; Stieglitz, Mirbabaie, Fromm, et al., 2018). In summary, HCI, CS, CSCW and IS have brought up different user-oriented approaches such as user-centred design, participatory design and end-user development (Lieberman et al., 2006; Rohde et al., 2017). Thus, most of this dissertation's work is published in international scientific conferences and journals within the research fields of HCI, CS, CSCW, IS, crisis management and crisis informatics. Furthermore, the epistemological foundations of this dissertation are important for the selection of methods and interpretation of results. Using the epistemological framework of Becker and Niehaves (2007) as depicted in Table 3.1, it follows *Kantianism* in assuming that there are entities independent and dependent of the human mind and that both experience and intellect are sources of cognition. Despite acknowledging the importance of "spirit, mind or language", which corresponds to ontological idealism and the examination of human perception within this dissertation, it does not "negate the existence of a real world which is independent of human thinking and speech" (Becker & Niehaves, 2007, p. 203).

Table 3.1 Epistemological framework, from Becker and Niehaves (2007); the epistemological foundations of this thesis are marked in bold

What is the object of cognition?	*Ontological realism.* A world exists independently of human cognition, for instance, independent of thought and speech processes.	*Ontological idealism.* The 'world' is a construct depending on human consciousness.	**Kantianism. There exist entities that are independent from (*noumena*) as well as dependent on human mind (*phenomena*).**
What is the relationship between cognition and the object of cognition?	*Epistemological realism.* Objective cognition of an independent reality is possible.	**Constructivism. The relationship of cognition and the object of cognition is determined by the subject.**	
What is true cognition?	*Correspondence theory of truth.* True statements are those which correspond with 'real world facts'.	**Consensus theory of truth. A statement is true (for a group), if it is acceptable to the group.**	*Semantic theory of truth.* A condition for truth is the differentiation of an object and a meta-language.
Where does cognition originate?	*Empiricism.* Cognition originates from the sense. Such experience-based knowledge is called *a posteriori* or *empirical knowledge.*	*Rationalism.* Cognition originates from the intellect. Such non-experience-based knowledge is referred to as *a priori knowledge.*	**Kantianism. Both experience and intellect are sources of cognition. Thoughts are meaningless without content, cognitions are blind without being linked to terms.**
By what means can cognition be achieved?	*Inductivism.* Induction is understood as the extension from individual cases to universal phases, the generalization	*Deductivism.* Deduction is the derivation of the individual from the universal.	**Hermeneutic. The understanding of a certain phenomenon is influenced by the pre-understanding of the entire/context.**

In favour of *constructivism* and the *consensus theory of truth*, it follows the assumptions of Rohde et al. (2009, p. 4) that "our experienced reality is being socially constructed [...] rather than 'discovered' in an independently existing world" and as "the relationship between cognition and the object of cognition is determined by subjective conceptualization and the way humans interact with each other, statements are seen as 'true' or valid by a social community, if they are accepted within this community (consensus theory of truth)". These foundations are important for crisis informatics as the actual situation may differ from emergency services' and citizens' evidence or perception of a crisis, which can lead to the dissemination of inaccurate information, such as fake news or rumours (Arif et al., 2017; Reuter, Hartwig, et al., 2019). This has the potential to threaten citizens' health and emergency services' public reputation as well as increasing their task complexity. Acknowledging the unique characteristics of a crisis situation, gaining knowledge is subject to *hermeneutics*, "where the understanding of a certain phenomenon depends on previous understanding of its context [...]" (Rohde et al., 2009). Therefore, the applied method should be capable of capturing the context and social practices of the involved crisis management stakeholders. The design, deployment and appropriation of ICT can be understood as an intervention into social practice (Rohde et al., 2009). However, as the changes in social practice cannot be anticipated thoroughly, evaluations of the appropriation of ICT are required to understand these changes and can furthermore serve as the starting point of iteratively re-designing ICT (Wulf et al., 2011).

3.2   Research Approach

In order to establish a rich understanding of social context and practice, Wulf et al. (2011) propose *design case studies* as a research framework which ideally comprises the steps of (1) empirical pre-study, (2) prototyping/(participatory) IT design, and (3) evaluation/appropriation study. The empirical pre-study should offer "[...] microlevel descriptions of the social practices before any intervention takes place" and "describe already existing tools, media, and their usage" (Wulf et al., 2015, p. 119). It aims for an understanding of the existing practice from technological, organizational and social perspectives and utilizes the application of empirical research methods to identify problems, needs, or opportunities for IT design (Stevens et al., 2018). The design of an innovative IT artefact should then tackle identified problems, needs, or opportunities and include the description of "the specific design process, the involved stakeholders, the applied design methods, and the emerging design concepts" (Wulf et al., 2015, p. 119). Using a participatory approach, the design may be continued in several iterations after the technology has been introduced to potential future users. Finally, evaluation or appropriation studies should document "the introduction, appropriation, and potential redesign of the IT artefact in its respective domain of practice" (Wulf et al., 2015, p. 119). To capture long-term changes in social practice based on the introduction of IT artefacts, repeated reflection with practitioners is required. Although there is a natural order of starting points of the phases, Wulf et al. (2015) do not understand the phases as being strictly consecutive, but as continuing: "once an analysis of existing practices has started, it does not make sense to stop reflecting upon the momentum of the existing practice; rather, it continues throughout the design and the study of appropriation" (Wulf et al., 2015, p. 120) (Figure 3.1).

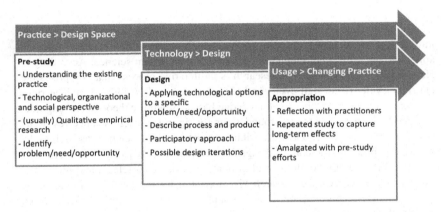

Figure 3.1 Schematic display of the structure of a design case study, illustration reproduced from Wulf et al. (2015)

Design case studies are inspired by *action research* which is "comparative research on the conditions and effects of various forms of social action and research leading to sociation action" (Lewin, 1948, pp. 202–203) and that uses "a spiral of steps, each of which is composed of a circle of planning, action, and fact-finding about the result of the action" (Lewin, 1958, p. 201). This dissertation integrates Reuter's (2014b, p. 22) understanding of a design case study, whereby "'planning' [refers to] the empirical analysis of the given practices, 'action' is the design and use of suggested ICT artefacts, and the evaluation leads towards 'fact-finding about the results of the action'". Furthermore, as design case studies centre around the design of ICT, they are linked to *design science* research, which "seeks to extend the boundaries of human and organizational capability by creating new and innovative artifacts" (Hevner et al., 2004, p. 75).

Figure 3.2 Three cycle view of design science research, informed by Hevner (2007)

Although Hevner and Chatterjee (2010) suggest integrating action research into design research, design case studies, which are an action research oriented approach, put a different emphasis on research than design science. According to Hevner (2007), design science research is an iterative process that comprises the (1) rigor cycle, (2) design cycle, and (3) relevance cycle (Figure 3.2). The rigor cycle aims to integrate "the vast knowledge base of scientific theories and engineering methods that provides the foundations for rigorous design science research" (Hevner, 2007, p. 89), and also contribute results of design science research to the knowledge base. As the core of design science research projects, the design cycle comprises the iterative building and evaluation of design artefacts and processes. Finally, as design science research "is motivated by the desire to improve the environment by the introduction of new and innovative artifacts" (Hevner, 2007, p. 88), the relevance cycle seeks to elicit requirements from the environment, comprising the application domain (people, organizational and technical systems) with its inherent problems and opportunities, and conduct field testing with the designed IT artefacts. Despite differences between the epistemological foundations of research from Hevner et al. (2004) and Rohde et al. (2009), the three phases of design case studies loosely correspond with the design and relevance cycles of design science research, while the interaction with the knowledge base is less elaborated in the framework of design case studies.

In this dissertation, an extended and modified view of design case studies is applied (Figure 3.3). First, in accordance with the rigor cycle of Hevner (2007), the foundations of and contributions to the knowledge base are elaborated within this research approach. Second, while Wulf et al. (2015, p. 119) propose "microlevel descriptions of the social practices" focusing on qualitative ethnographic methods, Ludwig et al. (2017) suggests, based on extensive crisis informatics

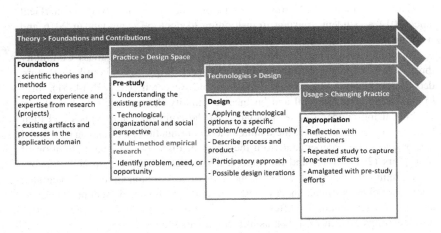

Figure 3.3   Extended and modified research design based on design case studies

research using design cases studies, the application of macrolevel research, including quantitative methods such as questionnaire-based surveys or social media analytics. Hence, this dissertation applies multi-method empirical research.

3.3   Research Context

My thesis comprises a theoretical review, empirical studies with citizens, and the design and evaluation of supportive ICT, which were embedded into the three research projects of EmerGent, KontiKat, and MAKI. Furthermore, the finalisation of my thesis, including the information refinement framework (chapter 12), and further research publications (Haunschild et al., 2020; Kaufhold, Haunschild, et al., 2020) were realised within the scope of ATHENE.

The research project EmerGent (Emergency Management in Social Media Generation; https://www.fp7-emergent.eu), funded by the European Union Framework Programme for Research and Technological Development (No. 608352), examined the positive and negative impact of social media during crisis situations. Based on the understanding of the impact of social media, the safety and security of citizens before, during and after emergencies should be enhanced. It had a duration of three years (2014–2017) and involved several stakeholders form the domain of crisis management in the European Union such as fire departments or the European Emergency Number Association. I have been involved in EmerGent since 2014 as student assistant, transitioning to doctoral candidate in 2016 and working in the project until its finalization in 2017.

The junior research group KontiKat (Civic-Societal and Business Continuity through Socio-Technical Networking in Disasters; https://www.kontikat.de), funded by the Federal Ministry of Education and Research (No. 13N14351), aims to encourage civic, societal and business continuity through socio-technical networking using cooperative technologies. At the core, not catastrophes and their handling are considered but the maintenance, continuation, and recovery of social and business life independent from the cause of disruption. Is has a duration of four years (2017–2021) and the superordinate research goals aim at the theory of continuity management, accurate methods of self-organization, and integrated findings of civic, societal, and business continuity through socio-technical networking in disaster situations. After the end of EmerGent in 2017 and until 2020, I have been involved in KontiKat as a doctoral candidate.

The collaborative research centre MAKI (Multi-Mechanisms Adaptation for the Future Internet; https://www.maki.tu-darmstadt.de), funded by Deutsche Forschungsgemeinschaft (SFB 1053), examines communication systems of the future. For instance, this could facilitate the ability to stream video on a smartphone in high quality without interruptions in spite of busy or overloaded mobile networks. The Internet has rapidly evolved into an integral part of our everyday life. Consequently, the necessary communication mechanisms and equipment are changing on a constant basis. The individual solutions resulting from this rapid evolution are widely considered to be problematic. Currently, as many as three standards exist for Bluetooth, Wi-Fi, and LTE alone. The result is a multitude of services often based on different technologies. MAKI (2013–2020) considers this diversity as an opportunity by utilizing the individual attributes of particular mechanisms to meet the desired high-quality objectives. In parallel to KontiKat, I have been involved in MAKI between 2018 and 2019 as doctoral candidate before moving to ATHENE in 2020.

The national research centre ATHENE (National Research Center for Applied Cybersecurity; https://www.athene-center.de), formerly known as the research centre CRISP (Center for Research in Security and Privacy, 2015–2019) is funded by the Federal Ministry of Education and Research (BMBF) and the Hessian Ministry of Science and Art (HMWK). It is an alliance of the Fraunhofer Institutes SIT and IGD as well as the universities Technical University of Darmstadt and Darmstadt University of Applied Sciences. ATHENE is the largest research centre for IT security in Europe and conducts cutting-edge research for the benefit of business, society, and government and strives for academic leadership. ATHENE

(2019–2026) examines automotive security, biometrics, cloud security, cryptography, cybersecurity analytics and defences, digital forensics, economics of IT security, industry 4.0, internet and infrastructure security, media security, mobile and cyber-physical system security, privacy and trust, secure software systems, security management, social and ethical aspects of IT security, threat modelling and security evaluation, as well as usable security. The goal of ATHENE's mission SecUrban is to provide secure and reliable ICT for critical infrastructures (CI) to achieve understandable and actionable security solutions, using the Digitalstadt Darmstadt as an empirical case. I have been involved in ATHENE since 2020.

3.4   Methods

Due to the collaborative and complex nature of the underlying research projects EmerGent, KontiKat, and MAKI, my research only contributes to a subset of their findings and is based on further works conducted in their contexts. In EmerGent, for instance, the basic literature review (Reuter, Ludwig, Friberg, et al., 2014), initial empirical studies (Reuter et al., 2017), two rounds of requirement elicitation (Akerkar et al., 2014, 2016) based on workshops with emergency services (O'Brien et al., 2016), as well as a first round of implementation and evaluation (Reuter & Amelunxen, 2016) were conducted before I entered the project as doctoral candidate. In order to contextualise my findings, an overview of all relevant research that informed the work of this dissertation is given in Table 3.2. Overall, the research was conducted in three cycles of technology development whereof most of my research contributions were achieved in the third phase, which are highlighted in grey. The first two development cycles were mostly conducted in the EmerGent project, while the third cycle was embedded in the first half of the KontiKat project and the second phase of the collaborative research centre MAKI. In compliance with design case studies, every cycle comprises findings of empirical, design and evaluative origin (Wulf et al., 2011). Furthermore, at the start of EmerGent (first cycle) and KontiKat (third cycle), theoretical foundations based on literature reviews informed the overall research designs of the projects. The following subchapters present methodological details on the primary studies in this dissertation.

Table 3.2 Applied methods, target audiences (A for authorities, C for citizens), content focuses, years of application and count (of literature or requirement documents, empiricism or evaluation participants, or designed or redesigned systems) with different research outputs (D for project deliverables in EmerGent, P for scientific publications in conferences or journals) across three development cycles and three research projects, whereof the research contributions of this dissertation are marked in italics

	Type	A	C	Focus	Year	Count	Project	Output
First Cycle	Theory (review)	x	x	Use, media, and role patterns	2014	1	EmerGent	D3.1
	Empiricism (interviews)	x		Social media in emergencies	2014	16	EmerGent	D3.1, P
	Empiricism (workshop)	x		End-user advisory board	2014	16	EmerGent	D2.6
	Empiricism (survey)	x		Social media perceptions	2015	761	EmerGent	D2.5, P
	Empiricism (workshop)	x		End-user advisory board	2015	18	EmerGent	D2.6
	Empiricism (survey)		*x*	*Social media perceptions*	*2016*	*1034*	*EmerGent*	*P (ch. 6)*
	Design (requirements)	x	x	Initial requirements	2014	1	EmerGent	D3.4
	Design (implementation)	x	x	System design (112.social, ESI, SMO)	2014–2016	3	EmerGent	D5.5, D6.2
	Evaluation	x		System evaluation (ESI)	2016	12	EmerGent	D3.7, P
Second cycle	Theory (contribution)	x	x	End-user based social media analytics	2017	1	EmerGent	P
	Empiricism (workshop)	x		End-user advisory board	2017	15	EmerGent	D2.7

(continued)

Table 3.2 (continued)

	Type	A	C	Focus	Year	Count	Project	Output
	Empiricism (survey)	x		Social media perceptions	2017	473	EmerGent KontiKat	P
	Design (requirements)	x	x	Revised requirements	2016	1	EmerGent	D3.5
	Design (implementation)	*x*	*x*	*System redesign (112.social, ESI, SMO)*	*2016–2017*	*3*	*EmerGent*	*D5.6, D6.3, P (ch. 9, 11)*
	Evaluation	*x*	*x*	*System evaluation (112.social, ESI, SMO)*	*2016–2017*	*14, 21, 12*	*EmerGent KontiKat*	*D6.4, P (ch. 9, 11)*
Third cycle	*Theory (review)*	*x*	*x*	*Use, role, and perception patterns*	*2017*	*1*	*KontiKat*	*P (ch. 4)*
	Empiricism (survey)		*x*	*Apps, guidelines, and infrastructure*	*2017*	*1069*	*KontiKat MAKI*	*P (ch. 5, 7, 8)*
	Design (implementation)		*x*	*System redesign (SMO)*	*2018*	*1*	*KontiKat MAKI*	*P (ch. 10)*
	Evaluation		*x*	*System evaluation (SMO)*	*2019*	*2*	*KontiKat MAKI*	*P (ch. 10)*

3.4.1 Theoretical Review

In order to review existing work and the state of the art on crisis informatics and social media, a literature survey was conducted. The review was specifically designed to identify major cases of social media use in emergencies, to extend research on usage patterns, to establish a role typology for the real and virtual realm, and to examine perception patterns of both authorities, such as emergency services, and citizens. Details on methods and results of this review are presented in chapter 4.

3.4.2   Empirical Pre-Study

Two representative surveys with the German population were conducted to get both qualitative and quantitative insights into their expectations, perceptions and use of infrastructure in crises, especially crisis apps and social media. The first representative survey was conducted in October 2016 using a panel of GapFish (Berlin), which is an ISO-certificated panel provider (Table 3.3). They guarantee panel quality, data quality and security as well as survey quality through various (segmentation) measurements for each survey within their panel of over 280,000 active participants. The survey comprised six questions on social media and three questions on crisis apps. The results of the survey were representative for the German population with regard to gender and age, and a wide spread of the survey sample in terms of region, education, and income was ensured. Quantitative questions were examined using statistical analysis and qualitative questions using qualitative coding. Details on methods and results of this empirical study are presented in chapter 6.

Table 3.3 Representative survey on social media and crisis apps

Panel	GapFish EntscheiderClub
Inquiry	October 2016
Sample Size	1.069
Population	Germany
Representativeness	Gender, age, region*, education*, income*
Topics	Social media (five quantitative and one qualitative questions) Crisis apps (three quantitative questions)
Analysis	Statistical analysis and qualitative coding

The second representative survey was conducted in July 2017 using a panel of GapFish (Table 3.4). The aim was not only to get additional insights on crisis apps, but also to examine infrastructure use in general and further evaluate previously developed social media guidelines. Thus, the survey comprised four questions on infrastructure, three questions on crisis apps and five questions for the evaluation of social media guidelines. The results of the survey were representative for the German population with regard to gender, age, region, education, and income. Again, quantitative questions were examined using statistical analysis and qualitative questions using qualitative coding. Details on the methods and results of this empirical study are presented in chapters 5, 7, and 8.

Table 3.4 Representative survey on infrastructure, crisis apps, and social media guidelines

Panel	GapFish EntscheiderClub
Inquiry	July 2017
Sample Size	1.024
Population	Germany
Representativeness	Gender, age, region, education, income
Topics	Infrastructure (three quantitative and one qualitative questions) Crisis apps (two quantitative and one qualitative questions) Social media guidelines (four quantitative and one qualitative questions)
Analysis	Statistical analysis and qualitative coding

3.4.3 Design of Artefacts

In summary, one non-technical and three technical artefacts were designed within the scope of this dissertation (Table 3.5, Figure 3.4). For the design of the Social Media Guidelines (SMG) as a non-technical artefact, we conducted literature reviews on the potentials and challenges of social media from the perspective of citizens and then reviewed guidelines on the use of social media in general and during emergencies. To design our SMG, their content was presented as input during an end-user advisory board (EAB) workshop, comprising 18 participants from emergency services. Based on these guidelines, EAB members and practitioners prioritized the most important information that should be communicated to citizens with the intention to keep the guidelines concise. Under consideration of their input, further literature, and existing guidelines, a first draft of the SMG was created. It was presented at another EAB workshop, comprising 15 participating emergency services, to gather insights for the design of the final guideline versions. The design of the SMG was primarily realised by the European Emergency Number Association (EENA). Details on the methods and results of the design are presented in chapter 8.

In terms of technical artefacts, the Emergency Mobile App (112.social) and Emergency Service Interface (ESI) were designed jointly within two development cycles in the EmerGent project. To identify requirements for both artefacts, a requirements analysis process was employed. Firstly, scenarios and use cases from real-life operations were chosen to be illustrated and analysed. They were then presented in workshops to experts, developers, and end users to discuss different approaches, establish a common understanding and allow interactions with

Table 3.5   Non-technical (SMG) and technical artefacts (112.social, ESI, SMO) within this thesis

Artefact	Short Description
Social Media Guidelines (SMG)	Concise graphical and text-based guidelines for citizens' social media use in general and in the preparation, response and recovery phases of an emergency.
Emergency Mobile App (112.social)	Mobile app for bidirectional communication between citizens (e.g., for reporting incidents) and emergency services (e.g., for broadcasting information and requesting additional information from citizens).
Emergency Service Interface (ESI)	Web interface for emergency services which displays app alerts (from 112.social) and aggregated social media alerts (from Facebook, Google+, Instagram, Twitter, YouTube) and allows alerts to be filtered by keywords, relevancy and information quality.
Social Media Observatory (SMO)	Web interface for citizens and emergency services which facilitates the cross-platform collection, export, filtering, visual analysis, relevance classification and dissemination of social media data.

each other. Furthermore, to involve a broader community, online surveys were conducted to collect data from involved actors. In order to consider the state of the art and to inform the application of appropriate methodologies, all interventions were supported by literature reviews. The design of 112.social was primarily realized by IES Solutions (IES). While the ESI frontend was primarily designed by the Oxford Computer Consultants (OCC) and Gasper Bizjak, the backend comprised components from the University of Siegen (USI), University of Paderborn, OCC, and IES. As a former employee of USI, I led the implementation of the information gathering component. Details on methods and results of the design are presented in chapters 9 and 11.

The Social Media Observatory (SMO), a web interface for the cross-platform collection and analysis of data, was subject to multiple design iterations involving user-centred evaluations (Reuter, Ludwig, Kotthaus, et al., 2016). During the dissertation project, an additional iteration was conducted to implement identified design requirements, upgrade the SMO with modern libraries and technologies, and improve its hedonic and pragmatic quality. Besides its basic functionality,

Figure 3.4 Overview of artefacts: graphical version of the social media guidelines (SMG, top-left), the emergency service interface (ESI, top-right), the mobile crisis app (112.social, bottom-left), and the social media observatory (SMO, bottom-right)

a novel supervised machine learning algorithm for relevance classification was added, featuring data labelling with real-time prediction of classifier, active learning to reduce amount of training data required for a well-performing classifier, and feedback classification to correct misclassifications of the algorithm. In the backend, the SMO uses an updated version of the information gathering component that was also used for the ESI. While the first iterations of interface design were realised by students of the University of Siegen coordinated by Christian Reuter, the latest design iterations were realised under my coordination and supervision by students of the Technical University of Darmstadt. Details on the methods and results of the design are presented in chapter 10.

3.4.4   Evaluation

All four designed artefacts were evaluated in my thesis. The final evaluation of SMG was part of the representative survey and is described in Section 3.4.1. The evaluations of 112.social, ESI, and SMO took place in the second and third development cycle (Table 3.2). First, a combined evaluation of 112.social and ESI took place, which comprised semi-structured interviews with emergency services based on live and paper-based demonstrations, a workshop, and three field trials (Table 3.6). During a live system demonstration, based upon a short introduction of its functionalities, the participants could interact with the system, which was preconfigured with fire and flood scenario keywords, before and during the guideline-based inquiry. In a paper-based demonstration, the interviewer introduced prepared screenshots of the system and explained its functionality as a foundation for the inquiry.

Table 3.6   112.Social and ESI evaluation: personal details of participants

Category	Data
Roles	crew (10), head (1), incident commander (8), other (6), press (5), PSAP operator (1), PSAP supervisor (5), section leader (3)
Level	gold (3), silver (9), bronze (8), none (1)
Age	20–29 (2), 30–39 (10), 40–49 (6), 50–59 (3)
Gender	male (18), female (3)
Country	Germany (13), Poland (7), Slovenia (1)
Type	Field trial (10), live and paper-based demonstration (9), workshop exercise (2)

Furthermore, the integrated system was tested at a convention in Salzburg, Austria. During the live exercise, a video of an incident (a fire) was shown, and the audience was asked to participate by using their Facebook and Twitter accounts. The audience, representing the role of active citizens, used the video for taking pictures of the incident scene and sending them along other valuable information to a simulated command and control room. Amongst others, it was manned with an incident commander and a social media manager from fire departments

(FD) who used the ESI to get relevant incident information from social media and to broadcast relevant information to ESI users on Facebook and Twitter. Finally, for longer-lasting testing periods in real-world scenarios, three field trials were conducted. The field trial of FD Dortmund was in March/April 2017 and of FD Hamburg in April/May 2017, both lasting four weeks. Another one was conducted in July 2017 with FD Hamburg at the G20 event. During these trials, the system was used by different functions of the organisation alongside their regular duties: The public relations (PR) department, the head of the dispatchers and the department of strategic planning. Details on the methods and results of this evaluation are presented in chapter 9.

Table 3.7 112.Social evaluation: OSCE survey participants by occupation, experience (multiple selection) and device

Category	Data
Occupation	Fire Brigade (2), Paramedic (1), student (1), none (2)
Experience	Voluntary Fire Brigade (5), Accident Ambulance (3), Emergency Service (1), none (1)
Devices	Samsung Galaxy S7 (2), Apple iPhone 6S (1), Moto G1 (1) OnePlus One (1), Samsung Galaxy S5 (1)

Second, as the combined evaluation put a stronger emphasis on ESI, two further evaluations were conducted for 112.social. A survey was conducted during the 23rd ministerial council meeting of the Organisation for Security and Co-operation in Europe (OSCE) that took place in Hamburg, Germany. Thirteen members of a Virtual Operations Support Team (VOST) used 112.social to send information to the ESI operator of FD Hamburg. After the two days of use, a survey was conducted with six VOST members (Table 3.7). Furthermore, to cover the citizens perspective, a mixed group of students and researchers evaluated 112.social during a treasure hunt on the campus of the University of Paderborn. The treasure hunt consisted of eight "treasures" that contained answers to previously asked questions. These treasures were located on different spots of the campus and, after finding one, the participants had to report them via 112.social. One researcher posed as the public-safety answering point (PSAP) using the ESI. After reporting a found treasure, the PSAP sent requests for further information and gave new instructions for finding the next treasure. After the treasure hunt, the citizens were asked to answer a survey (Table 3.8). Details on the method and results of both evaluations are presented in chapter 11.

Table 3.8   112.Social evaluation: treasure hunt survey participants with occupation and device information

Category	Data
Occupation	Research assistant (4), mechanical engineer (1), student (2), none mentioned (1)
Devices	Moto G1 (2), Moto G4 (4), OnePlus 3T (1), Samsung Galaxy S8 (1)

Finally, a supervised machine learning approach using active, incremental, and online learning for rapid relevance classification of social media posts was integrated into SMO and evaluated subsequently. The evaluation was based on two annotated datasets of the 2013 European floods and 2016 BASF SE Incident (Table 3.9). In a first step, a Random Forest implementation was evaluated in terms of accuracy, precision, recall, and time using different classification feature configurations. In the second step, a comparison of three incremental classification methods and the learning success both with and without active learning was conducted on both datasets. Details on the method and results of this evaluation are presented in chapter 10.

Table 3.9   SMO dataset evaluation: comparison of characteristics

Dataset	2013 European Floods	2016 BASF SE Incident
Collection period	30th May to 28th June 2013	17th October to 20th October 2016
Source	Twitter Search API	Twitter Search API
Number of messages	3923	3790
Relevant messages	1626 (41%)	1816 (48%)

Part II
Theoretical and Empirical Findings

Retrospective Review and Future Directions for Crisis Informatics

4

Social media has been established in many larger emergencies and crises. This process has not started just a few years ago, but already 15 years ago in 2001 after the terrorist attacks of 9/11. In the following years, especially in the last 10, sometimes summarized under the term crisis informatics, a variety of studies focusing on the use of ICT and social media before, during or after nearly every crisis and emergency has arisen. This article aims to recapitulate 15 years of social media in emergencies and its research with a special emphasis on use patterns, role patterns and perception patterns that can be found across different cases in order to point out what has been achieved so far, and what future potentials exist.

This chapter has been published as the journal article "Fifteen Years of Social Media in Emergencies: A Retrospective Review and Future Directions for Crisis Informatics" in the Journal of Contingencies and Crisis Management (JCCM) by Christian Reuter and Marc-André Kaufhold (Reuter & Kaufhold, 2018).

4.1   Introduction and Brief History

Social media is nowadays a part of everyday life. The former so-called Web 2.0 (O'Reilly, 2007) was initially defined as an architecture for participation with new possibilities for social interaction. According to O'Reilly (2006), Web 2.0 "does not only represent content that has been provided by an individual for the purpose of distribution; it also represents interaction among people". Over the years this interaction has been summarized more and more frequently under the term social media: a "group of internet-based applications that build on the ideological and technological foundations of Web 2.0, and that allows the creation

M.-A. Kaufhold, *Information Refinement Technologies for Crisis Informatics*, https://doi.org/10.1007/978-3-658-33341-6_4

and exchange of user-generated content" (Kaplan & Haenlein, 2010). In this context user-generated content refers to "the various forms of media content that are publicly available and created by end-users" (Kaplan & Haenlein, 2010). Allen (2004) points out that the "core ideas of social software itself enjoy a much longer history, running back to Vannevar Bush's ideas about the storage-device Memex in 1945 through terms such as augmentation, groupware, and Computer Supported Cooperative Work (CSCW) in the 1960s, 70s, 80s, and 90s". Accordingly, Koch (2008) argued that "most of what currently is advertised as a revolution on the web has been there as CSCW applications years (or even decades) ago—however, not as nice and as usable as today". However, during the last 10 years these services were intensively used. Currently the most common types of social media include Facebook with about 1.7 billion active users monthly, YouTube (1 billion), WhatsApp (1 billion), Instagram (500 million), LinkedIn (433 million), Twitter (320 million) and Google+ (235 million) (Kroll, 2016).

Social media is not only part of everyday live but also appearing in critical situations: Already after the 9/11 attacks in 2001, citizens created wikis to collect information about missing people (Palen & Liu, 2007), and FEMA and the Red Cross used web-based technologies to inform the public and to provide status report internally and externally (Harrald et al., 2002). Starting in about 2006, social media use in emergencies has become a very big research field, sometimes summarized under the term *crisis informatics*. Coined by Hagar (2007) and later elaborated by Palen et al. (2009) it "views emergency response as an expanded social system where information is disseminated within and between official and public channels and entities". Today crisis informatics "is a multidisciplinary field combining computing and social science knowledge of disasters; its central tenet is that people use personal information and communication technology to respond to disaster in creative ways to cope with uncertainty" (Palen & Anderson, 2016). During the last years, various studies have arisen addressing emergencies and the use of social media. Also, various international journals have published special issues (Hiltz, Diaz, et al., 2011; Pipek et al., 2014; Reuter, Mentler, et al., 2015) as well as tracks at various conferences, such as *ISCRAM*. This trend was predicted some years ago: "the role held by members of the public in disasters [...] is becoming more visible, active, and in possession of greater reach than ever seen before" (Palen & Liu, 2007).

Many studies focus on the concrete use of social media during a specific emergency, such as the 2011 London riots (Denef et al., 2013), the 2012 hurricane Sandy (Hughes et al., 2014) or the 2013 European floods (Reuter, Ludwig, Kaufhold, et al., 2015). These studies demonstrate the specific ways in which social media responded to various crises. Across various studies of emergencies and disaster events, numerous positive and negative aspects of social media have been identified, groups of users have been defined and perceptions have been studied. However, after 15 years of social media in emergencies, it is time to summarize what has been achieved so far to derive what should be the next step. Based on an overview of cases on social media in emergencies (section 4.2), use patterns (section 4.3), role patterns (section 4.4) and perception patterns (section 4.5) are derived, followed by a discussion and conclusion on future directions (section 4.6). According to the Oxford Dictionary, with pattern we refer to "the regular way in which something happens or is done". Broken down to our case we refer to repetitive ways in which social media is used in different cases, what roles have been observed and how this use is perceived. Having said this, our article focuses less on technological patterns, which can be found in other papers (M. Imran et al., 2015).

4.2 Published Cases of Social Media in Emergencies

According to the World Disaster Report (IFRC, 2015) during the last ten years in average about 631 disasters happened per year with 83,934 people killed, 193,558 people affected and estimated damage of 162,203 million US dollars. While natural disasters killed 76,420 people, technological disasters caused 7,513 lives per year. According to Hiltz et al. (2011), disaster, crisis, catastrophe, and emergency management "are sometimes used synonymously and sometimes with slight differences, by scholars and practitioners". This is also the case while searching for articles on social media use in these contexts. However, the *internationally agreed glossary of basic terms related to disaster management* (United Nations Department of Humanitarian Affairs, 2000) defines an *emergency* as a "sudden and usually unforeseen event that calls for immediate measures to minimize its adverse consequences". A *disaster* is a "serious disruption of the functioning of a society, causing widespread human, material, or environmental losses which exceed the ability of affected society to cope using only its own resources". The term *crisis* has not been defined in that document. Yet, *crises* are situations that the normal structural and process organization cannot overcome (BSI, 2008). The Greek root word *krisis* (judgment, decision) shows the ambivalent possibilities. In the following we will talk about emergencies, but do not want to limit ourselves in the cases to take into consideration.

For 15 years, the public has used social media in emergencies (Reuter et al., 2012). After the terrorist attacks of September 11th 2001 for example, wikis created by citizens were used to collect information on missing people (Palen & Liu, 2007), while citizens used photo repository sites to exchange information following the 2004 Indian Ocean tsunami (S. Liu et al., 2008) or the 2007

Southern California wildfires (Shklovski et al., 2008). Another early study focused on hurricane Katrina in 2005 (Murphy & Jennex, 2006). Social media were quickly revealed as an emergent, significant, and often accurate form of public participation and backchannel communication (Palen, 2008).

About ten years ago, Palen and Liu (2007), anticipated a future where ICT-supported public participation would be regarded as both normal and valuable. Subsequently, analysis of social media in emergencies, mainly in the USA but also in other places, has become conventional. There are fewer studies covering the situation in Europe (Reuter et al., 2012). Most studies have focused on the use of Twitter so far, partly due to its frequency of use in the USA. However, looking at the current statistics, Facebook has 1,7 billion active users while Twitter "just" has 320 million, this might not be the main reason. The ease of data selection (e.g. to obtain a statistically sound sample) in Twitter (Reuter & Scholl, 2014) might be most influential for this bias. While comparing larger emergencies and studies about social media in emergencies it appeared that nearly no emergency exists without articles on the use of social media there.

Many published research papers have mainly focused on the use of Twitter. Many articles focus on social media use during various disasters in the USA (e.g., 2001 9/11, 2005 Hurricane Katrina, 2012 Hurricane Sandy, 2013 Colorado Flood). Other studies have provided a more international backdrop (e.g., 2004 Indian Ocean tsunami, 2008 Sichuan earthquake, 2011 Tunisian revolution, 2011 Norway attacks, 2013 European floods, 2015 Paris shootings). Existing studies focus on both natural hazards (tsunamis, hurricanes, earthquakes, floods) and human-induced disasters (shootings, terror attacks, uprisings).

Table 4.1, which extends an earlier version (Reuter, Ludwig, Friberg, et al., 2015), gives an overview about studies on social media in emergencies. We are aware that this list cannot be complete. Our focus was less to identify and include all research on this topic, but more to highlight the existence of scientific cases about nearly all events during the last 10 to 15 years. The cases and studies have been identified following the instructions of Brocke et al. (2015) and searching in Google Scholar for the keywords "social media", "web 2.0", "Twitter", "Facebook", "emergency", "disaster", "crisis" in singular and plural without any restrictions concerning the timeframe. Furthermore, backward and forward search has been applied. In addition, for recent larger emergencies, we explicitly searched for studies on the use of social media while using the search term of the case (e.g. Paris shootings 2015). The cases are presented regarding their reference, the related case or scenario, and a brief overview of the scientific contribution. They are sorted by the year of occurrence. Due to the amount of studies, only a limited number per case has been selected to provide an overview.

Table 4.1 Overview of cases in literature

Reference	Case	Contribution
(Palen & Liu, 2007)	2001 9/11	Use of wikis to collect information about missing people.
(Harrald et al., 2002)		FEMA and the Red Cross used web-technologies to inform the public and to provide status report.
(S. Liu et al., 2008)	2004 Indian Ocean tsunami	Citizens used photo repository sites to exchange information.
(Endsley et al., 2014)	2005 Hurricane Katrina, 2010 volcano Eyjafjallajökull in Iceland	Credibility of social media information is less than of printed, official online or televised news and information from family, relatives, or friends.
(Shklovski et al., 2008)	2007 Southern California wildfires	Citizens used photo repository sites to exchange information.
(Hughes & Palen, 2009)	2008 Hurricanes Gustav and Ike	Highlights differences between the use of Twitter in crises and the general use.
(Qu et al., 2009)	2008 Sichuan earthquake	Outlines that people gather and synthesize information.
(Sutton, 2010)	2008 Tennessee River technological failure	Outlines the phenomena of broadcasting emergency-relevant information via Twitter.
(Heverin & Zach, 2010)	2009 Lakewood attack on police officers	Shows the ability of Twitter to organize and disseminate crisis-related information.
(Latonero & Shklovski, 2011)	2009 Los Angeles fire Department	Public Information Officers highlight the importance of the information evangelist within organizations.
(Starbird & Palen, 2010)	2009 Oklahoma fires	Highlights the role of retweeting for information processing, especially filtering and recommendation.
(Vieweg et al., 2010)	2009 Red River floods	Highlights broadcasting by people on the ground as well as activities of directing, relaying, synthesizing, and redistributing.
(Mendoza et al., 2010)	2010 earthquake in Chile	Shows that the propagation of tweets that correspond to rumours differs from tweets that spread news because rumours tend to be questioned more than news by the Twitter community.
(Birkbak, 2012)	2010 Bornholm blizzard	Two Facebook groups show that the geographical location and self-selection into groups create different views of a crisis.

(continued)

Table 4.1 (continued)

Reference	Case	Contribution
(Muralidharan et al., 2011)	2010 Deepwater Horizon oil spill disaster	BP's corrective action as the dominant image restoration strategy caused high presence of negative emotion.
(Starbird & Palen, 2011)	2010 Haiti earthquake	Analyses the earthquake with the help of translators and reveals the phenomenon of "digital volunteers".
(Reuter et al., 2012)	2010 Love Parade mass panic in Germany, volcano Eyjafjallajökull in Iceland	Systematizes the communication between authorities and citizens during emergencies, outlining the need for duplex communication.
(Nagy et al., 2012)	2010 San Bruno Californian gas explosion and fire disaster	Illustrates that sentiment analysis (analysis for identifying and extracting subjective information) with emotions performed 27% better than Bayesian Networks alone.
(Helsloot & Groenendaal, 2013)	2011 large-scale fire in Moerdijk, the Netherlands	Most tweets do not contain new relevant information for governments; tweets posted by governments got buried under an avalanche of citizen tweets.
(Starbird & Palen, 2012)	2011 Egyptian uprising	Shows how the crowd expresses solidarity and does the work of information processing through recommendation and filtering.
(Wilensky, 2014)	2011 Great East Japan earthquake	Emphasizes the use of Twitter to provide emotional support and mentions the problem of widely publishing obsolete or inaccurate information and the unequal distribution of useful information.
(Perng et al., 2012)	2011 Norway attacks	The notion of peripheral response has been developed in relation to emergent forms of agile and dialogic emergency response.
(Jennex, 2012)	2011 San Diego / Southwest blackout	The availability of social media illustrates that "the cell phone system did not have the expected availability and users had a difficult time using social media to contact family and friends".
(St. Denis et al., 2012)	2011 Shadow Lake fire	Shows the deployment of trusted digital volunteers as a virtual team to support an incident management team.
(Reuter et al., 2013)	2011 Super Outbreak	Distinguishes groups of twitterers, such as helpers, reporters, retweeters, and repeaters.

(continued)

Table 4.1 (continued)

Reference	Case	Contribution
(Wulf et al., 2013)	2011 Tunisian revolution	Social media linked the young activists with actors in other cities and stimulated the participation in weekly demonstrations.
(Kuttschreuter et al., 2014)	2011 Escherichia coli contamination crisis	Social media can act as a complementary information channel for a particular segment, but it is neither a substitute for traditional nor for online media.
(S. Yang et al., 2013)	2012 hurricane Isaac	Leads to knowledge, which classification algorithms work best in each phase of emergency.
(Hughes et al., 2014)	2012 hurricane Sandy	Shows that few departments used online channels in their response efforts and that communication differed between fire and police departments and across media types.
(Medina & Diaz, 2016)	2012 Madrid Arena tragedy	Opportunities according to the main principles of the theory of Crisis Communication Management provided by Twitter.
(J. I. White & Palen, 2015)	2013 Colorado flood	Highlights the blending of online and offline expertise to evacuate horses from an isolated ranch.
(Kaufhold & Reuter, 2014)	2013 European flood in Germany	Identifies challenges of the public response among emergent groups and digital volunteers highlighting the role of moderators.
(de Albuquerque et al., 2015)	2013 European flood in Germany	Messages near to severely flooded areas have a much higher probability of being relevant.
(Burnap et al., 2014)	2013 Woolwich (London) terrorist attack	The sentiment expressed in tweets is significantly predictive of both size and survival of information flows.
(Wan & Paris, 2015)	2014 Sydney siege	System to analyse posts of a special topic and visualize the emotional pulse of a geographical region.
(Chaturvedi et al., 2015)	2015 cyclone Pam 2014 Kashmir floods, Indonesia landslide	Data collection via Twitter for exploration of the ICT infrastructure for disaster management.
(Fung et al., 2014)	2014 Ebola fear in the USA	Examines the amplified fear of the imported Ebola virus through social media.

(continued)

Table 4.1 (continued)

Reference	Case	Contribution
(Fichet et al., 2015)	2015 Amtrak derailment, Baltimore protests, hurricane Joaquin floods	Examines the use of the live-streaming application Periscope by both citizens and journalists for information sharing, crisis coverage and commentary.
(Soden & Palen, 2016)	2015 Nepal earthquake	Investigates the work of mapmakers working and outlines factors contributing to the emergence of infrastructure.
(Zipf, 2016)	2015 Nepal earthquake, 2013 Philippines typhoon, 2011 Japan tsunami	Help of "Ambient Geographic Information" via social media (Twitter and Flickr) at crisis management.
(An et al., 2016)	2015 Charlie Hebdo shooting	Examines sociological theories in terms of the social factors that contribute to online individual behaviour.
(Zeng et al., 2016)	2015 Tianjin blasts	Clustering analysis and time series analysis of social network Weibo's rumour management strategies.
(Wiegand & Middleton, 2016)	2015 Paris shootings	Examines the velocity of newsworthy content and its veracity with regard to trusted source attribution.
(Sagar, 2016)	2016 Roanu cyclone in Sri Lanka	Twitter and Facebook were used to help flood-affected victims with disaster warnings, relief information, and weather alerts.

4.3   Usage Patterns—Types of Interaction in Social Media

The range of different emergency situations and responses to them have produced attempts to categorize the use of social media. The aim is to both promote systematic analysis of behaviours and interactions and to facilitate the use and development of qualified technology: Reuter et al. (2012) derived a classification matrix for cooperation in crisis situations, depending on the sender (X-axis) and the recipient (Y-axis) of digital content. Considering citizens (C) and authorities (A), such as emergency services, Reuter et al. (2012)'s crisis communication matrix distinguishes between four observed information flows or patterns of social media use in emergencies (Figure 4.1): On the inter-organizational level organizations of crisis response communicate with each other (A2A). On the public level, citizens and volunteers communicate with each other in real or virtually via social media such as Twitter or Facebook (C2C). This citizen-generated content is also being analysed by crisis response organizations (C2A). Besides the communication among the citizens, organizations responsible for recovery work inform the public (A2C).

Moreover, the "categories of organizational behaviour" of Quarantelli (1988) describe five different categories for the flow of information in a crisis, which have similarities to this categorization. However, to describe different use patterns and because of the relevance of communication among citizens (C2C) in social media, we will use the crisis communication matrix.

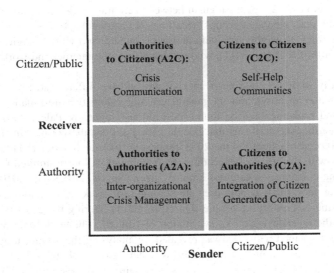

Figure 4.1 Crisis Communication Matrix (Reuter et al., 2012), adapted concerning the terminology

4.3.1 Citizens to Citizens (C2C)—Self-Coordination and Help

Not all activities of citizens in social media are intended for emergency services and we would argue that most of the activities aim to inform other citizens. People help each other and social media is a possible tool for this. However, this self-coordination and help has not been invented by social media: 40 years ago, Quarantelli and Dynes (1977) as well as Stallings and Quarantelli (1985) characterized these "emergent groups" as "private citizens who work together in pursuit of collective goals relevant to actual or potential disasters but whose organization has not yet become institutionalized". According to Quarantelli (1984), the essential conditions for the emergence of such groups are a legitimizing social environment, a perceived threat, a supporting social climate, a network of social relationships, and the availability of specific (immaterial) resources. According to some studies, citizens react in a largely rational way to crisis situations, rarely in panic, are not helpless, and do not loot (Helsloot & Ruitenberg, 2004). They are instead capable of taking part in a large amount of rescue and response work. Here, Reuter et al. (2013) distinguish between activities in the 'real' and the 'virtual' world: real "emergent groups" (Stallings & Quarantelli, 1985), which usually act in the form of neighbourly help and work on-site, and virtual "digital volunteers" (Starbird & Palen, 2011), who originate from the internet and work mainly online.

An early study on 2008 hurricanes Gustav and Ike differentiates between the use of Twitter in crises and its general use suggesting that information broadcasting and brokerage can be found more often (Hughes & Palen, 2009). 2008 Sichuan earthquake confirms that people gather and synthesize information (Qu et al., 2009); also a study on the 2008 Tennessee River technological failure confirms this with the effect of exceeding the boundaries of locally limited networks and raising emergency awareness among citizen (Sutton, 2010). The 2010 Yushu earthquake shows that people use microblogging to seek information about the status of the emergency or people (Qu et al., 2011). During the 2010 Haiti earthquake, the nowadays well-known notion of "digital volunteers" converging to strongly intertwined networks was revealed by analysing the Twitter usage with "Tweak the Tweet" translators (Starbird & Palen, 2011). Digital volunteers perform activities of relaying, amplifying, synthesizing, and structuring information in the wake of disaster events (Starbird, 2013). During the 2012 hurricane Sandy, citizens handled activities that are unlikely to be done by official emergency services such as recovering lost animals (J. I. White et al., 2014). Expressing solidarity, as seen during 2011 Egyptian uprising (Starbird & Palen, 2012) and providing emotional support, during 2011 Great East Japan Earthquake (Wilensky, 2014), are further tasks. A timeline analysis of the 2011 Super Tornado Outbreak (Reuter et al., 2013) indicates that during the preparedness and response phases highly retweeted warning and crisis tracking activities occur, while virtual self-help communities started their relief activities in the recovery phase along with a relatively increasing number of external resource links. Besides Twitter, also other media are used: The use of two Facebook groups during the 2010 Bornholm blizzard shows that the self-selection into groups create different views (Birkbak, 2012). Goolsby (2010) reports on ad-hoc crisis communities using social media to generate community crisis maps. Nowadays seven distinct crisis mapping practices in OpenStreetMap have been identified (Kogan et al., 2016).

To draw conclusions on a broader scope, Olteanu et al. (2015) investigated several crises in a systematic manner. They show that the average prevalence of different information types (32% other useful information, 20% sympathy and emotional support, 10% donations and volunteering, 10% caution and advice, and 7% infrastructure and utilities) and sources (42% traditional or internet media, 38% outsiders, 9% eyewitness accounts, 5% government, 4% NGOs, and 2% businesses) as well as their temporal distribution across a variety of crisis situations. Furthermore, Eismann et al. (2016) conducted a systematic literature review identifying that "sharing and obtaining factual information is the primary function of social media usage consistently across all disaster types, but the secondary functions vary".

However, there are also risks of widely publishing obsolete or inaccurate information and the unequal distribution of useful information (Wilensky, 2014). In cases of uncertainty, caused by redundant information and mistakes due to chaotic "unorganized" online behaviour of volunteers, "there will be a larger amount of collaboration on the platform" (Valecha et al., 2013). This might be addressed by cross-platform working moderators (Reuter, Ludwig, Kaufhold, et al., 2015) or attempts to install public displays for volunteer coordination (Ludwig et al., 2017). Furthermore, Purohit et al. (2014) propose a system for identifying seekers and suppliers in social media communities to support crisis coordination. However, besides all these achievements many aspects are still open: Cobb et al. (2014) suggest the coordination and integration of voluntary activities, the connection between different tools and tasks as well as the possibility to share own activities to generate learning effects for spontaneous and less experienced volunteers. Additionally, Kaufhold and Reuter (2014) identified challenges to support digital and real volunteers in achieving clarity and representation of relevant content, to facilitate processes of moderation and autonomous work as well as to promote feedback and updates in interaction relationships and to integrate technologies and interaction types.

4.3.2 Authorities to Citizens (A2C)—Crisis Communication and Public Alerting

Besides the use of social media among citizens, authorities nowadays and increasingly in the future integrate social media into their crisis communication efforts to share information with the public on how to avoid accidents or emergencies and how to behave during emergencies (Reuter, Ludwig, Kaufhold, et al., 2016). However, already the 2009 case study of Public Information Officers (PIO) of the Los Angeles Fire Department highlights the importance of the information evangelist, who promotes the use of new forms of media and technology within authorities to achieve an effective organizational utilization of social media (Latonero & Shklovski, 2011). Hughes and Palen (2012) argue that members of the public "have a changed relationship to the institution of emergency response" through the authorities' use of social media. A comparative study of police units in the 2011 London riots discusses the benefits and challenges of instrumental and public-including expressive communication approaches through Twitter (Denef et al., 2013), such as close relations and increased possible reach on the one hand and the requirement of high maintenance on the other hand. Another study about the 2011 Thailand flooding disaster describes the authorities' actions taken to correct the mistakes caused by the "emerging risks of the chaotic use of social media" (Chen et al., 2011). Therefore, Starbird and Stamberger (2010) recommend the use of structured crisis-specific Twitter hashtags to increase the utility of information generated during emergencies and to facilitate machine parsing, processing, and re-distribution for the proposed microsyntax "Tweak the Tweet". However, a study on 2012 hurricane Sandy also shows that communication differed between fire and police departments and across media types (Hughes et al., 2014). For this, they suggest new features and tools "to better track, respond to, and document public information". Furthermore, Veil et al. (2011) provide an overview of best practices, examples of social media tools and recommendations of practitioners. Time-series analyses reveal that relevant information became less prevalent as the crisis moved from the prodromal to acute phase and that information concerning specific remedial behaviours was absent (Spence et al., 2015).

Still, there are several barriers in the authorities' use of social media. The collaboration among humanitarian aid organizations and Volunteer and Technical Communities (V&TCs) was analysed in an exploratory study, categorizing the latter into software platform development communities, mapping collaborations, expert networks, and data aggregators (van Gorp, 2014). It further identifies six barriers of collaboration with aid organizations: limited resources, the management of volunteers, different levels of engagement, the level of commitment by V&TCs, different ways of working, and the aid for organizations' limited knowledge about the V&TCs' expertise. Plotnick and Hiltz (2016) show how social media is used by county-level US emergency managers and summarize barriers to effective social media use and recommendations to improve use: The lack of sufficient staff, a lack of guidance and policy documents, trustworthiness and information overload, and lack of skills.

4.3.3   Citizens to Authorities (C2A)—Integration of Citizen-Generated Content

In addition to the communication from authorities to citizens the use of citizen-generated content is important. The potential of benefitting from citizen-generated content lies within illustrating problematic situations through photographs taken with mobile phones. Thus, social media is used to estimate the citizen alertness (Johansson et al., 2012). The perceived unreliability of such information is a significant obstacle in exploring such opportunities (Mendoza et al., 2010). This could be alleviated by crowdsourcing strategies to confirm the trustworthiness of information visible in a picture (Reuter et al., 2012). In a comprehensive literature review regarding the integration of social media content, Hughes and Palen (Hughes & Palen, 2014) complement the challenges of verification, liability, credibility, information overload, and allocation of resources. Furthermore, a study on the 2010 Haiti earthquake showcases opportunities of social media for disaster relief in terms of donations towards the Red Cross (Gao et al., 2011). Akhgar et al. (2013) describe how public safety and security organizations are increasingly aware of social media's added value proposition in times of crisis. Another study suggests that volunteer groups in emergencies will in the future be challenged to mature and improve according to these enhanced possibilities, so that "professional responders will begin to rely on data and products produced by digital volunteers" (Hughes & Tapia, 2015).

During social media research, several applications and methods were examined to integrate citizen-generated content and support authorities in processing social media content. Ludwig et al. (2016) implemented a public display application with a robust communication infrastructure to encompass situated *crowdsourcing* mechanisms. Moreover, Castillo (2016) brings together computational methods (e.g., natural language processing, semantic technologies, data mining, etc.) to process social media messages under *time-critical constraints*. Several contributions aim on extracting *situational awareness* from social media: For instance, Vieweg et al. (2010) applied information extraction techniques to enhance situational awareness on Twitter. Based on the case of Japanese earthquakes in 2009, Sakaki et al. (2010) propose an algorithm that incorporates Twitter users as social sensors for real-time *event detection*, and Pohl et al. (2015) present clustering approaches for *sub-event detection* on Flickr and YouTube to automate the processing of data in social media. Furthermore, de Albuquerque et al. (2015) show that geographical approaches for quantitatively assessing social media messages can be useful to improve relevant content. Moi et al. (2015) propose a system to process and analyse social media data, transforming the high volume of noisy data into a low volume of rich content that is useful to emergency personnel. To achieve this goal, they identify the steps of information gathering and data preparation, data enrichment, information mining, semantic data modelling with ontologies, information quality assessment, alert detection and information visualization.

Also, in this area there are still unaddressed aspects: Imran et al. (2015) contribute with a comprehensive overview on processing social media messages to discuss challenges and future research directions including techniques for data characterization, acquisition, and preparation, event detection and tracking, clustering, classification, extraction, and summarization, and semantic technologies. Additionally, a study by Pohl (2013) summarizes existing frameworks and tools developed in the context of crisis related (e.g., Twitcident or "Tweak the Tweet") and non-crisis related (e.g., Twitinfo) research work to analyse social media or to include new functionalities into the social media usage for crisis management. The comparison reveals that there are systems for different applications, considering one or several social media platforms for monitoring, especially developed for crisis management, and performing different kinds of analysis: monitoring, event-detection, and sentiment analysis. However, at the same time, other studies have shown that not all emergency responders make use of such data during disasters, due to the difficulties of receiving and filtering particularly large amounts of data in emergencies (Hughes & Palen, 2012; Reuter, Amelunxen, et al., 2016).

4.3.4   Authorities to Authorities (A2A)—Inter- and Intra-Organizational Crisis Management

The inter- and intra-organizational collaboration (A2A) of authorities, as a last pattern, is often not supported by social media such as Facebook or Twitter. However, social media can help to improve inter-organizational awareness and informal processes. White et al. (2009) examined the potentials of online social networks with emergency management students: Sharing information, communication, and networking were the most popular features. They also show that possible concerns against those systems may be information integrity, user identification, privacy, and technology reliability. Experiences show that inter-organizational social networks for authorities might generate potentials (Pipek et al., 2013; Reuter, 2014b). Furthermore, authorities may use social media for internal communication. However, in this review, this pattern will not be explored in detail, as it does not directly involve citizens.

4.4   Role Patterns—Types of Users in Social Media

Within all those published cases (section 4.2) and detected usage patterns (section 4.3), different role patterns have been identified. Research regarding types of users active on social media began by identifying individual roles and proceeded with the development of role typologies. In their literature review, Eismann et al. (2016) state that different actor types make use of social media in similar ways, but perceive different conditions and restrictions for social media usage in disaster situations. These roles and role typologies take either a citizens' (public) or authorities' (organizational) perspective and are related to either the real or virtual realm. Based on the analysis of existing roles, this chapter proposes a role typology matrix for individual and collective roles.

4.4.1   Citizens; or Public Perspective

Citizens might be classified in various roles. Hughes and Palen (2009) initially identified information brokers who collect information from different sources to help affected citizens. For Starbird and Palen (2011), the second step was to recognize the actions of remote operators as digital volunteers who progress from simple internet-based activities like retweeting or translating tweets to more complex ones, e.g., verifying or routing information. To further differentiate potential user roles, Reuter et al. (2013) distinguish between activities in the 'real' world as opposed to the 'virtual' world: real emergent groups (Stallings & Quarantelli, 1985), whose involvement usually takes the form of neighbourly help and work on-site and virtual digital volunteers (Starbird & Palen, 2011), who originate from the internet and work mainly online. Ludwig et al. (2015) build on it and address these groups by enabling the detection of physical and digital activities and the assignment of specific tasks to citizens. Based on a timeline and qualitative analysis of information and help activities during the 2011 Super Outbreak, Reuter et al. (2013) suggest a more specific classification of Twitter users in different roles: helper, reporter, re-tweeter, repeater, and reader. Kaufhold and Reuter (2014) additionally suggested the role of the moderator.

Furthermore, according to Blum et al. (2014), three roles contribute to collective sensemaking in social media: The inspectors who define the boundaries of events; the contributors who provide media and witness statements and construct rich but agnostic grounded evidence; and investigators who conduct sensemaking activities to arrive a broad consensus of event under-standing and promote situation awareness. Table 4.2 presents terms that authors have used to describe different (overlapping) social media users in crisis from the public perspective.

Table 4.2 Public perspective on social media roles

Reference	Role	Description
(Stallings & Quarantelli, 1985)	Emergent groups	"Private citizens who work together in pursuit of collective goals relevant to actual or potential disasters […]"—actually not a social media role but still important.
(van Gorp, 2014)	V&TC	Virtual & Technical Communities with expertise in data processing and technologies development, have potential to inform aid organizations.
(Starbird & Palen, 2011)	Digital volunteers	Element of the phenomena popularly known as crowdsourcing during crises. In the twitter sphere, they are called Voluntweeters.
(Wu et al., 2011)	Celebrities	Celebrities are among the most followed elite users.
(Reuter et al., 2013)	Helper	Provide emotional assistance and recommendations for action, offer and encourage help, are involved in virtual and real activities.
	Reporter	Integrate external sources of information, thus providing generative and synthetic information as a news channel or eyewitness.
	Retweeter	Distribute important derivative information to followers or users, correspond with the information broker (Hughes & Palen, 2009).
	Repeater	Generate, synthesize, repeat, and distribute a certain message to concrete recipients.
	Reader	Passive information-catching participants who are interested in or affected by the situation.
(Kaufhold & Reuter, 2014)	Moderator	Establishes supportive platforms, mediates offers and requests, mobilizes resources, and integrates information.

4.4.2 Authorities; or Organizational Perspective

While the previous role descriptions and models address the public use of social media, Bergstrand et al. (2013) examined the utilization of Twitter by authorities and suggest an account typology containing *high-level formal organizational accounts, accounts for formal functions and roles, formal personal accounts,* and *affiliated personal accounts.* Furthermore, Reuter et al. (2011) proposed community scouts as amateur "first informers" to the perceived unreliability of social media information for authorities and St. Denis et al. (2012) describe the use of trusted digital volunteers during the 2011 Shadow Lake fire in virtual teams to inform a Type I incident management team about social media activities. On a higher level, Ehnis et al. (2014) distinguish *media organizations, emergency management agencies (EMAs), commercial organizations, political groups, unions,* and *individuals.*

From an emergency services' perspective, the German Red Cross contributed with the definition of *unbound helpers* which are non-affected citizens that mobilize and coordinate their relief activities autonomously and event-related, especially via social media (DRK, 2013). Accordingly, Kirchner (2014) summarizes the helper groups by their organization form as well as their spatial and social affection to the catastrophic event into the four categories *self-helpers and neighbourhood helpers* (I), *unbound helpers, ad-hoc helpers, and spontaneous helpers* (II), *preregistered helpers and first responders* (III), and *honorary office and full-time helpers in disaster management* (IV). Detjen et al. (2016) further specify the characteristics of these helper groups. Hence, unbound helpers (I, II) conduct reactive and (partially-)bound helpers (III, IV) proactive activities. From I to IV, the prosocial behaviour evolves from spontaneous to sustainable characteristics; the helping process grows in terms of long-term, continuous, plannable, involved, professional, and formal engagement; and the helper properties increase in awareness, commitment, experience, and professionalism. Table 4.3 presents terms that authors have used to describe different (overlapping) social media users in crisis from the organizational perspective.

Table 4.3   Organizational perspective on social media roles

Reference	Role	Description
(Olteanu et al., 2015)	Media organizations	Traditional or internet media have a large presence on Twitter, in many cases more than 30% of the tweets.
(Ehnis et al., 2014)	Commercial organizations	Publish rather small number of messages, e.g., containing humorous marketing messages.
(Olteanu et al., 2015)	Government	A relatively small fraction of tweets source from government officials and agencies, because they must verify information.

(continued)

Table 4.3 (continued)

Reference	Role	Description
(Reuter et al., 2011)	Community scouts	Proposed as amateur "first informers" to overcome the perceived unreliability of social media information for authorities.
(St. Denis et al., 2012)	Trusted digital volunteers	Used during the 2011 Shadow Lake fire in virtual teams to inform a Type I incident management team about social media activities.
(Bergstrand et al., 2013)	High-level formal organizational accounts	Used to formally inform the public about ongoing events in a unidirectional way of communication.
	Accounts for formal functions and roles	Distribute information about certain entities, retweet other civil security actors, and maintain a bidirectional communication.
	Formal personal accounts	Disseminate role-specific information and references of official work or actual topics.
	Affiliated personal accounts	Used for an expressive dissemination of information, personal opinions, reflections, and social conversation.
(Kircher, 2014)	Self-helpers and neighbourhood helpers	Directly affected by the event and work on overcoming it with or without organizational forces.
	Unbound, ad-hoc, and spontaneous helpers	Come from areas, which are not directly affected, are motivated by news and media, and work self-organized or in an organization.
	Preregistered helpers and first responders	Have registered themselves before the event and contribute with personal but no special disaster control qualifications.
	Honorary office and full-time helpers	Trained in specific tasks for disaster control.

4.4.3   Towards a Classification of Roles Related to Social Media Use

The literature review on roles and role typologies reveals two constant dimensions upon which a classification of roles seems suitable. Identified roles either (a) affiliate to the citizens' (public) or authorities' domain (Reuter et al., 2012) or (b) perform their activities in the real (Stallings & Quarantelli, 1985) or virtual realm (Reuter et al., 2013). Adopting the matrix style, four different role patterns may be distinguished considering the realm of the role's action (X-axis) and the affiliation of the role (Y-axis). The idea of the role typology matrix (Figure 4.2) is to provide an overview, to encourage systematic analysis and development of role patterns, and to promote the successful implementation of roles in public and organizational domains. However, there are further criteria to be considered in the classification of role patterns, for instance: In literature, roles are often defined according to the research interest or unit of analysis, e.g., collective sensemaking (Blum et al., 2014) or self-help activities (Reuter et al., 2013). Further criteria are types of activities (e.g., information processing, the status of the user (elite or ordinary), administrative autonomy (unbound or (partially-)bound), coordination (instructed or self-coordinated), or personal skills (none, personal, or disaster-specific skills)).

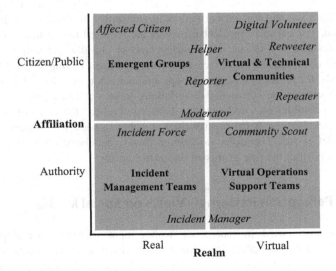

Figure 4.2   Role Typology Matrix

Emergent groups who include people "whose organization has not yet become institutionalized" (Stallings & Quarantelli, 1985) represent the public-real response. Typical roles of this pattern are affected citizens, self-helpers and neighbourhood helpers. Beyond, the public-virtual response is best characterized with *Virtual and Technical Communities (V&TCs)* who "provide disaster support with expertise in geographic information systems, database management, social media, and online campaigns" (van Gorp, 2014). Roles like celebrities, digital volunteers, readers, repeaters, and retweeters fit in this pattern. However, because emergent groups and V&TC's potentially (horizontally) collaborate in the course of an emergency (Kaufhold & Reuter, 2016), there are roles performing activities in both realms, e.g., different types of helpers, media, or reporters. Additionally, moderators even seek a direct collaboration with authorities.

Regarding the real-authority response, *Incident Management Teams* perform on-the-ground operations aiming "to save human lives, mitigate the effect of accidents, prevent damages, and restore the situation to the normal order" (Chrpa & Thórisson, 2013). In order to integrate the virtual-authority response, emergency services deploy *Virtual Operations Support Teams (VOST)* adapting "to the need for emergency management participation in social media channels during a crisis, while also having that activity support but not interfere with on-the-ground operations" (St. Denis et al., 2012). For this activity, official personnel or roles like *community scouts* or *trusted digital volunteers* are considered. Furthermore, to cover both the real and virtual realms in authorities, horizontal collaboration is required. For instance, incident managers are required to synthesize real and virtual information in the decision-making process. Besides that, different kinds of vertical collaboration take place during emergencies. During the 2013 European floods, for instance, emergent groups and incident teams worked together in order to overcome the emergency (Kaufhold & Reuter, 2016). However, because virtual communities on Facebook and Twitter influenced the work of emergent groups, a collaboration between authorities and citizens became necessary to coordinate relief efforts. Therefore, *moderators* closely collaborated with authorities to eventually fulfil the role of trusted digital volunteers.

4.5   Perception Patterns—Views on Social Media

Social media is used in emergency management. Different usage patterns (section 4.3) and role patterns (section 4.4) showed this. However, the question arises how these activities are perceived—both by the pubic (e.g., citizens) and by authorities. In the following, the results of larger surveys on authorities' and citizens' perception on social media are summarized.

4.5.1   Authorities' Perception of Social Media

There are a few (quantitative) studies on authorities' perception of social media (Reuter & Spielhofer, 2017), although most are from North America. First, San, Wardell III and Thorkildsen (2013) analysed the results of a survey of a comparative study conducted in 2012 by the American National Emergency Management Association (NEMA) among members of emergency services from all 50 Federal States of the US about social media use in emergency management. Second, Plotnick, Hiltz, Kushma and Tapia (2015) conducted a survey of 241 US emergency managers at county level in 2014 about use patterns, barriers, and improvement recommendations for the use of social media during emergencies. Third, the annual study of the International Association of Chiefs of Police (IACP) reports about law enforcement's use of social media on "the current state of practice and the issues agencies are facing in regard to social media" (International Association of Chiefs of Police, 2015). Last but not least, Reuter, Ludwig, Kaufhold and Spielhofer (2016) published their findings of the survey conducted with 761 emergency service staff across Europe in 2014 about current attitudes and influencing factors towards the use of social media in emergencies.

On the one hand, there is a positive attitude towards the use of social media in general (San et al., 2013), including private and organizational use (Reuter, Ludwig, Kaufhold, et al., 2016). The majority of US authorities already use social media as they value its suitability for information dissemination (San et al., 2013). This includes warnings, advice, and guidance on how to cope with or prevent emergencies or disasters, hints and advice on how to behave during an emergency, coordination of the help of volunteers, summary information after an emergency, and coordination of clean-up activities (Reuter, Ludwig, Kaufhold, et al., 2016). Currently, agencies' use of social media has already increased from 81% (77% Facebook, 37% Twitter, 16% YouTube) to 96% (94% Facebook, 71% Twitter, 40% YouTube) during the last five years (International Association of Chiefs of Police, 2010, 2015). Additionally, the number of social media policies has also increased from 35% to 78% (International Association of Chiefs of Police, 2010, 2015). Further increase of social media use is expected (74%), even more for organizations already using it (Reuter, Ludwig, Kaufhold, et al., 2016).

On the other hand, there are several restrictions within the use of social media: First, there is a huge gap between rhetoric and reality (Reuter, Ludwig, Kaufhold, et al., 2016). Despite the overall positive attitude towards social media for obtaining an overview of the situation and for raising situational awareness, in fact only a few agencies have often or sometimes used social media sites for this purpose (Reuter, Ludwig, Kaufhold, et al., 2016). As the predominant use of social media is more to share information (Reuter, Ludwig, Kaufhold, et al., 2016; San et al., 2013) than to receive messages (Reuter, Ludwig, Kaufhold, et al., 2016), so that only a modest use of social media can be observed, ground-breaking crowdsourcing and crisis-mapping activities are neglected (San et al., 2013). In addition, about 20% of the local and about 30% of the county agencies surveyed "had not identified a goal for social media operations" at all (San et al., 2013). Also, only about half of the observed emergency agencies at county level in the study of Plotnick et al. (2015) use social media at all.

Identified barriers for the use were despite a lack of dedicated personnel (San et al., 2013), doubts about its credibility and reliability (Reuter, Ludwig, Kaufhold, et al., 2016; San et al., 2013), concerns about privacy issues (Reuter, Ludwig, Kaufhold, et al., 2016), and still a lack of formal policies to guide the use of social media (Plotnick et al., 2015). But even for those emergency agencies who do have formal policies, prohibitions for the use of social media still exist (Plotnick et al., 2015). Reasons for limited success in the use of social media could be stated for limited reach and insufficient resources as data collection and analysis capabilities (San et al., 2013). For this, enabling conditions for the use of social media could be identified within the organizational culture and skills (Reuter, Ludwig, Kaufhold, et al., 2016) and the verification of citizen-generated content (San et al., 2013).

4.5.2   Citizens' Perception on Social Media

Very few quantitative studies have been conducted where citizens have been asked about their perception of using social media in emergencies. In particular, four are worth mentioning. These include a comparative study with over 1,000 participants conducted by the Canadian Red Cross (2012), which aimed to identify to what extent Canadian citizens use social media and mobile devices in crisis communication and what they expect from the emergency services both currently and in future. Secondly, the American Red Cross (2012) studied citizens' use of social media during emergencies, with 1,017 online and 1,018 telephone survey respondents. In the third study worth mentioning, Flizikowski et al. (2014) present a survey within Europe, conducted among citizens (317 respondents) and emergency services (130 respondents), which identified the possibilities and challenges of social media integration into crisis response management. Finally, Reuter and Spielhofer (2016) analysed the findings of a survey of 1,034 citizens across Europe conducted in 2015 to explore citizens' attitudes towards the use of social media for private purposes and in emergency situations.

In principal, the participants' attitude towards the use of social media was largely positive (Flizikowski et al., 2014). Benefits in using social media during emergencies can be seen in the reassurance for citizens, in providing situational information, and monitoring (Canadian Red Cross, 2012). Due to these benefits, social media is seen as a support for existing channels, however, it cannot replace them (Canadian Red Cross, 2012). Especially friends, family, news media (or reporters), and local emergency officials are seen as the most trusted sources (American Red Cross, 2012). Therefore, the Canadian Red Cross employs "trusted volunteers" to support official response via social media (Canadian Red Cross, 2012). In contrast to authorities' use, citizens use social media rather to search (43%) than to share information (27%) (Reuter & Spielhofer, 2017). Most likely, users seek information about weather, traffic, damage caused, and information on how other people were coping (American Red Cross, 2012). If they do provide information, users not only share weather information, safety reassurances, and their feelings about the emergency but also their location and eyewitness information (American Red Cross, 2012).

4.5.3 Towards Comprehensive Perception Patterns

Even if the attitude towards social media is positive, only 12% of the general public, but still 22% of high school graduates, conducted in the study of the American Red Cross (2012), have used social media to share or obtain information during emergencies and disasters or in severe weather conditions. Because of this, challenges in the use of social media during emergencies were identified within all studies. Both, citizens and emergency services identify the same challenges (Flizikowski et al., 2014). Possible barriers for the use of social media are especially credibility doubts of citizen-generated content (Canadian Red Cross, 2012; Flizikowski et al., 2014; Reuter & Spielhofer, 2017), a lack of knowledge and personnel issues (Canadian Red Cross, 2012; Flizikowski et al., 2014), lacking uniform terms of use (Flizikowski et al., 2014), and difficult accessibility for older generations (Flizikowski et al., 2014). Regarding the trustworthiness, unknown people in the general vicinity of the emergency are the least trusted (American Red Cross, 2012) so that emergency services are expected to monitor social media

(Reuter & Spielhofer, 2017). Beyond, unfortunately, there is only very little awareness of social media Safety Services and Emergency Apps (Reuter & Spielhofer, 2017).

4.6   The Past and the Future: Discussion and Conclusion

4.6.1   Achievements of 15 Years of Research

Social media become more and more mature. However, not only social media in general—but also the use of social media in emergencies. 15 years ago, the first documented case of disaster support with social media was found and for about 10 years, social media has been used in a more and more intensive way. Research has tried to examine different cases, different users, different methods, practices, and tools; trying to support all actors involved in crises, disasters, and emergencies of different type and size. Summarized under the term crisis informatics (Hagar, 2007; Palen et al., 2009) much has been achieved so far—and much is still to do. This article contributed to the development in providing a compilation about existing cases of social media use in emergencies—knowing that this list cannot be complete (section 4.2). Furthermore, the article analyses the state of the art regarding different use patterns (section 4.3). Based on this, different role patterns that have been identified across various studies are elaborated and synthesized (section 4.4). Additionally, perception patterns of both authorities and citizens are elaborated (section 4.5). Finally, in this chapter, we discuss the past and especially the future.

During nearly every larger emergency of the last 10 years and during many of the last 15 years, we found studies highlighting the use of social media. While many had a focus on the USA initially, studies from other continents are catching up allowing more comparative and systematic analysis across different circumstances and types of emergencies. Still, most of the studies focus on Twitter; we suggest this is based on the ease of data selection there (section 4.2). The analysis focused on different *usage patterns*, including the communication among citizen (C2C), with concepts of self-coordination and help, emergent groups, and (digital) volunteers; the communication from authorities to citizens (A2C), including concepts of crisis communication; from citizens to authorities (C2A), including concepts like big data- or social media analysis, crowdsourcing, and crowd tasking; and among authorities (A2A), including inter-organizational social networks (section 4.3). Within the two basic affiliations of authorities and citizen as well as the real and virtual realms, various *role patterns* become apparent. Affected citizens and helpers form emergent groups to overcome the emergency on-site; digital volunteers self-organize in virtual and technical communities (V&TC) to remotely support amateur and professional response; emergency services deploy incident management teams (IMT) for professional on-site emergency response; and trusted digital volunteers are organized in Virtual Operations Support Teams (VOST) to assist professional response in the virtual realm (section 4.4). Also, social media is perceived in different ways by authorities and citizens while the challenge is to match different expectations and to address current barriers. Both groups see the same challenges, like a lack of trust, lack of knowledge—however, citizens expect authorities to monitor social media (section 4.5).

4.6.2   Future Practice and Research Potentials

For future practice and research, many issues are still open: *Self-coordination and help (C2C)* have been proven to be of high importance, however, chaos is a characteristic pattern detected. Here, the question arises how this can be addressed. The automatic cross-media suggesting of relevant posts according to crises dynamics (Kaufhold & Reuter, 2016) of interest or the matching of needs and offers (Purohit, Hampton, et al., 2014) might help to structure communication, while flexibility is required as well. The granularity of citizen activities—are single citizens supporting or rather groups of citizens, like clubs—is also important to determine appropriate organization and work practices. Furthermore, currently many different tools are used in an opportunistic way. The visibility of different practices that have shown to work seems important to facilitate appropriation amongst citizens and, in the long term, to improve disaster preparedness and overcoming. In *crisis communication (A2C)* it is still a challenge to apply "perfect" crisis communication. According to some studies, many citizens expect responses to messages in social media from authorities within one hour (Reuter & Spielhofer, 2017). However, not all emergency services might be able to act in that speed, sometimes caused by a lack of personal, or a lack of skills, as some studies suggest (section 4.5). Press officers must adapt to a new role including more dynamics compared to pre-social media times. The verification and careful creation of own posts is necessary which conflicts with the need of quick response. Therefore, types of communication, like instrumental or public-including expressive communication approaches (Denef et al., 2013) should be further elaborated, to also suggest smaller authorities ways of crisis communication.

For analysing and *integrating citizen-generated content (C2A)* from social media, research applied various algorithmic approaches (M. Imran et al., 2015).

They, on the one hand, intend to detect or predict critical events and to transform the high volume of big and noisy data, which cannot be processed by emergency mangers in a limited amount of time before or during large-scale emergencies, into a low volume of rich and thick content (Moi et al., 2015). On the other hand, algorithms aim to detect underlying patterns (e.g., mood or geospatial correlations) using statistical approaches or visual analytics (Brynielsson et al., 2014; Fuchs et al., 2013). Social bots and fake news challenge these attempts. However, not only large-scale emergencies but also small incidents with suitable algorithms and different granularities and thresholds must be considered. While emergency managers are sometimes sceptical about the quality of citizen-generated content and social media (Hughes & Tapia, 2015; Reuter, Ludwig, Kaufhold, et al., 2016), it must be ensured equally that they trust in the quality of algorithms as an additional filtering layer, e.g., by providing a certain degree of customizability and transparency (white-box approach). Furthermore, research examined crowdsensing approaches to sharpen the authorities' awareness about citizens' activities (Ludwig, Reuter, Siebigteroth, et al., 2015; Sakaki et al., 2010). Concerning *inter- and intra-organizational crisis management (A2A)*, social media can be used for the coordination of crisis communication and more informal networking among authorities and employees. Here, structures of social media could support the development of collaborative ICT or inform encapsulated social networks. The latter have the benefit of trust because the usage group is limited and controlled.

The systematization of *role patterns* and role properties potentially supports the interaction among authorities and citizens. For instance, a semi-automatic identification of role patterns (Reuter et al., 2013) and their display in social media may improve role awareness, the self-finding process, and guidance for citizens to take a role. Due to the chaotic organization in such emergencies (Valecha et al., 2013), well-defined role properties could furthermore improve capacity planning for authorities and among citizens, for instance, to crowdsource tasks to the situationally correct audience. Considering these opportunities, the role typology matrix may be used to systematically optimize collaboration and communication structures among different crisis actors in the real and virtual realm, e.g., to improve the communication between first responders and digital volunteers, or the incident manager's awareness of the activities and scope of VOST. Then, from an IT perspective, role patterns should inform the tailoring of ICT to support role-specific activities. This, however, is problematic because users tend to use general software they are familiar with, like Facebook, during emergencies and not always specific and maybe better tools. Here, applications that are embedded into the social media ecosystems, e.g., Facebook apps, may allow a smooth appropriation of emergency-specific tools (Reuter, Ludwig, Kaufhold, et al., 2015).

The *perception* of social media is both a result and a starting point of the aforementioned aspects. It depends on own experiences and media coverage. Online rubbernecking is widely reported during crises (Bruns, 2014). However, while looking at the published studies (section 4.5), it seems that there is a gap between reported cases in academia, looking more at potentials, but also at risks, and the coverage by mass media, where negative aspects are more present. They include rumour propagation, dissemination of false or misleading information, ethical dilemmas (Alexander, 2014), and propaganda or social bots (Reuter, Pätsch, & Runft, 2017b). Risks and bad sides of media usage cannot be controlled comprehensively, but research may try both to foster good sides and to guide bad aspects to the right direction. In addition to that, some studies about the perception of social media exist, however they are not representative. This is also important as long as social media is used in different ways among countries. Trust is the main issue, so future work might focus on the key enablers, like positive examples of social media use. The feeling to be part of a movement that productively works together to overcome crises and emergencies, is the intended result of this.

In sum, *crisis informatics* has achieved a lot. It is only sometimes named alike, but the use of social media in crisis management has been established as an important research area. This article could—as a limitation—just look at a part of it. The article tried to summarize some selected aspects to give a current overview and to suggest at least some aspects for the upcoming years.

Survey on Media Perception and Use During Infrastructure Failure

5

Security and conflict research mainly focuses on issues of vulnerability, reinforcement of resilience and preservation or recovery of critical infrastructures. Additionally, the importance of social media and crisis apps is being increasingly acknowledged. To what degree is German civil society effectively prepared for the case a crisis occurs? Which kinds of ICT are being used in everyday life and in a potential infrastructure failure? Our contribution presents the results of a representative survey with 1,024 participants in Germany, which demonstrate that threat awareness, preparation, effective crisis management and distribution of crisis apps are still relatively low in Germany, while traditional communication and information channels as well as informal information networks are preferred. The results additionally point out the significant potential for support which ICT have for increasing risk awareness, facilitating information transfer and improving communication between civil society, critical infrastructure operators and authorities.

This chapter has been published as the conference article "Potentiale von IKT beim Ausfall kritischer Infrastrukturen: Erwartungen, Informationsgewinnung und Mediennutzung der Zivilbevölkerung in Deutschland" at the International Conference on Wirtschaftsinformatik (WI) by Marc-André Kaufhold, Margarita Grinko, Christian Reuter, Marén Schorch, Amanda Langer, Sascha Skudelny, and Matthias Hollick (Kaufhold, Grinko, et al., 2019).

M.-A. Kaufhold, *Information Refinement Technologies for Crisis Informatics*,
https://doi.org/10.1007/978-3-658-33341-6_5

5.1   Introduction

Even though extreme natural events like floods and storms occur particularly often in Europe (Guha-Sapir et al., 2004), in Germany, collective crises and long-term infrastructure failures of power and telecommunication, for instance, are rather uncommon (Bundesnetzagentur und Bundeskartellamt, 2016). A low threat awareness in civil society, however, increases the risk as the sense for the own potential vulnerability and a corresponding prevention of failures seem to be missing (Holenstein & Küng, 2008). In a crisis situation, the concerned parties refer to different sources for retrieving and offering information, preserving civil continuity and cooperating, also by using ICT (Austin et al., 2012; Hughes et al., 2008). While the application of ICT in crises has become a separate research field—crisis informatics—especially since the attacks on September 11th, 2001 (Hughes et al., 2008), there is a lack of representative studies on information expectations, preparations for and behaviour in crises—especially in Central Europe. Our study attempts to address this issue and answer the following four research questions:

RQ1: How do German citizens estimate their own behaviour as well as their degree of preparedness and knowledge in an infrastructure failure?

RQ2: What kind of media channels do they use for communicating and gathering information in crisis situations and why?

RQ3: Which information types and channels do they expect from critical infrastructure operators?

RQ4: How can ICT increase awareness, instruct behaviour and support communication as well as civil continuity in such situations?

In our contribution, we first analyse the current state of research on the socio-technical character of critical infrastructures and crisis communication (section 5.2). Subsequently, the methodology (section 5.3) and results of a representative survey in Germany (section 5.4) will be presented which aim to answer the research questions above. In the following discussion and conclusion (section 5.5), the results will be reflected regarding the four research questions and we will draw implications for further research as well as ICT design.

The results indicate a low awareness of and preparation for the risk of a critical infrastructure failure. Information gathering and communication via different media channels create a high demand for specific, timely and reliable information and instruction. Also, this contribution emphasises the potential of ICT usage in CI failures and design requirements considering users' needs and characteristics.

5.2   State of the Art

5.2.1   Characteristics of Socio-technical Critical Infrastructures

An infrastructure can be defined as an underlying framework which allows a group, organisation or society to function and which consists of material, institutional and personal sub-aspects (Jochimsen, 1966). In our case, we are considering the material category, especially the supply of resources like energy and (tele-) communication, which are also called critical infrastructures (CI). The Council of the European Union defines CI as "an asset, system or part thereof [...] which is essential for the maintenance of vital societal functions, health, safety, security, economic or social well-being of people, and the disruption or destruction of which would have a significant impact" (The Council of the European Union, 2008), namely as a disruption of civil and operational continuity (Boin & McConnell, 2007).

Even though infrastructure failures are not fully avoidable, power supply in Germany is relatively stable; power outages are rare and quite short (Bundesnetzagentur und Bundeskartellamt, 2016). As a consequence, many German citizens have never experienced such an emergency and are accordingly unprepared, e.g., concerning risks and security measures, as well as communication and information channels. Citizens expect authorities and organisations with safety tasks to take care of potential crises (Reuter, Kaufhold, Spielhofer, et al., 2017). Even in cases citizens have experienced a crisis, the probability for them to take precautionary measures is rather low, as they consider it unlikely to happen again.

The preparation for a crisis and the stability of a CI can only lower, but not eliminate the risk of failures, meaning that reactions of citizens and authorities are crucial for emergency management, system continuity and recovery. This "social resilience" contains not only citizens' preparations for a potential crisis, but also the enabling of an effective training, framework agreements and communication between experts, authorities, CI operators, media, emergency forces and citizens (Boin & McConnell, 2007). Here, a "human" or "social infrastructure" (O'Sullivan et al., 2013) emerges, which means organisations, teams or groups of people who act together, support each other, communicate and exchange information. Nowadays, these unions are increasingly established via social media and contribute to maintaining and efficiently recovering certain infrastructure functions as well as public security (Semaan & Mark, 2011).

5.2.2   Crisis Communication and Requirements for ICT

Crisis communication commonly refers to the one-to-many communication and public image of an organisation (Seeger, 2006)—in our case, the CI operator. Combined with official information sources, informal media have been used increasingly by citizens and emergency managers alike—especially when central information is lacking—to gather and exchange current information, keep track of the events, organise help, stay in contact with friends and family and offer active and moral support (Austin et al., 2012). Apart from social media, which are being used more and more and analysed in crises ever since their emergence, multi-channel systems support the dissemination of warnings, e.g., via text message, email and RSS (Klafft, 2013). Furthermore, a current study found that 16% of citizens in Germany are using crisis apps (Reuter, Kaufhold, Leopold, et al., 2017b).

However, with the growing number of media channels and a faster dissemination of contents, there is the threat of information overload and fake news, unreliable sources and incomplete information (Austin et al., 2012; Huang et al., 2015). A communication channel's usability and credibility have a great influence on its usage (Austin et al., 2012): In earlier studies, concerned parties have primarily used traditional media such as TV, radio and newspapers, as these are commonly regarded as the most trustworthy sources (Austin et al., 2012; Petersen et al., 2017). Also, direct communication via personal conversations, phone calls and text messages are regarded as important (Austin et al., 2012).

To initiate the desired reaction during an infrastructure failure, CI operators have to keep people up to date on current developments, establish a dialogue and appear honest, accessible and credible. Furthermore, their messages should be consistent with trustworthy references (Seeger, 2006). The information amount, medium and source significantly influence the perception of a crisis and the organisation behind it as well as citizens' response (Austin et al., 2012; Huang et al., 2015). During a crisis, status updates, particularly on the duration of a failure, should occur frequently to avoid emotional stress, negative reactions, misunderstandings and unpredictable behaviour—especially in case of a discrepancy between the sources (Reynolds & W Seeger, 2005). A survey in four European countries has shown that the more participants use social media, the higher their expectation of CI operators to use these platforms to share information concerning the ongoing infrastructure failure (Petersen et al., 2017). The study could determine a significant difference in age regarding the usage of social, but not traditional media.

5.2.3   Research Gap

When CI break down, research mostly focuses on the complex effects and failure management, but also on the continuity and resilience of the affected population. While there are studies on how emergency institutions such as police and fire-fighters communicate and react to infrastructure failures, the maintenance of civil communication and continuity has not been widely researched (Petersen et al., 2017). For example, it is not sufficiently examined to which degree the public in these areas is informed about their local infrastructure and crisis measures—particularly alternative infrastructures. Also, what kind of information the public expects from the CI operator and via which channels it is communicated requires a more extensive analysis. Furthermore, many relevant studies date back more than a decade, which increases the demand for research in the context of current technological developments and corresponding user attitudes. One study which has aimed at answering similar questions does not represent the European population demographically (Petersen et al., 2017); we would like to complement this with a representative study by surveying a wide spectrum of participants from different age and income groups, educational backgrounds and German federal states.

5.3   Methodology

The data for this paper originates from a representative online survey which we conducted in July 2017 in Germany using the ISO-certified panel provider Gap-Fish (Berlin). To answer the research questions, we used three closed and one open question (see appendix) which deal with the behaviour and media usage during infrastructure failures. Questions on power outages were based on Reuter (Reuter, 2014a) and those on expectations towards infrastructure providers were partly taken from Petersen (Petersen et al., 2017). The participants were asked about their usage of different media channels in crisis situations (Q1) and which kind of communication they expect of CI operators in case of a failure (Q2). Furthermore, they had to indicate their knowledge concerning measures taken by authorities should the emergency call fail (Q3). The last, open question gave the opportunity to describe the own behaviour in a crisis (Q4). Since infrastructure failures are uncommon, the questions could also be answered hypothetically—regardless of whether participants had experienced such a situation or not.

5.3.1 Study Participants

Our sample (N = 1.024) has been adapted to the distribution of age, region, education and income with respect to the general German population (Bundeszentrale für politische Bildung (bpb), 2016; Statista, 2016; Statistisches Bundesamt, 2016). According to this statistic, our sample consisted of 49.5% female and 50.5% male respondents between 18 and 64 years, of whom one half was 45 years and older (48%). We interrogated participants from all federal states, while most participants originated from North Rhine-Westphalia (22%) and Bavaria (16%). Only 1% of all participants had not graduated from school; 15% had obtained a higher degree. The majority earned between 1.500€ and 3.500€ (gross) a month. Also, we gathered data on technology usage habits via a five-point scale (from "hourly" to "never"). For the smartphone, almost half of all participants indicated to use it daily (49%). A similar result could be found for social media, e.g., Facebook (46%), messengers (43%) and YouTube (28%). In total, 33% stated that they used smartphone and messengers daily.

5.3.2 Data Analysis

For our analysis, we first eliminated incomplete datasets and reduced the sample size from N = 1.069 to N = 1.024. Also, we clustered demographic variables like age and income in categories to allow an easier comparison. Subsequently, we calculated the absolute frequency and percentage of answers to the closed survey questions in Microsoft Excel. For statistical analysis of data, we used the software package IBM SPSS Statistics 25. Non-parametric tests were chosen based on ordinal data. Chi-squared tests served to examine significant differences between the demographic variables as well as media usage habits. Correlations between variables were analysed using Spearman's rho. For the qualitative analysis of the open question, we used the method of Open Coding (Glaser & Strauss, 1967). The resulting insights are a valuable addition to the results from the quantitative analysis. In the following, we will present the results of our analyses with regard to the research questions.

5.4 Results

5.4.1 Media Usage During Crises

One of our questions focused on which information sources the respondents had used in crisis situations they had experienced, and which had been especially useful to them (see Table 5.1). For one third of the participants (34%), the TV had been a very helpful source during a crisis, followed by the radio (30%), social media (20%), face-to-face communication (17%) and emergency services (15%). In total, most individuals had used the TV (82%) and personal conversations with friends, family and neighbours (74%). Further 74% had listened to the radio and 61% made phone calls. On the fifth place we can find social media, which had been used by 55% of the respondents, closely followed by newspapers and magazines (54%). Other media had been referred to by less than half of our participants. Crisis apps were the least used media channel (25%).

Table 5.1 Results for Q1: Which of the following options have you used in an acute crisis situation (when you were concerned or a helper) to obtain information about the events, and which was especially useful for you?

	Yes, very helpful	Yes, I used it	No
TV	34%	48%	18%
Face-to-face conversations	17%	57%	26%
Radio	30%	44%	27%
Phone conversations	14%	47%	38%
Social media	20%	35%	46%
Newspapers/magazines	12%	42%	46%
Other internet offers	12%	31%	58%
Emergency services	15%	22%	63%
Local information	9%	25%	66%
Crisis app	10%	15%	74%

All sources significantly correlated with each other ($p < .01$). Also concerning demographic and socio-technical factors, we could find significant influences on the answers. Here, we clustered the answer options in three groups: traditional media (TV, radio and newspapers/magazines), digital media (social media, crisis apps and websites) as well as local, personal communication (on-site information,

face-to-face conversations, phone calls and emergency services, see Table 5.2).
Gender and region did not have any significant influence on the usage of information channels, however, age and education did: in particular, older participants
used the radio to a higher degree (r = .104), but at the same time displayed a
lower tendency to refer to social media (r = .335), websites (r = .205) or personal conversations with friends and family (r = .181). The factors of young age
(r = .175), higher education (r = .053) and lower income (r = −.0.58) led to a
greater tendency to rely on local sources. Furthermore, the usage of smartphones
had a positive effect on using local information (r = .129) and digital media (r
= .296). Frequent social media users more often relied on local (r = .283) and
traditional media (r = .115). The frequency of posting on social media influenced
the usage of traditional (r = .152), digital (r = .365) as well as local media (r =
.260).

Table 5.2 Significant results of chi-squared tests on media usage in crisis situations (*p < .05,
**p < .01, N = 1,024)

	Traditional media	Digital media	Local communication
Age		$\chi^2(30,N) = 121.64^{**}$	$\chi^2(40,N) = 70.32^{**}$
Education (ED)		$\chi^2(30,N) = 53.16^{**}$	$\chi^2(40,N) = 64.81^{**}$
Income (IC)			$\chi^2(24,N) = 37.62^*$
Smartphone usage (SP)		$\chi^2(24,N) = 111.75^{**}$	$\chi^2(32,N) = 49.00^*$
Social media usage (SM)	$\chi^2(180,N) = 274.15^{**}$		$\chi^2(240,N) = 424.12^{**}$
Posting behaviour (PB)	$\chi^2(24,N) = 63.69^{**}$	$\chi^2(24,N) = 171.76^{**}$	$\chi^2(32,N) = 126.68^{**}$

5.4.2   Expectations Towards Critical Infrastructure Operators in Crises

Further, we wanted to know which information the participants expect from CI
operators in case of an infrastructure failure (see Table 5.3). Most respondents
named messages in traditional media (72%) and the CI operator's website (70%).

Apart from this, the majority expected a text message (60%), a social media post (59%), a phone call (58%) or a personal address on-site (57%). Two-way communication was regarded as neutral or unnecessary by 60%. The central expectations of the content of information included instructions (83%), the reason for the CI breakdown (80%) as well as the expected duration of the failure (78%). For 70%, the source of the information was relevant.

All answers significantly correlated between each other as well as with the previous questions (p < .0001). The usage of social media during crises (r = .388, p < .0001) significantly influenced the expectation of a message on the CI operator's social channels. The same result could be found concerning traditional media, e.g. newspapers (r = .070, p < .05), TV (r = .105, p < .0001) and radio (r = .159, p < .002). Furthermore, we could determine that a younger age increases the expectation of a phone call (r = −.098) and a message in social media (r = −.114), while the opposite is true for a message in traditional media (r = .160). The older the participants, the more they expected the CI operator to publish instructions (r = .088), indicate the information source (r = .089) and provide information on the duration of the failure (r = .102). We also found influences of income and education status: there was specifically a positive influence on messages in traditional media (r = .099 und r = .085) and websites (r = .116 und r = 0.94). A higher income additionally increased the information expectation especially on the breakdown duration (r = .115). All significant influences of the chi-squared tests can be found in Table 5.4.

5.4.3 Measures During an Emergency Call Failure

Our third question served to find out which kind of knowledge the participants had regarding the measures taken during a CI breakdown (see Table 5.5). The majority were aware of where the next police station (70%) and other administration buildings (68%) are and knew the position of the nearest ambulance station (55%) and firehouse (55%). More than half were informed about the fact that media report alternative measures (56%) and the police enforce their activity in the affected area (51%). In contrast, the location of the nearest building of the Federal Agency for Technical Relief (THW, German abbreviation for "Technisches Hilfswerk") was only known to 27% of the respondents, and about a quarter (23%) knew that a fire vehicle is placed on a central square in areas without a fire station. Only 19% were aware of the fact that in a crisis situation, the fire stations are occupied by authorities and organisation with safety tasks (BOS, German abbreviation for "Behörden und Organisationen mit Sicherheitsaufgaben").

Table 5.3 Results for Q2: What do you expect on part of the operator (e.g. Telekom, energy suppliers) in case of infrastructure failures (e.g. electricity, gas or telecommunications)?

	Strongly agree	Agree	Neutral	Disagree	Strongly disagree
Instructions	45%	38%	15%	2%	1%
Information on reason	45%	35%	17%	3%	1%
Information on duration	45%	33%	17%	3%	1%
Message in trad. media	35%	37%	21%	5%	2%
Message on website	35%	35%	21%	6%	3%
Information source	32%	38%	26%	3%	1%
Text message	25%	35%	29%	8%	4%
Social media post	25%	34%	25%	9%	6%
Phone call	22%	36%	31%	7%	3%
Address on-site	22%	35%	33%	8%	2%
Two-way communication	13%	28%	51%	6%	3%

The answers for all measures significantly correlated with each other and with the usage of all communication channels in crises ($p < .0001$). The reliance on phone calls, personal contacts and emergency services displayed the highest correlation ($r = .289$, $r = .282$ and $r = .273$). To analyse the potential influence of demographic factors as well as social media usage on the answers, we clustered them in one value and carried out chi-squared tests based on these values. Here, we observed a significant influence of education ($\chi^2(180, 1{,}024) = 225.51$; $p < .05$; $r = .006$) and income ($\chi^2(108, 1{,}024) = 1407.23$; $p < .0001$; $r = .047$), even though we could not determine a distinct trend due to weak Spearman correlations. The usage of smartphones ($\chi^2(144, 1{,}024) = 178.25$; $p < .05$; $r = .037$) and social media ($\chi^2(1080, 1{,}024) = 225.51$; $p < .05$; $r = .202$) as well as posting frequency in social networks ($\chi^2(144, 1{,}024) = 224.53$; $p < .0001$; $r = .141$), however, did prove to be significantly correlated to the knowledge.

Around a third of all participants ($n = 344$, 33,59%) had a low average knowledge of the measures (neutral or less). Compared to the total sample, the

Table 5.4 Significant chi-squared results on communication expectations in crises (*p<.05, **p<.01, N = 1,024), see table 5.2 for explanations of abbreviations

	Phone call	Text message	Social media post	Website
Age	$\chi^2(20,N) =$ 43.13**			
State		$\chi^2(56,N) =$ 77.95*		
ED	$\chi^2(20,N) =$ 41.95**			
IC				$\chi^2(12,N) =$ 24.96*
SP		$\chi^2(16,N) =$ 35.01**	$\chi^2(16,N) =$ 44.65**	
SM	$\chi^2(120,N) =$ 175.96**		$\chi^2(1080,N) =$ 1407.23**	
PB	$\chi^2(16,N) =$ 31.15*	$\chi^2(16,N) =$ 28.54*	$\chi^2(16, N) =$ 144.91**	$\chi^2(16,N) =$ 35.18**

	On-site	Trad. media	Two-way	Duration	Reason
Age		$\chi^2(20,N) =$ 52.73**		$\chi^2(20,N) =$ 41.20**	
ED	$\chi^2(20,N) =$ 43.07**	$\chi^2(20,N) =$ 46.52**			
IC		$\chi^2(12,N) =$ 21.48*			
SM			$\chi^2(120,N) =$ 159.77**		
PB	$\chi^2(16,N) =$ 34.92**		$\chi^2(16,N) =$ 32.63**		$\chi^2(16,N) =$ 30.48*

distribution demographic characteristics is very similar for this group, while a t test for independent samples showed some significant differences: This group was older (t(1022) = 2.47, p = .014), indicated a lower usage of social media (t(1022) = 5.09, p<.000) and posted less frequently in them (t(1022) = 2.39, p = .017).

Table 5.5 Results for Q3: In the case of a 112 emergency call failure, various measures are taken to enable the population to report an emergency. To what extent do you know the following measures or locations? (THW = "Technisches Hilfswerk", Federal Agency for Technical Relief; BOS = "Behörden und Organisationen mit Sicherheitsaufgaben", authorities and organisations with safety tasks)

	Strongly agree	Agree	Neutral	Disagree	Strongly disagree
Police station	29%	41%	14%	10%	7%
Administration	31%	37%	16%	10%	7%
Alternatives in media	13%	43%	22%	14%	8%
Firehouse	24%	31%	18%	16%	10%
Ambulance station	22%	33%	21%	15%	10%
Enforced police presence	13%	38%	20%	18%	12%
THW	10%	17%	21%	34%	17%
Fire vehicles in areas without fire station	6%	17%	17%	35%	25%
Fire stations occupied by BOS	7%	12%	15%	36%	30%

5.4.4   Anticipated Behaviour in Crisis Situations

In order to find solutions for a better ICT support and effective communication strategies for CI operators, we asked our participants to describe in detail their behaviour in crisis situations in an open question Q4: describe in as much detail as possible how you would proceed in an extraordinary situation (crisis, catastrophe)? Which activities would you carry out? Via Open Coding (Glaser & Strauss, 1967), we attached each answer to one or more codes describing the statement, which were iteratively refined. By the end of the coding, we derived 41 individual codes which we sorted into seven categories, indicated in italics in the following. The total numbers of statements in the categories are as follows: *Communication*: 692; *information*: 646; *personal safety*: 490; *no answer*: 209; *material safety*: 145; *problems*: 140; *passive behaviour*: 112. Regarding *personal safety*, about a fifth (n = 178) indicated that they would seek shelter or stay at home if it was safe. 10% of respondents also mentioned family members of whom they would take care. Apart from this, 158 participants would also offer help to others outside their family, like neighbours, and look out for people in need or danger. Also, warning others—be it personally, via phone call, message or in social networks—was a common reaction. Concerning *material safety*, the supply with food and other resources played an important role for 78 individuals. In total, 13 of respondents were aware of the necessity of preparation. The second aspect of this category consisted of personal belongings, such as documents and keys (n = 67).

In the *communication* category, the focus lay on warning close ones and forwarding information. Second was contacting other people in general (n = 63). Family (n = 176), friends (n = 81) and neighbours (n = 65) were the most frequently indicated recipients, especially to exchange each other's safety status. Asking for advice, offering cooperation and exchanging information (n = 47) were further reasons to establish contact. In the *information* category, we found that more than half of the participants (n = 447) would gather information on the respective crisis. Of those participants who mentioned a specific media channel, 189 people indicated to use traditional, while slightly less named digital media (n = 136). Additionally, 69 participants said they would ask passers-by, helpers on-site or listen to loudspeaker announcements. Others would directly address emergency institutions, government representatives or CI operators (n = 44). Meanwhile, 83 participants explicitly stated that they would wait for and follow instructions communicated by the media and emergency services.

We comprised codes not containing any actions the participants would carry out in case of a crisis in the last three categories. Empty answers and statements like "I don't know" were assigned to the category *no answer*. As *passive behaviour*, we counted doing nothing (n = 32) and staying calm (n = 77). Moreover, we encountered *problems* our respondents experienced with answering the open question. For instance, they indicated their behaviour would depend on the type of crisis (n = 106). 22 participants explicitly stated they had never before experienced a crisis and therefore could not give a qualified answer. Similarly, nine individuals pointed out that they would react differently in an actual crisis situation and could not predict their reactions.

5.5 Discussion and Conclusion

The breakdown of critical infrastructures affects many resource and communication channels, hinders the transfer of necessary information and causes economic damage (Pipek & Wulf, 2009). Even though research has examined the characteristics, resilience as well as the cooperative and technical aspects of CI (Huang et al., 2015), citizens' attitude towards CI failures and ICT usage for compensation in Central Europe has not been extensively analysed. Our study therefore contributes with specific aspects of awareness, behaviour and communication of the German population in CI breakdowns. Although our sample is representative only based on demographic aspects, we can derive current results in an underrepresented field which we have analysed under different points of view and which can help draw a big picture on media usage in CI failures in Germany, as well as instruct further research.

Awareness and behaviour in CI breakdowns (RQ1). Earlier studies have already shown that the majority of citizens have never experienced a severe CI failure (Holenstein & Küng, 2008; Reuter, Backfried, et al., 2018). Around 10% of our participants did not have an answer to what they would do in an emergency, while others believed the measures to be dependent on the type of crisis and therefore not predictable. However, we can observe an increasing threat awareness, particularly due to terrorist attacks. When asked about their crisis behaviour, most participants stated to rely on known media channels to fill information gaps and receive recommendations for action. In case of a crisis, the own safety and that of close ones, the exchange of information and mutual help were of highest priority. Because in Germany, effective crisis preparation and management lies in the hands of authorities and emergency services, their primary task is to sensitise the population for potential threats, guide their behaviour and avoid a greater damage through lack of preparation and risk awareness (Holenstein & Küng, 2008). On the other hand, many of our respondents knew about buildings and measures they can use in case of a CI breakdown. This even correlated with the usage of social media and other behaviour like contacting emergency services and communication with others via phone. We can see that media can be effectively used for crisis preparation.

Communication and information via media channels (RQ2). Compared to social media and websites, traditional media like TV and radio were the most frequently mentioned information sources (Austin et al., 2012). This is probably due to their perception as official, up to date and trustworthy, aspects which were important to some participants. In accordance with a qualitative study (Huang et al., 2015), many respondents tended to trust witnesses on-site and 70% practiced personal communication. However, the latter was rarely named as very helpful source—other than social media. In comparison to a 2017 study, the participants even more often resorted to social media (55%), while the usage of crisis apps increased from 16% to 25% (Reuter, Kaufhold, Spielhofer, et al., 2017). Although they are not yet widespread, 40% of their users perceived them as very helpful. Therefore, CI operators could use such apps to inform about institutions, measures and potential risks as well as point out how to prepare and behave. Influences of demography and ICT usage in crisis situations found in prior works (Petersen et al., 2017) could be replicated for this and the following research question. However, this was not the strongest influence: The more participants used specific ICT in their daily life as well as in crises, the more they perceived their benefit in crisis situations. ICT primarily served to gather information, contact friends and family, as well as exchange information and safety status.

Expectations for CI operators (RQ3). With regard to communication with CI operators, our participants appreciated almost every kind of media and over half of them expected the CI operators to use social media in crisis situations. Providing information, especially recommendations for action, as well as the duration of and reason for the crisis, was perceived as more important than establishing a two-way communication (Reuter, Kaufhold, Spielhofer, et al., 2017).

Implications for design and application of ICT (RQ4). Our results show that there is a high demand for information on and sensibilisation for crises. Traditional and social media are considered helpful and reliable channels for disseminating this information. However, in specific infrastructure failures (e.g., power), the battery life of smartphones allows a limited usage of crisis apps, which, among others, contain check lists and behaviour instructions before, during and after different types of emergencies (Reuter, Kaufhold, Leopold, et al., 2017b). Therefore, a diffusion of suitable ICT can increase social resilience and awareness for behaviour measures (Boin & McConnell, 2007). Additionally, this could increase the general understanding of infrastructures as an important basis for many aspects of cooperation (Pipek & Wulf, 2009; Semaan & Mark, 2011). Since demographic factors and technology usage had an influence on the answers, communication channels and information should be adapted to their respective target group (O'Sullivan et al., 2013). At the same time, fast, accessible, correct, complete and proven information should be provided to establish a more effective crisis management and cooperation between CI operators, authorities and citizens via ICT.

Concluding remarks. To a high extent, our results support the statements of earlier studies. Compared to established research, however, we provide current insights into the poorly examined area of attitude towards, expectations and usage of media in failures of stable CI based on a representative sample from the whole of Germany. Under consideration of diverse factors such as daily technology usage, our research offers a basis for comparison, among others, with further target audiences, types of crises, spheres of influence and countries. In case of the breakdown of certain channels, the relevance of others can increase or temporarily replace them. Meanwhile, it is important to cover as many channels as possible and adapt instructions depending on the kind of crisis. Here, it became evident that ICT can be effectively used for crisis communication in the following areas:

- Information on potential threats (risk communication) and improvement of civil resilience (recommendations for action for personal and material safety, emergency stockage, etc.)
- During a CI breakdown: Communication support between actors (authorities, CI operator, civil society) on multiple channels
- During and after the infrastructure failure: Supply of current, local, specific, trustworthy and target group-oriented information by authorities and CI operator.

The study has some limitations: Since crises like infrastructure failures are rather uncommon in Germany, our questions were placed hypothetically, so that it has to be empirically examined to what extent they correspond to actual crisis behaviour. In existing studies on actual infrastructure breakdowns, a lack of preparation by citizens and an increased demand for information has been proven (Nestler, 2017). An answer option for participants who had never been affected by a crisis should be added to achieve more specific and valid results. A focus on specific events, in this case, CI failures, is also necessary to define behaviour and expectations in different situations. Moreover, definitions and a clear distinction of scales would contribute to a better understanding of the questionnaire. The fact that we surveyed our respondents online further limits representativity. Therefore, the study should be carried out offline and with participants of different affinity towards technology. Furthermore, the reasons behind the behaviour and media usage indicated in the open answer could be examined in detail to better understand intentions and needs of affected individuals.

5.6   Appendix: Survey Questions

Q1: In an acute crisis situation (when you yourself were affected / when you were a volunteer), which of the following possibilities did you use to inform yourself about the events and which were very helpful for you? (Yes, it was very helpful; Yes, I have used it; No): 1. newspapers and magazines | 2. television | 3. radio | 4. Face-to-face conversations (e.g. with family, friends and neighbours) | 5. Phone calls (e.g. with family, friends and neighbours) | 6. contacting emergency services, fire brigade, police or hospital | 7. On-site information opportunities (e.g. notices, flyers and loudspeaker announcements) | 8. social media (e.g. Facebook, Twitter, Instagram and YouTube) | 9. other Internet services | 10. a crisis app (e.g. Katwarn or NINA) | 11. none | 12. don't know

Q2: What do you expect in the event of infrastructure (e.g. electricity, gas or telecommunications) failures on the part of the operator (e.g. Telekom, energy supplier)? (Strong approval; approval; neutral; rejection; strong rejection): 1. personal contact on site | 2. Phone call | 3. text message | 4. message on social media | 5. message on the website | 6. message in traditional media | 7. two-way communication | 8. information on duration | 9. information on reason | 10. information on what to do | 11. source of information

Q3: In the event of a failure of the emergency number 112, various measures are taken to ensure that the population continues to have the possibility of reporting an emergency. To what extent are the following measures or locations known to you? (Is very well known to me; Is known to me; Neutral; Is not known to me; Is not known to me at all): 1. the BOS will occupy fire equipment stores in the affected area. | 2. in villages without a fire equipment shed, a fire brigade vehicle is positioned on a central square. | 3. the police increase their activity in the affected area. | 4. further alternatives are reported by the media. | 5. location of the nearest rescue station (location of ambulance, etc.) | 6. location of the nearest fire station | 7. location of the nearest building of the German Federal Agency for Technical Relief (THW) | 8. location of the nearest police station | 9. location of the nearest administrative buildings (e.g. town hall, district hall)

Q4: Describe in as much detail as possible how you would proceed in an extraordinary situation (crisis, catastrophe)? What activities do you perform, or would you perform?

Survey on Citizens' Perception and Use of Social Media

The value of social media in crises, disasters, and emergencies across different events (e.g. floods, storms, terroristic attacks), countries, and for heterogeneous participants (e.g. citizens, emergency services) is now well-attested. Existing work has examined the potentials and weaknesses of its use during specific events. Fewer studies, however, have focused on citizens' perceptions of social media in emergencies, and none have deployed a representative sample to examine this. We present the results of the first representative study on citizens' perception of social media in emergencies that we have conducted in Germany. Our study highlights, for example, that around half (45%) of people have used social media during an emergency to share and / or look for information. In contrast, false rumours on social media (74%) are perceived as a threat. Moreover, only a minority of people have downloaded a smartphone app for emergencies (16%), with the most popular ones' weather and first aid apps.

This chapter has been published as the conference article "Social Media in Emergencies: A Representative Study on Citizens' Perception in Germany" in the Proceedings of the ACM: Human Computer Interaction (PACM): Computer-Supported Cooperative Work and Social Computing by Christian Reuter, Marc-André Kaufhold, Thomas Spielhofer, and Anna Sophie Hahne (Reuter, Kaufhold, Spielhofer, et al., 2017).

6.1 Introduction

Research into crisis management in Computer Supported Cooperative Work (CSCW) and Human Computer Interaction (HCI) has become more common as it has become evident that the work of professional bodies, volunteers and other

M.-A. Kaufhold, *Information Refinement Technologies for Crisis Informatics*, https://doi.org/10.1007/978-3-658-33341-6_6

organizations is increasingly mediated by computer technology and, more specifically, by the use of social media. A series of disasters and other crisis events have been examined in the light of this fact (e.g. Hughes et al., 2014; Palen & Anderson, 2016; Reuter, Ludwig, Kaufhold, et al., 2015; Reuter & Kaufhold, 2018). Such events are of an eclectic, and to a degree unforeseeable, character. For example, the attack at the Christmas market in Berlin in December 2016, the shootings in and around the Olympia shopping mall in Munich in July 2016 or the attacks at the Champs-Elysées in France and in London in June 2017 were both shocking and largely unanticipated. However, also small occasions, such as a car accident, or a fire are emergencies that are considered in this context. The United Nations Department of Humanitarian Affairs define emergencies like those mentioned above as "[a] sudden and usually unforeseen event that calls for immediate measures to minimize its adverse consequences" (United Nations Department of Humanitarian Affairs, 2000). Likewise, we will consider emergencies such as an accident, power cut, severe weather, flood or earthquake in this paper.

For our purposes, the use of social media in critical situations is especially interesting: Already after the 9/11 events in 2001, citizens were creating wikis to collect information about missing people (Palen & Liu, 2007), and FEMA and the Red Cross used web-based technologies to inform the public and to provide status reports internally and externally (Harrald et al., 2002). This burgeoning research field, sometimes summarised under the term *crisis informatics*, has revealed interesting and important real-world uses for the social media. Coined by Hagar (Hagar, 2007), crisis informatics is "a multidisciplinary field combining computing and social science knowledge of disasters; its central tenet is that people use personal information and communication technology to respond to disasters in creative ways to cope with uncertainty" (Palen & Anderson, 2016).

The so-called Web 2.0 (O'Reilly, 2007) was initially defined, in fact, as an architecture for supporting new possibilities for social interaction. Over the years, this interaction has been increasingly defined in terms of social media as a "group of Internet-based applications that build on the ideological and technological foundations of Web 2.0, and that allows the creation and exchange of user-generated content" (Kaplan & Haenlein, 2010). In this context, user-generated content refers to "the various forms of media content that are publicly available and created by end-users" (Kaplan & Haenlein, 2010). Currently [in August 2017], the most common types include Facebook with about 2.0 billion active users monthly, WhatsApp (1.2 billion), YouTube (1 billion), Instagram (700 million), Twitter (328 million), and LinkedIn (106 million) (Statista, 2017b).

Current research, e.g. during the last 5 years, in CSCW focuses on how (mobile) applications and technologies are capable of supporting authorities'

(Alcaidinho et al., 2017) and citizens' (Dittus et al., 2017) collaborative efforts before, during, and after crises or emergencies. In terms of social media, many available studies focus on the concrete use of social media during a specific emergency, such as the 2011 London riots (Denef et al., 2013), the 2012 hurricane Sandy (Hughes et al., 2014), the 2013 European floods (Reuter, Ludwig, Kaufhold, et al., 2015), or the 2014 Oso landslide (Dailey & Starbird, 2017). These studies demonstrate the specific ways in which social media have responded to various crises. Across various studies of emergencies and disaster events, numerous positive and negative aspects of social media have been identified (Alexander, 2014), and different groups of users' contextually specific attitudes and behaviour have been studied (Reuter & Spielhofer, 2017). Such qualitative studies provide for an in-depth approach to the practices of various parties in relation to specific events. The very fact of the eclectic nature of these events, however, means that generalisation about overall attitudes to social media use is difficult. Here, we provide data of a quantitative kind in order to complement the important qualitative work that has already been done, so as to provide a more general understanding of attitudinal tendencies with respect to how much social media are used, and for what purposes.

To this end, our study aims to research citizens' perception of social media in emergencies with a view to some more generalised conclusions. The sample we engage with concentrates on the situation in Germany, and subsequently compares selected findings with a further survey in the UK. This paper contributes findings on the perception of social media use in the German population with a representative survey of 1,069 participants. Our findings show that social media is used in emergencies more to search for than to share; emergency services are clearly expected to monitor social media and to respond within an hour as well; the main barriers to use social media are rumours and unreliable information.

6.2   Perception of the Use of Social Media in Emergencies

During the last few years, several studies in HCI, CSCW, and other disciplines on social media use in emergencies have emerged (Reuter & Kaufhold, 2018). The qualitative research we refer to, almost by definition, has focused on local practices and rather less on general attitudinal matters. In this paper, then, we explicitly focus on the findings of quantitative studies of perceptions of the use of social media in emergencies of which, hitherto, there have been relatively few.

6.2.1   Citizens' Perception of Social Media

In particular, four quantitative studies examining attitudes and perceptions are worth mentioning. Firstly, a comparative study with over 1,000 participants was conducted by the Canadian Red Cross in 2012 (Canadian Red Cross, 2012). This aimed to identify the extent to which Canadian citizens use social media and mobile devices in crisis communication and what they expect from the emergency services, both currently and in the future. Secondly, the American Red Cross (2012) studied citizens' use of social media during emergencies, with 1,017 online and 1,018 telephone survey respondents in 2012. In the third study worth mentioning, Flizikowski et al. (2014) presented a survey within Europe in 2014, conducted among citizens (n = 317) and emergency services (n = 130), which identified the opportunities and challenges of social media integration into crisis response management. Finally, Reuter and Spielhofer (Reuter & Spielhofer, 2017) analysed the findings of a survey of 1,034 citizens across Europe conducted in 2015 to explore citizens' attitudes towards the use of social media for private purposes and in emergency situations.

Flizikowski et al. (2014) identified issues such as lack of knowledge, non-uniform terms of use, personnel issues, credibility of citizen-generated content, and accessibility from both citizens' and emergency services' points of view. Personnel issues and credibility issues were also identified in the study of the Canadian Red Cross (2012). According to the American Red Cross (2012), 22% of high school graduates and 12% of general citizens stated that they had already used social media sites during emergencies to get information about damage caused, information on how other people were coping, weather, and traffic or to share personal feelings, safety reassurances, or eyewitness information.

In principle, participants' attitudes towards the use of social media were positive overall (Flizikowski et al., 2014). Benefits of using social media during emergencies can be seen in reassurance for citizens, providing situational information, and monitoring (Canadian Red Cross, 2012). Accordingly, social media is seen as a support for, adjunct to, existing channels (Canadian Red Cross, 2012). Friends, family, news media (or reporters), and local emergency officials especially are seen as the most trusted sources (American Red Cross, 2012). Therefore, the Canadian Red Cross employs "trusted volunteers" to support the official response via social media (Canadian Red Cross, 2012). Moreover, citizens' use social media to search (43%) rather than to share information (27%) (Reuter & Spielhofer, 2017). Users are most likely to seek information about weather, traffic, damage caused, and information on how other people were coping (American Red Cross, 2012). If they do provide information, users not only share weather information, safety reassurances, and their feelings about the emergency, but also their location and eyewitness information (American Red Cross, 2012). Even if the attitude towards social media is positive, only a few people have used it to share or obtain information during emergencies and disasters or in severe weather conditions [14].

In conclusion, the difference between the apparently positive attitude manifested and the actual usage suggests that there are barriers to the use of social media. Indeed, within all studies, challenges in the use of social media were identified. Both, citizens and emergency services identify the same challenges, especially concerns about the credibility of citizen-generated content (Canadian Red Cross, 2012; Flizikowski et al., 2014; Reuter & Spielhofer, 2017), a lack of knowledge and personnel issues (Canadian Red Cross, 2012; Flizikowski et al., 2014), lack of uniform terms of use (Flizikowski et al., 2014), and difficult accessibility for older generations (Flizikowski et al., 2014). Regarding trustworthiness, unknown people in the general vicinity of the emergency are the least trusted (American Red Cross, 2012). It follows that, from the point of view of the emergency services, the monitoring of social media is expected and anticipated (Reuter & Spielhofer, 2017).

In addition to citizen's perception of social media, some research focuses on the perceptions of the authorities. For example, Reuter et al. (2016) published the findings of a survey in 2016 conducted with 761 emergency service staff across Europe about current attitudes influencing the use of social media in emergencies.

6.2.2 Research Gap

These studies we have cited have been conducted in different countries, largely based on opportunity-based samples. This means that the number of answers might be high, however it is not guaranteed that they cover the population in terms of gender, age, region, education and income in an appropriate form (e.g. 51% female, 49% male). From a methodological perspective, it seems that none of the existing studies was based on a representative sample that guarantees this and thus we have little information about how often and with what regularity social media are used. It is not obvious whether those samples are representative of a cross-section of the population. This of course limits the reliability of these studies in terms of generalizable statements to some extent and provides an opportunity for a representative and therefore more robust and reliable study.

Of course, representativeness can only be assessed in relation to a target population. We focus on a representative sample in Germany. Given the fact of existing work in the US, Canada, and Europe, it was interesting from our point of view to conduct a study in Germany to verify or disconfirm existing findings, thus providing a basis for comparability. The transferability of our results to other countries obviously is uncertain but for those with similar statistics on social media use, socio-economic wealth, technical development, and infrastructure, our study design might be applicable.

According to our *definition of emergency*, based on the United Nations Department of Humanitarian Affairs (2000), it comprises everything such as an accident, power cut, severe weather, flood, or earthquake. This was mentioned in Q2 for the participants as a basic interpretation of that term.

6.3   Methodology

6.3.1   Survey Design

The survey conducted consisted of nine questions, most of them in closed answer format except for the last question, which was open format (see appendix for the questions). First, participants were asked about the frequency of their social media (Q1) and their previous usage of social media in emergencies (Q2). For participants who answered with the options "Yes, I have used it to find out and share information" or "Yes, I have used it just to share some information", the kinds of information shared was collected (Q3). Then, for all participants, we requested participants to assess the responsiveness of emergency services to messages posted via social media (Q4) and elicited opinions about factors which might discourage social media use in emergencies (Q5). We also considered previously downloaded apps for emergencies (Q6) and specifications concerning the kinds of emergency-related apps the participants had already downloaded (Q7). We also wanted to know, regarding all participants, what possible future usage of apps in emergencies for exemplified purposes were anticipated (Q8). Finally, the last question was open format and covered further details about experiences with social media in emergencies (Q9).

6.3.2   Characteristics of Survey Participants

6.3.2.1 Gender, Age, Region, Education, Income

The survey was conducted using a panel of GapFish (Berlin) in October 2016. GapFish is an ISO-certificated panel provider. They guarantee panel quality, data quality and security as well as survey quality through various (segmentation) measurements for each survey within their panel of 180,000 active participants.

We conducted a representative survey of the adult German population stratified for **gender** and **age**, meaning that gender and age of the sample matched the population as a whole. There were no significant differences between the population statistics and the survey statistics in relation to gender and age (χ^2 (1, N = 1,069) = .000, p = .994 and χ^2 (5, N = 1,069) = .426, p = .995, respectively).

In addition, we ensured a wide spread of the survey sample in terms of **region, education,** and **income**. Significant differences were found between population statistics and the survey statistics for region and income (χ^2 (3, N = 1,069) = 175.92 and χ^2 (3, n = 802) = 225.36, p < .001, respectively). Some respondents did not provide information about their income (n = 802). Several categories for each variable were combined in order to perform the chi-squared tests. States in Germany were combined to the regions 'North', 'South', 'East', and 'City States' and income categories were combined to 'Less than 1,500', '1,500–2,500', '2,500–4,500', and 'Over 4,500'. It was not possible to test for significant differences for the variable 'highest education' between the survey and the population as there were no national statistics available that matched the categories used in the survey.

The **gender** distribution of our sample (N = 1,069) with 50.9% female and 49.1% male participants was representative in relation to Germany (51% female, 49% male) (Bundeszentrale für politische Bildung (bpb), 2016).

All participants were older than 18 years and the majority (60%) were older than 45 years, which corresponds to the age distribution in Germany (57.2% over 40 years) (Statista, 2015). Our sample consists of participants from all states in Germany, whereby the majority was of North Rhine-Westphalia (14.1%), the state with the highest population in Germany (Bundeszentrale für politische Bildung (bpb), 2016). About 34% of the survey respondents stated that they completed 'A-Levels' as their highest education (combines the responses 'A-Levels' and 'university-degree'). In Germany as a whole, the proportion is slightly lower with 29% (Statistisches Bundesamt, 2016). 22.7% of Germans had a secondary school certificate in 2015, whereas in our sample 33.2% stated this as their highest educational qualification. Nevertheless, our sample comprises participants of all educational levels. About 21.9% of our participants stated that they earn about 2,500€ to 3,500€ per month after tax deductions, which corresponds to the income distribution in Germany (Bundeszentrale für politische Bildung (bpb), 2016). Again, our sample consists of participants with income from under 1,000€ to over 4,500€ per month to ensure a wide spread of household income.

6.3.2.2 **Smartphone and Social Media Use**

Our first question concerned the frequency of smartphone and social media use, whereby 50% of our German participants use a smartphone daily, and 29% even use it hourly (Figure 6.1). However, there are also nearly 8 million German citizens (13%, as according to OECD the population of Germany over the age of 18 is 64,160,544 in 2013 (https://stats.oecd.org/)) who never use a smartphone and therefore have limited access to their social media accounts. Looking at this 13% group in detail, they are appreciably more likely to be aged 65 and over. Of all respondents, aged 65 and over, 28% have never used a smartphone; in comparison, all respondents aged between 18 and 24 years old use a smartphone. This equates to a significant relationship between age and frequency of smartphone use, χ^2 (20, N = 1,069) = 363.59, p <0.001. Those numbers correspond with the representative ARD/ZDF study (W. Koch & Frees, 2016). 49% of participants overall use a smartphone daily to get access to the internet, whereas that figure is 86% of under thirty-year-olds.

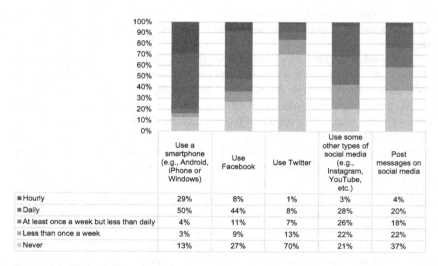

	Use a smartphone (e.g., Android, iPhone or Windows)	Use Facebook	Use Twitter	Use some other types of social media (e.g., Instagram, YouTube, etc.)	Post messages on social media
▪ Hourly	29%	8%	1%	3%	4%
▪ Daily	50%	44%	8%	28%	20%
▪ At least once a week but less than daily	4%	11%	7%	26%	18%
▪ Less than once a week	3%	9%	13%	22%	22%
▪ Never	13%	27%	70%	21%	37%

Figure 6.1   Please indicate how often, on average, you do the following things (Q1)

Overall, 44% indicated use of Facebook and 20% indicated use of Twitter daily, whereas 28% used other types of social media daily. These results are consistent with another German study on social media use—even though the questioning differed a little between both surveys. Among the 589 participants in that study, 51.1% used social media several times a day (Statista, 2017a). Daily usage of Facebook has increased from 23% in 2015 to 26% in 2016 (W. Koch & Frees, 2016). In 2016, 2% of the participants used Twitter daily. Hourly usage was indicated for Facebook by 8% and for Twitter by 4%.

- The proportion of people who use Facebook at least daily declines with age, starting with 76% in the youngest age group down to 33% in the oldest age group, χ^2 (20, N = 1,069) = 175.45, p<0.001.
- Apart from age, gender was shown to have a significant relationship with the frequency of Facebook use, χ^2 (4, N = 1,069) = 13.19, p = 0.01. More than half (57%) of women use it at least daily compared with 47% of men.
- Similar to Facebook use, younger citizens are more likely to use Twitter than older citizens. The relationship between age and frequency of Twitter use was significant, χ^2 (20, N = 1,069) = 38.25, p = 0.008. In particular, while 13% of 18 to 44-year-olds use it at least daily, only 6% of those aged 45 or older do so.
- Furthermore, the frequency of Twitter use differed significantly between men and women, χ^2 (4, N = 1,069) = 11.02, p = 0.026. Of all men, 10% use Twitter at least daily whereas it is only 8% for women. Nevertheless, 70% stated that they never post messages on social media.

6.3.3 Quantitative Analysis

For the quantitative analysis, the survey data was extracted and analysed using IBM SPSS Statistics 23, a software package for analysing quantitative data (IBM, 2014). Microsoft Excel was used for qualitative coding and for the design of the displayed figures. The analysis consisted of three key steps: (1) *Preparing* the data including assignment of missing values and data values, and combination of categories of demographic background variables. (2) *Exploring basic frequencies* for each question. (3) *Using cross-tabulations* with chi-squared tests to explore any significant differences across different types of respondents in relation to gender, age, region, income, and highest education level. Unless otherwise stated response categories from the questions (e.g., Q1: hourly, daily, at least once a week but less than daily, less than once a week, never) were utilized for tests of significance. For the variable age, the following categories were used: 18–24, 25–34, 35–44, 45–54, 55–64, 65+. In general, significant differences were only found for age and gender whereas region, education, and income did not show any significant differences in social media use nor attitudes towards the social media (Q1–Q8). Kendall's Tau was used to determine correlations between ordinal items.

6.3.4 Qualitative Analysis

The analysis of our free-text survey question was based on the inductive approach of *grounded theory* (Strauss, 1987). We used *open coding* associated with grounded theory to derive categories from the more qualitative free-text answers by carefully reading and aggregating categories. Categories were jointly checked and agreed. Each open-ended response was then assigned to one or multiple categories to achieve a quick overview of the interesting and relevant topics. The previously acquired knowledge from the literature review and quantitative analysis was used to increase theoretical sensitivity. Each quotation is referenced with the participants' response identifier.

6.4 Empirical Results

6.4.1 Use of Social Media in Emergencies (Q2+Q3): More Searching, Less Sharing

Nearly half of the survey sample (44%) indicated that they had used some form of social media during an emergency (Figure 6.2). For this purpose, 19% have used social media to find out and share information and at least 20% have already used it merely to gather information about emergencies. Only 5% of the participants have used social media channels specifically to share information in emergencies. Of the 1,040 citizens that made a clear statement, in relation to gender, women are more likely to find and share information (23%) as well as find information (22%) than men (17% and 19%, respectively), χ^2 (3, n = 1,040) = 8.05, p = .045. In relation to age, there was a significant relationship between the use of social media in emergencies and age, χ^2 (15, n = 1,040) = 142.62, p <0.001. The frequency for all forms of use decreases with age. While 27% of 18 to 24-year-olds reported having looked for and shared information on social media during an emergency, only 10% of the respondents in the age group 65+ said they had done so.

For those who shared information on social media (n = 259), the most shared information is weather conditions or warnings (63%), road or traffic conditions (59%), feelings or emotions about what was happening (46%) and one's own location (37%) (Figure 6.3). However, some significant differences were revealed

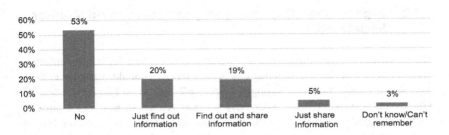

Figure 6.2 Have you ever used social media (such as Facebook, Twitter, Instagram) to find out or share information in an emergency such as an accident, power cut, severe weather, flood, or earthquake close to you? (Q2)

depending on the age and gender of citizens. Female respondents (52%) were more likely to share emotions or feelings than male respondents (39%; χ^2 (1, N = 259) = 4.80, p <.028), while men (15%) were more likely to share what they were doing in order to stay safe than women (6%; χ^2 (1, N = 259) = 6.03, p <.014). Moreover, the older age groups, 45–54 (60%), 55–64 (56%), and 65+ (58%) are more likely to have shared emotions and feelings via social media during emergencies than the younger age groups, 18–24 (29%), 25–34 (39%), and 35–44 (45%), χ^2 (5, N = 259) = 11.50, p <.042.

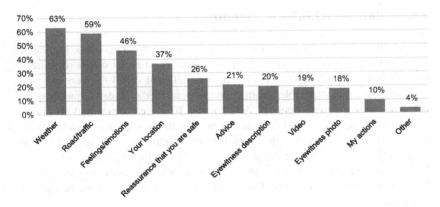

Figure 6.3 What types of information did you share? (Select as many as apply) (Q3)

6.4.2   Expectations (Q4): Emergency Services Should Monitor Social Media

The most stated expectation when posing an urgent request for help or information on a social media site was that emergency services should regularly monitor
their social media use (31% strongly agree; 36% agreed) (Figure 6.4). Second,
47% expect a response from emergency services within an hour (17% strongly
agree; 30% agree). In contrast, 43% think that emergency services are too busy to
monitor social media during an emergency (17% strongly agree; 26% agree). Interestingly, only the statement 'emergency services should regularly monitor SM'
and the statement 'expect a response from emergency services within an hour' are
correlated ($r_\tau = .527$, $p < .001$) whereas neither of these are correlated with the
statement 'emergency services are too busy to monitor SM during an emergency'
($r_\tau = -.01$, $p = .7$ and $r_\tau = -.003$, $p = .911$, respectively). This indicates, perhaps unsurprisingly, that respondents who expect emergency services to monitor
social media also expect them to respond within an hour.

Looking at the differences, it was shown that the three younger age groups,
18–24 (57%), 25–34 (53%), and 35–44 (53%) are more likely to agree to the
statement that emergency services are too busy to monitor social media during an
emergency than the group of respondents aged 45–54 (37%) and 65+ (31%), F
(5, 1,063) = 8.65, $p < .001$.

6.4.3   Barriers (Q5): False Rumours

Despite all the positive aspects of social media usage for emergencies, barriers to
its use remain (Figure 6.5). The many false rumours on social media (definitely)
put off participants the most (34% were definitely put off; 40% were put off).
43% suggested that unreliable information on social media tended to put them off
and 22% were 'definitely' put off. Additionally, 62% were concerned about data
privacy such that this would (definitely) prevent them from using social media
in emergencies. The chance that social media might not work properly in an
emergency would also prevent 33% (additional 27% definitely) from using it. For
22%, it was definitely better to call 112 than to post messages on social media
and 14% stated that they were definitely not confident using social media for such
purposes.

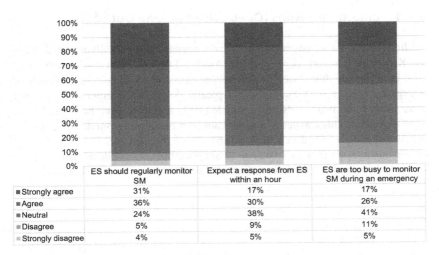

	ES should regularly monitor SM	Expect a response from ES within an hour	ES are too busy to monitor SM during an emergency
▪ Strongly agree	31%	17%	17%
▪ Agree	36%	30%	26%
▪ Neutral	24%	38%	41%
▪ Disagree	5%	9%	11%
▪ Strongly disagree	4%	5%	5%

Figure 6.4   Imagine that you posted an urgent request for help or information on a social media site of a local emergency service, such as your local police, coastguard, fire, or medical emergency service. To what extent do you agree with the following statements (Q4)

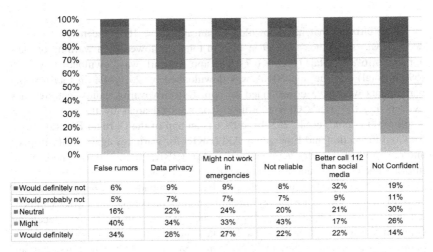

	False rumors	Data privacy	Might not work in emergencies	Not reliable	Better call 112 than social media	Not Confident
▪ Would definitely not	6%	9%	9%	8%	32%	19%
▪ Would probably not	5%	7%	7%	7%	9%	11%
▪ Neutral	16%	22%	24%	20%	21%	30%
▪ Might	40%	34%	33%	43%	17%	26%
▪ Would definitely	34%	28%	27%	22%	22%	14%

Figure 6.5   What might put you off using social media during an emergency? (Q5)

6.4.4   Emergency Apps (Q6–8): Used by One Out of Six

Findings on the use of emergency apps have already been partially reported (Reuter, Kaufhold, Leopold, et al., 2017b): Among 1,069 German participants, only 16% have already downloaded a smartphone app that could help in a disaster or emergency (Figure 6.6). People who typically download an emergency app are more likely to be male (60%) than female (40%) and to be aged between 25 and 54 years (65%). This equates to a significant relationship between downloading an app and gender, χ^2 (3, n = 1,012) = 9.67, p = 0.002 and between downloading an app and age, χ^2 (5, n = 1,012) = 42.88, p < 0.001. Respondents that were not sure whether that have downloaded an app were excluded from this analysis.

Figure 6.6 Have you ever downloaded a smartphone app that could help in a disaster or emergency? (Q6)

Of all citizens that have downloaded an app (n = 166), the most downloaded app was a weather app (69%; 11% in total), followed by a warning or alert app (42%, 6% in total) (Figure 6.7). The app Katwarn was downloaded by 37% (6% in total) and the app NINA was downloaded by 25% (4% in total) of the participants who have downloaded an app. Katwarn warns against catastrophes by knowing the user's GPS coordinates or selected areas by the user (Reuter, Kaufhold, Leopold, et al., 2017b). NINA is an app of the German Federal Office for Civil Protection and Emergency Aid (BKK) for warnings against catastrophes and recommendations for action as well as tips (BBK, 2015). Affected people can inform contact points about their state of affectedness.

While the proportion of people that have downloaded an emergency app is currently quite low (16%), a significantly higher percentage of citizens indicated that they were likely to use apps for such purposes in the future. Receiving emergency warnings is the most likely stated future usage purpose of such smartphone apps in an emergency (57%) (Figure 6.8). About 51% stated that they are quite/very likely to receive tips about how to stay safe or find out information about the emergency in this way. Connecting with other citizens to help others affected by the emergency was indicated by 46% as at least quite likely and another 42% might share information about the emergency with an emergency

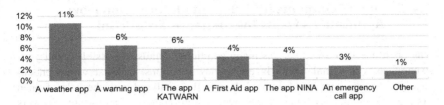

Figure 6.7 What type of app did you download? (Q7)

service. To contact an emergency service instead of making a 112 call is the least likely usage purpose as 32% agreed and 42% stated this as not very likely and not likely at all.

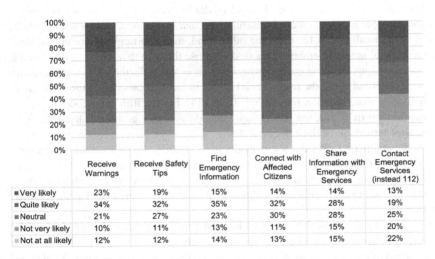

	Receive Warnings	Receive Safety Tips	Find Emergency Information	Connect with Affected Citizens	Share Information with Emergency Services	Contact Emergency Services (instead 112)
▪ Very likely	23%	19%	15%	14%	14%	13%
▪ Quite likely	34%	32%	35%	32%	28%	19%
▪ Neutral	21%	27%	23%	30%	28%	25%
▪ Not very likely	10%	11%	13%	11%	15%	20%
▪ Not at all likely	12%	12%	14%	13%	15%	22%

Figure 6.8 Please indicate how likely you are in future to use a smartphone app for each of the following purposes as a result of an emergency? (Q8)

6.4.5   Open Comments (Q9): Benefits, Information Flow, Applications Fields, Inexperience, Scepticism

The open question asked for "as many additional details as possible about your experience with social media in emergency situations, or what would encourage you to use them in the future". It was also analysed with the help of open coding. We created five meta-codes, which were coded for 890 usable answers (not empty) of the participants: *benefits*, *information flow*, *ranges of application*, *scepticism*, and *inexperience*. Each answer could be coded with a maximum of two meta-codes.

6.4.5.1 Benefits: Faster Than Other Channels

With respect to the meta-code *benefits* (n = 151), it became apparent that several people rate social media as useful or quick to use, for instance, and support, at least to some extent, the use of social media in emergency situations. Moreover, some people find it particularly useful to *get and spread information* with the help of social media (n = 135). On the other hand, many participants are not familiar with social media in emergencies or do not use them so that their answers can be classified into the meta-code *inexperience* (n = 487). Further, a few participants were *sceptical* about the use of social media in emergencies due to rumours, alternatives, or data security, for example (n = 114). For the fifth meta-code, *ranges of application* (n = 50), the participants named cases as e.g. natural disasters, fight against terrorism, or the safety-checks.

Several participants use social media in emergency situations as they think they are more immediate than other media and reach many people (n = 45): *"Social Media are more current and faster than other news channels in such situations"* (#943). Additionally, it is assumed that help can be better organised at short notice (#86).

Concerning the usefulness of social media in emergencies, few participants (n = 32) believe that social media are particularly practical for all purposes: *"At a local flood in my city, the help was organised via Facebook. That was very helpful and productive – during the coup in Turkey, I could contact friends and make sure that they are okay. I would like official warnings via Twitter or Facebook, for example. These would be seen directly in the timeline. All this, however, should only be an addition to normal communication in case of an emergency (radio, TV, etc.)"* (#333).

According to some participants, the authorities could also use social media in case of a catastrophe to prevent panic. One participant points to the 2016 Munich shooting, as an example as there was a very professional spokesperson (#1210). Another sub-code of the meta-code productivity is connectivity (n = 27). That means that some participants assume that social media encourages the exchange of information as social media makes it easy to communicate with each other. In emergencies, people can contact each other more readily and are therefore better connected. One participant sees even more advantages in terms of emotional processing: *"You can exchange with several people at the same time; You can hear more opinions and thus act/behave better; You can see what others do; You are not "alone" or do not feel lonely"* (#788).

6.4.5.2 Information Flow: Spreading and Obtaining Up-To-Date Information

In this context, information flow is also one significant advantage of social media for some participants. Because of the degree of connectivity, people have more opportunities to spread and obtain information. Several people (n = 89) use social media for this purpose, based on the enormous capacity for information dissemination and therefore users always feel well informed. They perceive social media to be always up-to-date, especially in unusual situations: *"Social media are useful to gather information about events. In the meantime, the technology has advanced so much that information is contemporary. The example of Halloween: With the help of Facebook we could see where horror clowns were spotted and knew what places we had to avoid"* (#739).

This example shows that many people use social media to obtain information as they think that social media usually provides the latest news and are quicker and more up-to-date than traditional media. For instance, one participant mentions that the police can publish recommendations in emergencies in social media as in the case in Munich, where the police warned people via Facebook not to leave the house (#716).

Further, Facebook, for example, offers features such as the 'Safety-Check'. In the case of an emergency, the telephone network is often overloaded so that you cannot reach your relatives or friends directly. With the given feature, people can confirm that they are okay if they stay near a catastrophe. In this context, one participant cited the Munich shooting again because in this case, he obtained information about status with the help of social media since he thought that traditional media were not that quick (#955). Additionally, social media are used for warnings (n = 23) and the dissemination of information (n = 23), which also counts for the meta-code information flow.

6.4.5.3 Ranges of Application: Natural Disasters, Safety Checks

If people use social media in emergencies, they have different areas of application (n = 50). The most frequently mentioned instances where people used social media were natural disasters (n = 15). In the case of a natural disaster, which cannot be influenced by humans, many people may be in danger. In this occasion, it is felt that social media may be helpful to obtain information about the status or the extent of the catastrophe. Further, people may be warned, or help can be organised. One participant used social media in an emergency and found it very useful: *"Example of a flood: By shared postings and information I could estimate whether the region where I live or the residences of relatives are at risk or rather warn others, who do not have access to social media (e.g. elder generations)"* (288).

The above-named feature 'Safety-Check' was also used by a few other participants (n = 12). They were very thankful for this feature as social media had a calming effect. One participant sees even more advantages: *"I like the possibility of stating 'safe' (on Facebook) in a natural disaster or the like. At various floods in Germany, emergency appeals on Facebook could be posted in a group which gave me, as a volunteer, possibilities to help systematically"* (#382).

Another participant used Facebook during the shooting in Munich: *"[…] I found it very helpful and reassuring to see on Facebook that friends on-site are safe (with the help of the quite new 'I am safe'-function)"* (#208). Another tool, which was named at least three times, was Katwarn (n = 3) (see Q6–8). Further, people use social media for the search for a missing person (n = 7).

6.4.5.4 Scepticism: General Sceptics and Rumours

Several participants, even so, take a sceptical view towards the use of social media in emergencies, thus their answers are classified into the meta-code scepticism (n = 114). As a reminder, Q5 (N = 1,069) asked about barriers, highlighting rumours (74%), data privacy (62%), or that it is simply not working (60%) or not reliable (65%). From these 114 sceptical participants, some (n = 48) are sceptical in general and therefore do not want to trust social media in dangerous situations. One reason is the spreading of rumours, which unsettles a few respondents: *"I am sceptical. It is not reliable; there is too much misinformation, users share news, search calls, etc. without checking their truthfulness"* (#277). Information flow may, therefore, lead to unnecessary panic (#639).

Moreover, some sceptical participants do not completely reject the use of social media but prefer alternatives (n = 30). In everyday life, they use social media, but in emergencies, they still switch to traditional media as they find it difficult to rely on the internet. As one participant states: *"I would rather rely on a phone call; afterwards, you can look and read on social media. We find it difficult to rely on the internet completely"* (#156). Another respondent suggested he does not totally trust social media and therefore used other media in emergencies: *"In emergencies, the news channels would run day and night. (On that basis that there is*

electricity, if not, an app would also be useless at some point due to empty battery).
Since there was too much mischief on Facebook and co, I would probably not totally
rely on it [...]; news channels are more competent [...]" (#988).

6.4.5.5 Inexperience

Of course, several participants do not use social media at all and mentioned this
in the comments (n = 88), or are not familiar with the use of social media in
general and especially not in emergencies (n = 339) and therefore belong to the
meta-code inexperience. On the one hand, they do not use social media since they
think it is unnecessary as one participant confirms: *"I do not have a smartphone,*
and I will not get one. For this reason, I do not need social media, which I overall
regard as superfluous as they only have very limitedly use for social development
[...]" (#323).

On the other hand, they still prefer other media, for example, if they are at
an advanced age: *"I belong to the elder generation, and I do not know about*
technologies. I rely on traditional news" (#631). Others are inexperienced but
would not exclude the use of social media in emergencies: *"No experience with*
social media in case of an emergency. Still, I would be encouraged if particular
information disseminates quicker" (#402).

6.5   Discussion and Conclusion

Social media may be used in crises, disasters, and emergencies across different
types of events, countries, and continents as well as there being various types
of participant involvement. It is important to examine the attitudes of involved
participants to understand the appropriation of social media, to identify existing
barriers, and to develop solutions, for instance via mobile applications that promote an efficient use before, during, and after safety-critical events. A number of
existing studies have already researched the citizens' (American Red Cross, 2012;
Canadian Red Cross, 2012; Flizikowski et al., 2014; Reuter & Spielhofer, 2017)
and emergency services' attitudes towards social media (International Association
of Chiefs of Police, 2015; Plotnick et al., 2015; Reuter, Ludwig, Kaufhold, et al.,
2016; San et al., 2013). Focusing on the citizens' perspective, however, few of
the above studies involved a large scale and statistically reliable sample.

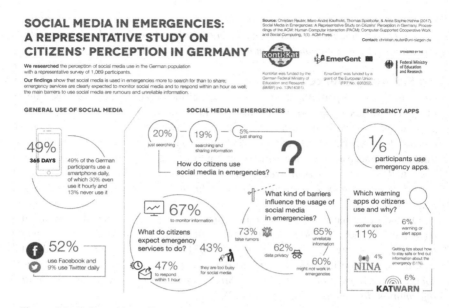

Figure 6.9   Infographic on Citizens' Perception of Social Media in Emergencies in Germany (2017)

6.5.1   The Perception of Social Media in Emergencies by Citizens' in Germany

This paper makes a contribution in the following two ways to crisis informatics, both of which we consider to be important: (a) providing data which is agnostic to a particular disaster, and (b) examining a large population of people who have both used and not used social media in an emergency. In order to understand our findings appropriately, the results are to be interpreted along our broadly chosen definition of emergency in section 6.1.

Our study revealed that 79% of the adult German population use a smartphone daily, while 13% never use a smartphone; half of those are 65 years old or over (Figure 6.9). More than half of the population use Facebook daily with younger citizens using it more often but also a third of those aged 65 or older use it daily. Twitter is less frequently used with 70% of the population never using it (Q1).

The main results of our study are listed below:

- **Current Use:** Social media is used in emergencies more to **search for** (20%) than to **share** (5%) information and with the 19% who use it both to find out and share information, there are significant relationships with gender (as well as content-related) and with age.
- **Expectations:** On the one hand, emergency services are clearly expected to **monitor social media** (67% (strongly) agree) and to **respond within an hour** (47%), and these two points correlate significantly. On the other hand, they are perceived as **too busy to monitor social media** (43%), with a significant bias towards younger participants.
- **Main barriers** of using social media are false rumours (73% were (definitely) put off), unreliable information (65%), data privacy (62%), and the possibility that social media might not work properly in an emergency (60%).
- **Emergency apps** are only used by one out of six (16%), with a significant relationship with gender towards male participants, and age towards younger (25–54 years) participants. The most downloaded apps are weather apps (69%) and warning or alert apps (42%). For future use, receiving emergency warnings (57%) and tips about how to stay safe or find out information about the emergency (50%) are the most likely perceived opportunities (see Reuter, Kaufhold, Leopold, et al., 2017b for details).

Focusing on the open question (Q9), the following results can be summarised in five meta-codes:

1) **Benefits** (n = 151): Social media are perceived as faster than other channels (n = 45), practical for all purposes (n = 32) and connectivity (n = 27).
2) **Information flow** (n = 135): Spreading and obtaining up-to-date information (n = 89), receiving warnings (n = 23) and disseminating information (n = 23) are appreciated within the social media usage.
3) **Ranges of application** (n = 50): In case of natural disasters (n = 15) and for safety checks (n = 12), social media is used the most.
4) **Scepticism** (n = 114): General scepticism (n = 48) due to rumours and the preference of alternatives (n = 30), especially for older generations, could be identified.
5) **Inexperience** (n = 487): People that do not use social media at all (n = 88) and that are not familiar with the use, especially in emergencies (n = 339), could be found in our study.

Moreover, our study revealed that some characteristics of citizens are more important than others to predict their attitudes towards and actual use of social media in emergencies:

- Especially **age** was shown to be strongly correlated with the general use of social media (Q1, e.g., Facebook ($p < .001$), Twitter ($p < .01$)), the use during emergencies (Q2, $p < .001$) and the availability of an emergency app (Q6, $p < .001$).
- Apart from age, **gender** was correlated with some of the questions, but we did not find any significant effects for income and location.
- However, as **income** is correlated with most of the background variables it is worth to continue including it in further surveys.
- We found numerous correlations with **education** level, although in most cases there was no clear direction. One exception is the item 'Emergency services are too busy to monitor social media' (Q4)—younger ($p < .001$) and more educated people ($p < .05$) were more likely to agree with this statement. The influence of education level on the attitudes towards use of social media in emergencies should be therefore explored in further studies, potentially alongside working status and social grade, which were not included in this study but in surveys that we conducted in other European countries.

6.5.2   Directions for Future Research

The results indicate implications for further research:

1) Firstly, the propagation of emergency-related apps needs to be promoted. As users tend to use platforms that they are familiar with, the augmentation of general apps or platforms with disaster capability, such as Facebook Safety check, seems a promising approach to improve social resilience during crises and emergencies. Also, work on the distribution of these apps with sharing and other collaborative functionality is needed (Reuter, Kaufhold, Leopold, et al., 2017b).
2) Secondly, expectations towards emergency services lead to two conclusions: monitoring is needed, but monitoring should not cost time during emergencies. Obviously, small emergency services cannot handle this task manually, a problem which has also been identified in earlier research (Reuter, Ludwig, Kaufhold, et al., 2016). VOST teams are a possibility (Cobb et al., 2014), however, further cost effective and convenient ways of monitoring (Ludwig,

Reuter, Siebigteroth, et al., 2015) are needed, e.g. semi-automatic generation of alerts and notifications based on social media (Reuter, Amelunxen, et al., 2016) that can also be transferred to control station systems. Another aspect concerns the perception of social media as entailing negative aspects such as rumour propagation, dissemination of false or misleading information, ethical dilemmas (Alexander, 2014), and propaganda or social bots (Reuter, Pätsch, & Runft, 2017a).

3) Moreover, although researchers consider citizens to act rationally during crises in western societies (Helsloot & Ruitenberg, 2004), participants reflected on the nature of panic. These problems have become even more apparent, especially in CSCW, dealing with uncertainty (Starbird et al., 2016) and mechanisms of self-correction (Arif et al., 2017) related to false rumours that are hot topics in current research. However, more work is evidently needed.

6.5.3   Relationship With Prior Work

Our work contains similarities to existing studies, but also offers some distinctive conclusions. Compared to the study of the Canadian Red Cross (2012), more people in our study used social media to either get or share information (25% vs. 12%), which might be due to the fact that the Canadian study was published in 2012. Meanwhile the overall social media usage in general has developed (Kroll, 2016). In comparison to Reuter and Spielhofer (2017), the percentages for receiving (39% vs. 43%) and sharing (24% vs. 27%) information during crises are slightly lower (Q2). In both studies, the sharing of weather conditions and warnings, road or traffic conditions, as well as feelings and emotions about what was happening were most prominent (Q3). However, there is a big difference in sharing eyewitness photos (18% vs. 59%). According to our study, slightly fewer people expect emergency services to monitor social media (67% vs. 69%), and fewer expect emergency services to be too busy (43% vs. 56%) and thus more people demand a response within one hour (47% vs. 41%) (Q4). However, it must be noted that we are comparing studies which have different degrees of representativeness, were conducted in different timeframes, and have either a single- or multi-country focus. Regardless, the fact of these variations suggests there is a need for a systematic and large-scale approach to studies of attitudinal work, which can be used to supplement the more site-specific studies that constitute the main body of work in this area.

Besides, qualitative studies offered insights into citizens' activities, phenomena and types of information in social media during emergencies. A body of case studies examined the dissemination of situational updates (Vieweg et al., 2010), the expression of solidarity (Starbird & Palen, 2012) or emotional support (Wilensky, 2014), reassurance of safety (Acar & Muraki, 2011), sharing of eyewitness information (Dailey & Starbird, 2017) or general advice (Sreenivasan et al., 2011). To draw conclusions on a broader scope, Olteanu et al. (2015) investigated several crises in a systematic manner showing the average prevalence of different information types in Twitter, such as 32% other useful information (incl. weather and road conditions), 20% sympathy and emotional support, 20% affected people, and 10% caution and advice. Similarly, our findings reflect situational updates, such as information on weather (63%) or roads/traffic (59%), to be the largest category, followed by emotional content (46%) and affection-based information, such as own location (37%) or reassurance of own safety (26%), but much less general advice (21%) (Q3). In contrast, plenty of qualitative and mixed method studies identified misinformation (Reuter, Ludwig, Ritzkatis, et al., 2015), unreliable information (Reuter, Ludwig, Kaufhold, et al., 2015) and rumours (Oh et al., 2013; Wilson et al., 2017) as barriers of social media use; our study confirms false rumours (74%) followed by the reliability of information (65%), enriched by 114 open-ended responses expressing a sceptical view on social media use during emergencies (Q9), as the strongest barriers (Q5).

This study revealed that only 21% of the participants downloaded a smartphone app that could help in a disaster and emergency (Q6); a qualitative study with 22 interviewees suggests that there is a weak motivation due to the low probability of (natural) disasters in Germany, and identifies warning messages, recommendations for action, sending warnings and all-clears, chat and organisation and tailorable behaviour as desired functionality of emergency apps (Reuter, Kaufhold, Leopold, et al., 2017b). The wider distribution of weather apps (11%) over warning (6%), first aid (4%) or emergency call apps (3%) could be related to a higher relevance in daily life (Q7). On the other hand, participants indicate that they likely will use smartphone apps to receive warnings (57%) or safety tips (51%) in future (Q8), confirming the need for and potentials of providing instructing information (Coombs, 2009) and orientation information (Nilges et al., 2009) during emergencies. A mixed method approach could be applied to gather qualitative in-depth feedback about motivational and functional requirements, e.g. by considering a persuasive system design approach (Kotthaus et al., 2016), of emergency apps and test findings qualitatively against a larger sample.

6.6   Limitations

Of course, this study has limitations. Firstly, we study many aspects that have already been studied in the past. However, they have not been studied in a representative way and not with regard to all the aspects we deal with. Secondly, the study was conducted using an online survey which still might provide representative results with regard to some factors, but just covers people how are willing to do online-surveys; therefore, they are most likely more familiar with internet and social media. Thirdly, we only study citizens' perceptions in Germany. This gives us the possibility to conduct a representative study on a wider basis. In future research, we want to conduct and analyse further representative studies in other countries in Europe (UK, Italy, and the Netherlands). The representative study in the UK has already been conducted and provides some interesting preliminary findings about existing differences: Accordingly, in the UK social media is less used during emergencies (not used by 53% in GER vs. 66% in UK; Q2). In the UK, emergency services are more often seen as being too busy to use social media in emergencies (62% UK vs. 43% GER, Q4) and citizens expect less that social media is monitored by emergency services (67% GER vs. 37% UK, Q4). In addition, the distribution of emergency related apps is much lower in the UK (7%) than in Germany (15.5%). These findings are very preliminary and at present nonparametric. Detailed comparisons and analysis about the reasons of those findings will be conducted in future work.

6.7   Appendix: Survey Questions

Q1: Please indicate how often, on average, you do the following things (Hourly, Daily, At least once a week but less than daily, Less than once a week, Never): *Use a smartphone (e.g. Android, iPhone or Windows) | Use Facebook | Use Twitter | Use some other types of social media (e.g. Instagram, YouTube, etc.) | Post messages on social media*
Q2: Have you ever used social media such as Facebook, Twitter, Instagram etc. to find out or share information in an emergency such as an accident, power cut, severe weather, flood or earthquake close to you? *Yes, I have used it to find out and share information | Yes, I have used it just to share some information | Yes, I have used it just to find out some information | No, I have not used it in this way | Don't know/Can't remember*
Q3: What types of information did you share? (Select as many as apply) Weather conditions or warnings | Road or traffic conditions | Reassurance that you are safe

| Your feelings or emotions about what was happening | Your location | What actions you were taking to stay safe | An eyewitness description of something you experienced | Advice about what actions others should take to stay safe | An eyewitness photo | A video | Other (please specify)

Q4: Imagine that you posted an urgent request for help or information on a social media site of a local emergency service, such as your local police, coastguard, fire or medical emergency service. To what extent do you agree with the following statements (Strongly agree, Agree, Neutral, Disagree, Strongly disagree) *Emergency services should regularly monitor their social media | I would expect to get a response from them within an hour | Emergency services are too busy to monitor social media during an emergency*

Q5: What might put you off using social media during an emergency? (This would definitely put me off; might put me off; Neutral; would probably not put me off; would definitely not put me off) *Information on social media is not reliable | There are many false rumours on social media | I am concerned about data privacy | It is better to call 112 than to post messages on social media | I am not confident using social media | Social media might not work properly in an emergency*

[By the term 'confident', we relate to people who are conscious about handling their social media tools, respectively by 'not confident' it is obvious that this refers mostly to people with few social media experiences.]

Q6: Have you ever downloaded a smartphone app that could help in a disaster or emergency? (Yes; No; do not know/Not sure)

Q7: What type of app did you download? A weather app | A warning app | A First Aid app | An emergency call app | The app Katwarn | The app NINA | Another type of app (please specify)

Q8: Please indicate how likely you are in future to use a smartphone app for each of the following purposes as a result of an emergency? (Very likely, Quite likely, Neutral, Not very likely, Not at all likely). *To receive emergency warnings | To receive tips about how to stay safe | To contact an emergency service instead of making a 112 call | To share information about the emergency with an emergency service | To find out information about the emergency | To connect with other citizens to help others affected by the emergency*

Q9: Please provide any additional details of your experience of using social media in emergencies or what might encourage you to do so in future.

We asked two additional questions that were not analysed in the scope of this paper.

Survey on the Adoption, Use and Diffusion of Crisis Apps

7

Although media and ICT play an increasingly large role in crisis response and management, in-depth studies on crisis apps and similar technology in the context of an emergency have been missing. Based on responses by 1,024 participants in Germany, we examine the diffusion, usage, perception and adoption of mobile crisis apps as well as required functions and improvements. We conclude that crisis apps are still a little-known form of disaster ICT, but have potential for enhancing communication, keeping users up to date and providing a more effective crisis management as supplement to other media channels dependent on different underlying infrastructures. However, they should be adaptable to user characteristics, consider privacy, allow communication and offer valuable information to raise awareness of potential disasters without creating an overload. Also, the familiarity with and trust in crisis apps should be addressed to maximize their beneficial impact on crisis communication and management. We discuss further implications as well as directions for future research with larger target groups and specific usage scenarios.

This chapter has been published as the conference article "Adoption, Use and Diffusion of Crisis Apps in Germany: A Representative Survey: Design and evaluation of citizens' guidelines" at the Mensch und Computer 2019 (MuC) by Margarita Grinko, Marc-André Kaufhold, and Christian Reuter (Grinko et al., 2019).

7.1 Introduction

Germany is a typical example of a Central European country, where storms and floods are most likely natural disasters, causing human and material damage (Guha-Sapir et al., 2004; Höppe, 2015). At the same time, most German citizens

© The Author(s), under exclusive license to Springer Fachmedien Wiesbaden GmbH, part of Springer Nature 2021
M.-A. Kaufhold, *Information Refinement Technologies for Crisis Informatics*,
https://doi.org/10.1007/978-3-658-33341-6_7

119

have never experienced a disaster and therefore do not consider the risk very high, which is different to other countries in the Middle East or Pacific region (Guha-Sapir et al., 2017). The lack of preparation enhances the potential damage inflicted by a crisis which means locals' threat awareness should be raised and according measures for a higher crisis risk should be supported (Bundesministerium des Inneren, 2009). One topic which has raised the attention for crises is terrorism in Europe in the recent years (Giuliani, 2016). However, such human-induced disasters are only one part of the crisis spectrum which can also be natural incidents, bio-medical or chemical emergencies, or accidents (Boin & McConnell, 2007), whose potential is rising with issues such as global warming and international conflict.

Information and communication technologies (ICT) have been researched as a tool to help manage these crises under the term of crisis informatics (Hagar, 2010; Reuter, Hughes, et al., 2018). However, in past studies, crisis informatics heavily focused on the utilization of social media by diverse actors (Reuter & Kaufhold, 2018) and current research suggests a broadening of its scope (Soden & Palen, 2018). Among other crisis-related ICT, we can find mobile crisis apps, such as Katwarn or NINA (Kotthaus et al., 2016), which support different crisis types, warning and communication functionality, and different degrees of configurability. Their advantage, among others, lies in resilience against infrastructure breakdowns such as power outages, providing an additional channel for crisis communication, allowing for ubiquitous usage, utilizing their battery life and providing recommendations for action even offline. However, especially due to the mentioned lack of threat awareness, their diffusion is still low: According to a recent representative survey, only a sixth of all participants are using crisis apps in general, where particularly weather information is being retrieved (Reuter, Kaufhold, Spielhofer, et al., 2017).

For this paper, we are reviewing literature concerning crisis communication and its demands, specifically referring to crisis apps as a form of ICT (section 7.2). The literature review revealed that there is a low coverage of crisis apps in research, while existing research focuses on factors which need to be addressed to increase their dissemination and the benefit they offer for crisis support. Therefore, our aim is to answer the following research questions on the diffusion, use and adoption (Choudrie & Dwivedi, 2005) of mobile crisis apps among the German population:

- **RQ1**: What is the awareness and diffusion of mobile crisis apps among the German population?
- **RQ2**: What are the desired behaviours and features for mobile crisis apps?

- **RQ3**: What are characteristics and requirements for the adoption of mobile crisis apps?

We address these research questions by conducting a mixed-method study via a survey with a representative sample of German citizens (N = 1,024), concerning their usage of and attitude on crisis apps (section 7.3). We also ask for reasons and seek for improvement suggestions to offer implications for further design. Our results indicate that Facebook Safety Check, Katwarn and NINA are the most diffused crisis apps in Germany (section 7.4). We then discuss implications in section 7.5.

7.2   Background and Related Work

In 2017, about 2.32 billion people were using smartphones, a number which is estimated to increase to 2.87 billion by 2020 (Statista, 2017c). Since the introduction of smartphones, according apps have been popular worldwide and in Germany: Facebook is currently the most used social network with over two billion users, followed by YouTube and WhatsApp (Statista, 2018). Among other purposes like messaging, blogging and sharing media, these networks have also been used to seek information in crisis situations and communicate this information to others (Austin et al., 2012; Eismann et al., 2016; Hughes et al., 2008; Lin et al., 2016; Palen & Anderson, 2016). Besides general-purpose apps mostly for social media, specific mobile crisis apps have been developed to supplement crisis management efforts (Karl et al., 2015). While crisis communication has been previously understood as being bound to an organization (B. F. Liu et al., 2011), citizens can now play an active role in emergency management and offer valuable support via active and digital participation (Kaufhold et al., 2018; Mirbabaie & Zapatka, 2017; Reuter, Ludwig, Kaufhold, et al., 2016; Reuter & Kaufhold, 2018). However, information demands vary throughout crisis phases and there are certain prerequisites for and barriers to an effective crisis communication and management via ICT.

7.2.1   Citizens' Information Demand During Crises

A crisis seldom occurs at once, but usually is preceded by prerequisites or countermeasures and often has lasting effects. Therefore, there are frameworks dividing such events into different stages. Based on earlier theoretical works, Coombs

(2010) distinguishes between the pre-crisis, crisis response and post-crisis phases. Similarly, according to Fischer, Posegga and Fischbach (Fischer et al., 2016), a crisis situation can be divided into four stages: The mitigation phase, where measures are taken to prevent a crisis or to decrease its potential impact; the preparation phase, which precedes a disaster; the response phase immediately after the incident; and the recovery phase, where material and social structures are reconstructed. Citizens' information expectations differ throughout these stages and equally depend on the crisis type (natural or man-made) and the associated predictability. While preparation and response have been widely researched, information demands in recovery and mitigation have been less considered academically (Austin et al., 2012).

In general, information from various sources and channels has to be managed by organizations and emergency services in order to deliver up-to-date, credible and relevant notifications to recipients potentially affected by the crisis (Hagar, 2010). During the **mitigation** phase, many citizens concerned by the crisis take precautionary measures (Thieken et al., 2007). Organizations inform others about risks as well as potential actions and take steps to reduce further risks (Fischer et al., 2016; Volgger et al., 2006). What is important here is not only the physical, but also community resilience (Boin & McConnell, 2007; Dawes et al., 2004; Norris et al., 2008). Measures to prevent man-made crises stretch further than the local communities, into politics and international relations.

To **prepare** the population for a disaster, threat awareness in combination with early and accessible warnings as well as information on potential threats, their consequences and required behaviour is needed (Coombs, 2009; Geenen, 2009; Volgger et al., 2006). In Germany, reasons for low threat awareness are poor local preparation, instruction and warning systems (Menski & Gardemann, 2008; Nestler, 2017). It is therefore necessary to provide guidelines (Kaufhold, Gizikis, et al., 2019) and utilize as many channels as possible: While authorities mostly use mass media like TV and radio or audio cues, ICT enable to reach a broad audience considering the high usage of smartphones in the population (Kaufhold, Grinko, et al., 2019; Klafft, 2013; Reuter, Kaufhold, Leopold, et al., 2017b; Statista, 2017c).

In the **response** phase, those concerned need targeted information on the origin, duration and consequence of the crisis, as well as instruction on behaviour and the safety status of close ones (Nilges et al., 2009; Ryan, 2011; Wade, 2012). In this stage, knowledge is interpreted, disseminated and discussed, which has a significant effect on citizens' behaviour (Kaufhold & Reuter, 2016; Mirbabaie & Zapatka, 2017). This means reliability, consistency and correctness are especially relevant and determine the usage of a certain information channel (Austin et al.,

2012; Geenen, 2009; Petersen et al., 2017; Seeger, 2006; Utz et al., 2013). Emergency services, friends and eye witnesses are the most trusted sources especially on social media (Austin et al., 2012; Huang et al., 2015; Kogan et al., 2015; Olteanu et al., 2015). While ICT play a role for active social media users, they should not replace, but add to traditional information sources (Huang et al., 2015; Petersen et al., 2017; Shklovski et al., 2008).

During crisis **recovery**, communication is equally important: Those who experienced the crisis are concerned with (re-) establishing contact with family and friends as well as asking for further help to enable the recovery (Kaufhold & Reuter, 2016; Semaan & Mark, 2012). ICT is mostly used for organization of volunteers and collaboration of organizations with each other and with citizens (Fischer et al., 2016; Reuter, Ludwig, Kaufhold, et al., 2015). Once the disaster has passed and the impacted community has recovered, the cycle begins again with the mitigation phase.

7.2.2   ICT and Apps for Crisis Management

Ever since September 11th, 2001, ICT have played a role in fulfilling these information demands and helping concerned parties communicate with each other (Hughes et al., 2008). Apart from acute warnings, ICT like crisis apps can be used to provide information, give behaviour advice and support communication and cooperation between citizens, authorities and emergency managers (Jagtman, 2010). However, ICT are not the only channel to be relied on for effective crisis management. Fischer, Posegga and Fischbach (2016) have identified social barriers for crisis communication via ICT like information overload, quality and reliability issues as well as inconsistency. "Social overload" (Maier et al., 2012) can keep users from seeing relevant posts or even from using social media altogether. Besides individual and organizational characteristics, technology with a poor configuration of aspects such as system breakdowns, usefulness, complexity, reliability, presenteeism, anonymity and pace of change may induce technostress, thus hindering the adoption and use of ICT (Agogo & Hess, 2018), which is likely to be a particular issue during stressful crisis situations. Furthermore, privacy and the threat of technical dysfunction keep citizens from using ICT in disasters (Reuter, Kaufhold, Spielhofer, et al., 2017).

When power outages occur, crisis apps can have an advantage over other ICT applications (Nestler, 2017), while one of the barriers for organizations to apply or develop ad-hoc apps for disaster management is a common data language (Shih et al., 2013). Concerning their usage, low familiarity with such apps and their benefits is currently the greatest barrier: As found in a representative survey, only 16% of Europeans have been using crisis apps (Reuter, Kaufhold, Leopold, et al., 2017b). Meanwhile, their functions can range from location-based warning of and information on disasters to instructions and support to eye witness reporting and information sharing as well as emergency calls (Karl et al., 2015). This way, they can support information and communication demands throughout several crisis stages.

Groneberg et al. (2017) have so far conducted the broadest international crisis app comparison based on categories of information, communication and preparation. Accordingly, the most frequent crisis app functionalities in the information category are warnings, followed by maps and general information or news. The first aspect is also the most expected one, while (potential) users also often wish to receive behaviour advice and to help emergency services by providing on-site information (Reuter, Kaufhold, Spielhofer, et al., 2017). However, communication and preparation functions are less widely spread in crisis apps (Groneberg et al., 2017). The medical aspect of emergencies has also become the focus of apps where information and an option to contact emergency services is provided (Bachmann et al., 2015). Of over 600 apps analysed by Bachmann et al. (2015), more than half are directed towards the public.

Among the most popular existing apps in Germany is Katwarn, which warns its users of crises either by knowing their GPS coordinates or by letting the users indicate a region they would like to be informed about (Fraunhofer FOKUS, 2018). For testing purposes, a test alarm can be set to see how the app works. Furthermore, NINA warns of crises based on GPS or Wi-Fi coordinates and offers recommendations for action and general tips (BBK, 2015). Affected people can inform contact points and tell them to what extent they are affected by the crisis or not (in terms of an all-clear signal). An overview and comparison of all free crisis apps currently available in the German app store *Google Play* is provided in Table 7.1, where we have motivated the categories in the table based on our pre-study (see section 7.3). We can see that most apps offer location-based warnings and according settings mostly apply to the location and notification type. Furthermore, some apps offer mostly occasional or static behaviour advice while communication is limited to sharing a warning on other media channels, but seldom includes contacting emergency services or authorities. The development of a crisis app supporting user needs across all crisis stages requires further insight into current grievances, expectations and wishes.

Table 7.1 Comparison of free crisis apps in Germany

App	Crisis types	Warnings	Communication	Instructions	Emergency Contacts	Settings
NINA (DE) (BBK, 2015)	Natural disasters	Push notifications based on entered and current location; map overview	/	Hints by BBK (Office of Civil Protection and Disaster Assistance)	/	GPS or custom locations, push message on/off and sound, types of crises
Katwarn (DE) (Fraunhofer FOKUS, 2018)	Natural disasters, crimes, missing persons	Location-based warning; public displays, websites; text, symbol and map; int. network	Sharing warnings over social media; sending own information	Based on preferred types of events	/	GPS or custom locations (worldwide); test warning available
Disaster Alert (INT) (Pacific Disaster Center, 2018)	Natural and bio-medical disasters	Worldwide map and list overview of current warnings	/	/	/	Map view and push notification settings
Safeture (INT) (GWS Production, 2018)	Political events, accidents, natural and human-induced crises	Location-based list overview of incidents	Personal tracking and sharing position with friends	Advice for each incident	Emergency numbers indicated for each country	Country

(continued)

Table 7.1 (continued)

App	Crisis types	Warnings	Communication	Instructions	Emergency Contacts	Settings
Facebook Safety Check (INT) (Facebook, 2018)	Natural disasters and terrorism	Indication if people in user's location used the service	Informing friends of one's own safety during an incident	Display of news and recommendations from different sources	Can be included in news	/
Sicher reisen (INT) (German Federal Foreign Office, 2018)	Natural and human-induced disasters; political unrest	Location-based travel warnings and push notification	Option to send an all-clear message to contacts	Information on travel preparation and behaviour in an emergency	/	Custom locations and notification type
BIWAPP (DE) (Marktplatz GmbH, 2018)	Natural, chemical, bio-medical disasters, power outage, terrorism and police reports	Push notification with location and incident information; overview of alerts and catastrophes	Option to share notifications in social media	Can be included in alert message	Option to call directly from app	Location and size of notification area; test warning available
Cell Broadcast (INT) (Wikipedia, 2018)	Mostly natural disasters	Warning message issued by broadcaster	/	Depending on the message	Can be included in the message	Dependent on mobile network

(continued)

Table 7.1 (continued)

App	Crisis types	Warnings	Communication	Instructions	Emergency Contacts	Settings
SoftAngel (DE) (Techno et Control GmbH & Co. KG, 2018)	Personal, child or pet emergency; Medical, transport or general assistance	Calls for help by and for app users	Send messages to friends and emergency contacts; play siren or SOS light	/	Entry of personal or general emergency numbers	Location, emergency numbers, notification, Bluetooth, sound and light
Deutsches Rotes Kreuz (DE) (DRK, 2018)	Medical emergencies, accidents	News concerning DRK (German Red Cross) activities; no warnings	/	First aid and behaviour in emergencies; view of own coordinates	Overview of national emergency numbers	/
safeREACH (AT, DE) (safeREACH, 2018)	Organizational emergencies	Push notification SMS to all concerned organization members	Users can issue an alarm	Included in the message	/	Target groups, communication channels, scenarios

7.2.3   Research Gap

As several authors have found, there are currently only few studies focusing on the usage of existing crisis apps (Al-Akkad & Raffelsberger, 2014; Groneberg et al., 2017). Especially how crisis apps should be designed in order to be widely used needs to be examined to increase their popularity and usefulness for crisis management, thus facilitating communication and creating a higher awareness for potential disasters in the German population, or even revealing other ways of usage. Reuter et al. (Reuter, Kaufhold, Spielhofer, et al., 2017) have already conducted a representative study on crisis apps in Germany to find out how many locals intend crisis apps for which purpose, especially to receive warnings and safety tips, and Fischer et al. (2019) identified that perceived risk, trust, and subjective norms positively influence usage intention and compliance intention. However, there is still a lack of extensive research on expectations and needs to increase crisis app diffusion, facilitate use and support adoption. In contrast to previous work and considering the potential of technostress (Agogo & Hess, 2018; Maier et al., 2012), our paper aims to broaden the focus by comparing more apps and by finding out which app functionalities are desired or, in contrast, not needed by German population and how usage barriers can be overcome. We focus on aspects such as how many functions a single app should provide, how much effort users are ready to invest as well as the attitude towards test warnings. Our findings can inform further, more context-related research and design of effective, efficient and satisfactory crisis apps which are well established and help raise awareness and citizen safety level for crisis situations.

7.3   Representative Study: Methodology

The presented questions are taken from a representative online survey, which we conducted in Germany in July 2017 using the ISO-certified panel provider Gap-Fish (Berlin). Our survey consisted of 30 questions in total. In this work, we are specifically examining three of these (Q4–6) which are related to crisis apps, while analysing all survey questions in-depth is outside the scope of this paper. We used two closed and one open-format question, which leads to both quantitative and qualitative results and thus characterizes our mixed-method study. Participants were asked about whether they have used or planned to use different crisis apps (Q4) and to indicate the reason for this, depending on their answer (Q5.1 and Q5.2). Also, they should rate their agreement with different settings and functionalities of such apps (Q6). Six options of Q6 based on a qualitative pre-study that was already published as part of a research paper (Anonymized, 2017). In Table 7.2, the most important crisis app features are displayed which have been found as a result of 22 individual interviews. Among the participants, whose age ranged between 19 and 32 years, were eight males and 14 females.

7.3.1   Characteristics of Survey Participants

The sample of survey respondents (N = 1,024) was adapted to the distribution of age, region, education and income according to the general German population (Bundeszentrale für politische Bildung (bpb), 2016; Statista, 2016; Statistisches Bundesamt, 2016). Our sample consisted of 49.5% female and 50.5% male respondents between 18 and 64 years, nearly half being 45 and older (48%). We recruited participants from every federal state of Germany, where the largest sample came from North Rhine-Westphalia (22%) and Bavaria (16%). Only 1% of participants did not graduate from a school, while 15% held a degree from a university or college. The majority (69%) earned between 1500€ and 3.500€. Almost half of participants indicated to use a smartphone daily (49%). Similar results could be found for daily usage of social media, namely Facebook (46%), instant messaging services (43%) and YouTube (29%). Another third even stated an hourly usage of smartphones (43%) and messengers (33%).

7.3.2   Data Analysis

Our data analysis was undertaken in several steps. First, we eliminated missing values, reducing the participant sample from 1,069 to N = 1,024. Also, we combined demographic variables to categories for a better comparison. We then computed frequencies and percentages of our closed survey question responses in Microsoft Excel. To statistically analyse quantitative data, we used IBM SPSS Statistics 25 (IBM, 2014). Non-parametric tests were chosen based on ordinal data. Chi-squared cross-tabulations served to explore significant differences between demographic factors, media use habits and attitudes and correlations between variables were determined using Spearman's Rho. Regarding the qualitative analysis of open-ended questions, we used open coding (Strauss, 1987) by carefully scanning the responses and establishing codes which were jointly verified and adapted. Each open-ended response was assigned to one or multiple codes to achieve an overview of the relevant topics, and in the last step, meta-codes or categories were derived. Previously acquired knowledge from the literature review and quantitative analysis was used to increase theoretical sensitivity.

Table 7.2 Main functions of the crisis app prototype

Functionality	Description
Warning messages	Overview about current and upcoming crises in a map view
Recommendations for action	Information on how people should behave before, whilst, and after a crisis with detailed explanations and pictures
Warning and all-clear	Inform private persons or official contact points via call or SMS
Chat and organization	Exchange with others and with relief organizations via chat and map of current helpers
Settings	Individual settings of crisis types, site, tone etc.
Emergency contacts	Setting personal contacts relevant during crises

7.4   Results

7.4.1   Distributions of Crisis App Usage: Little Awareness and Interest

In the first question, we asked whether the participants had used, were using or planning to use a crisis app (see Table 7.3). For every crisis app, we found a negation for 61–77% of all respondents. NINA and Katwarn are the apps relatively many participants have used in the past (6% each) and were planning to use in the future (7% each). Even greater was the readiness to use the Malteser app (9%), "Sicher reisen" ("Travelling safely", 11%) and Facebook Safety Check (13%). For the latter, 8% indicated they had already made use of it during a past crisis, making this the most well-known crisis app. The percentage of those who have not used a crisis app and do not intend to use one is over 50%. In another question in our survey (Q7), which was not analysed in-depth for this paper, 15% indicated they have used crisis apps and for additional 10%, they were even a very useful source of information.

Generally, the answers for all crisis apps significantly correlated (p<.01). For chi-squared tabulations to determine differences between demographic groups, we aggregated the responses over all items in Q4. Again, Spearman correlations were used to determine the direction of the trend. The overall usage of crisis apps was significantly positively influenced by everyday usage of social media (χ^2(1,200, 1,024) = 2,998.13, p<.0001; r = .248) and income (χ^2(120, 1,024) = 152.57, p<.05; r = .043), while negatively influenced by age (χ^2(200, 1,024) = 270.50, p<.005; r = − .099) and education (χ^2(200, 1,024) = 361.36, p<.0001; r = − .031). We did not find significant dependencies for other demographic variables.

7.4.2 Attitudes Towards Crisis Apps: Low Effort, High Functionality

To find out which features and characteristics a crisis app should have, we asked about participants' attitude towards different related statements (see Table 7.4). Most participants (68%) agreed that there should be a single crisis app for Germany, while only 21% would install several of them: 44% would like to have an app preinstalled on their phone, but 49% would like to have one with everyday utility. The attitude towards a combination with a weather app is almost equally split: 33% would embrace it, 32% would disagree and 35% are neutral.

It becomes clear what kind of functions are required in a crisis app: For 73%, it would be the possibility to configure the types of disasters they are warned of themselves, and recommendations for action are met with an acceptance rate of 71%. Two thirds (67%) of participants wished for support in their individual preparation for disaster and 57% would like the app to notify contacts of danger and resolution of a crisis. With only 42% of agreement, the chat function is least popular.

In general, only around one third (35%) of respondents would trust warnings they receive without prior setup or agreement. Concerning the frequency of mock warnings, the majority disagreed to receiving them daily (14% agreement), while the most popular frequency was monthly, accepted by 33%. More frequent test warnings would be rather annoying (66%) than accepted (46%). In contrast to that, only about a fifth (21%) would be bothered by too rare notifications.

As for the effort, over two thirds preferred only one crisis app and receive monthly test warnings rather than more frequent ones. Concerning functionalities, most of them are appreciated except for the chat, and an included feature which is useful outside of a crisis suggests that participants would rather have less applications to handle, while having as many possibilities of interaction as possible. Also, the option of configuration and individualization is strongly supported.

Table 7.3 Have you ever used one or more of the following crisis apps, are you currently using them or planning to use them in the future? (Q4)

	Yes, I used it	Yes, I am using it	Yes, I will use it	No, neither	Uncertain/ maybe
Facebook SC	8%	3%	13%	61%	14%
Katwarn	6%	6%	7%	67%	14%
NINA	6%	4%	7%	69%	13%
Sicher reisen	4%	3%	11%	66%	17%
Malteser	3%	2%	9%	68%	18%
Another crisis app	2%	1%	6%	68%	23%
Galileo-LawinenFon	2%	1%	6%	75%	16%
saip	2%	1%	5%	76%	17%
Cell Broadcast	2%	2%	3%	77%	16%
Safeture	1%	1%	4%	76%	17%
ANIKA	1%	1%	4%	77%	17%
BIWAPP	1%	1%	4%	77%	17%
SoftAngel 2.0	1%	1%	4%	77%	17%

For a better comparison of demographic data and the responses, we divided and clustered the sub-aspects from the second quantitative question (Q6) into further two groups based on related insights: Required effort to use the app (including using as few apps as possible) and desired functionalities. In the effort group, we included statements on how many apps participants would like to install and if they should be combined with other functions.

To verify whether the variables in the groups were interrelated, we correlated them using Spearman's Rho. For the sake of clustering, we reverted the answer coding (for example, giving the answer option Strongly agree the coding 1 instead of 5 for the statement I am willing to install several mobile crisis apps on my smartphone). Therefore, a lower code can be assigned to a lower readiness to put in effort. The answers to statements in the functionality cluster showed highly significant correlations ($p < .001$). Similarly, we also found significant correlations for all variables in the effort group ($p < .05$). The initially included statements about the attitude towards frequency of warnings did not correlate with all variables and have therefore been comprised in a separate group. For this purpose, we combined these statements and applied a corresponding weighting (the higher the result, the more frequent participants wish the warnings to be). Furthermore, we regarded general trust in warnings as a separate item. All in all, we arrived at four groups: Effort, functions, trust and warning frequency. With these groups, we aimed to determine whether demographic factors and media usage behaviour influenced attitude towards crisis apps. The results of chi-squared tests are displayed in Table 7.5.

Correlations show that the younger the participants, the greater the trust in warnings without a known source ($r = .154$, $p < .001$). A greater smartphone use led to a greater tendency to accept effort concerning crisis apps ($r = .137$), the need for more functions ($r = .185$) and a greater trust in warnings ($r = .126$). The latter was also positively influenced by social media use ($r = .234$). However, the more participants used social media, the less they agreed to more frequent warnings ($r = -.013$). Similarly, posting behaviour negatively influenced a need for more functions ($r = -.027$). Interestingly, greater use of crisis apps had a positive influence on function expectation ($r = .060$) and trust ($r = .125$), but a negative influence on agreement to frequent test warnings ($r = -.129$).

7.4.3   Reasons for (Non-)Usage of Crisis Apps: Ignorance Is the Primary Issue

In our open question, we asked participants to give the reason for their respective answer to whether they have used or would use crisis apps and indicate features they would like to see in a crisis app. Q5.1: *(If Yes): For what reasons have you used one or more crisis apps, are currently using them or planning to use them in the future?* and Q5.2: *(If No): Why have you not used a crisis app yet, which functions do you miss and/or do you not want to use one in the future?*

For Q5.1, we identified n = 486 valid answers by eliminating those who did not give an answer, made statements like "I would not use it" without stating a reason, or replied "I don't know". For Q5.2, the number of valid answers amounted to n = 538. As some of the answers for Q5.1 contained negative statements and vice versa, thus creating overlapping replies, we did not separate between the questions for the coding, and thus are analysing 1,024 answers in total.

By applying open coding, we divided the answer codes in four general categories: 1) *advantages*, 2) *features*, 3) *scenarios* and 4) *disadvantages*. An overview of the total number of responses for each category together with the five most frequent answers is provided in Table 7.6. For every citation in the following, the respondent ID is given in R[ID].

In the *advantages* category, we summarized all responses highlighting a reason for participants to use a crisis app. The most frequent reply was up-to-date

Table 7.4 What do you think of crisis apps that you can install on your smartphone? (Q6)

	Strongly agree	Agree	Neutral	Disagree	Strongly disagree
One crisis app	38%	30%	25%	4%	3%
Crisis app with everyday utility	16%	33%	34%	9%	8%
Preinstalled crisis app	19%	25%	30%	11%	15%
Crisis app in weather app	10%	23%	35%	18%	14%
Several crisis apps	7%	14%	27%	28%	24%
Warning configurator	35%	38%	19%	3%	4%
Recommendations for action	38%	33%	21%	3%	4%
Support of personal preparation	28%	39%	25%	3%	4%
Warning and all-clear function	24%	33%	32%	5%	6%
Chat function	16%	26%	39%	12%	7%
General trust in warnings	10%	25%	38%	18%	9%
Bothered by frequent test warnings	37%	29%	23%	7%	4%
Acceptance of frequent test warnings	18%	28%	35%	11%	8%
Monthly test warnings	9%	24%	32%	19%	16%
Quarterly test warnings	10%	20%	36%	17%	17%
Bothered by seldom test warnings	7%	16%	36%	25%	16%
Weekly test warnings	6%	15%	31%	26%	23%
Daily test warnings	5%	9%	29%	30%	27%

Table 7.5 Chi-squared results for crisis app expectations. Values which were not significant are indicated in grey. * p<.05, **p<.01

	Effort	Functions	Trust	Warning frequency
Gender	χ^2(25, 1,024) = 18,87	χ^2(19, 1,024) = 11,53	χ^2(4, 1,024) = 4,82	χ^2(45, 1,024) = 39,79
Age	χ^2(125, 1,024) = 120,87	χ^2(95, 1,024) = 96,09	χ^2(20, 1,024) = 49.63**	χ^2(225, 1,024) = 250,92
Region	χ^2(350, 1,024) = 420.91**	χ^2(266, 1,024) = 278,12	χ^2(56, 1,024) = 47,01	χ^2(630, 1,024) = 692.48*
Education	χ^2(125, 1,024) = 140,32	χ^2(95, 1,024) = 107,98	χ^2(20, 1,024) = 23,24	χ^2(225, 1,024) = 231,13
Income	χ^2(75, 1,024) = 74,66	χ^2(57, 1,024) = 34,86	χ^2(12, 1,024) = 12,78	χ^2(125, 1,024) = 132,95
Smartphone use	χ^2(100, 1,024) = 177.14**	χ^2(76, 1,024) = 158.93**	χ^2(16, 1,024) = 30.81*	χ^2(180, 1,024) = 202,70
Social media use	χ^2(750, 1,024) = 760,86	χ^2(570, 1,024) = 594,72	χ^2(120, 1,024) = 194.90**	χ^2(1350, 1,024) = 1949.08**
Posting rate	χ^2(100, 1,024) = 89,81	χ^2(76, 1,024) = 100.18*	χ^2(16, 1,024) = 10,66	χ^2(180, 1,024) = 181,85
Crisis app use	χ^2(1000, 1,024) = 931,80	χ^2(760, 1,024) = 831.66*	χ^2(160, 1,024) = 268.09**	χ^2(1800, 1,024) = 3058.88**

information (named by 129 participants). Related to this were fast and efficient warnings (26 responses): *"In case of an accident, it could be a matter of seconds, so it is helpful to have a direct connection"* (R409). Safety (mentioned 46 times), including being prepared for a crisis, ranked second in this category. Ten participants indicated that they would simply feel safer having a crisis app installed. An equal number referred to the growing need: *"Since recently, 'crises' have been occurring more often, I think it is important to be informed and to know what to do in specific situations"* (R273). This is especially the case for terrorist attacks where Facebook Safety Check has been used. A particularly important advantage for six individuals was that they can reach a large group of people: *"My circle of acquaintances stretches out all over the world, so it is helpful to know in what kind of situation someone is or what might lie ahead of myself"* (R965). In total, six participants already had a positive experience where a crisis app helped them in an emergency, and further six particularly pointed them out to be reliable. Three even view them as an alternative to other media and communication channels regarded as less reliable: *"Once I was very ill and was staved off by the emergency service on the phone, I used an app by an aid agency. Should I find myself in a situation where I can see that several people are in danger, I would immediately use a crisis app"* (R882).

For 19 respondents, the warnings issued by crisis apps provide a possibility to avoid threats and to *"prepare for crises and similar situations and react to them faster"* (R30). Six participants value the practical nature of the app: *"Since there are more and more bad things happening in the world, it makes sense to use these apps, because most people have their smartphone with them all the time anyway"* (R449). Finally, 39 respondents indicated they used or would use crisis apps out of interest or curiosity.

Although *features* were another frequent reason for usage, we looked at them distinctly from advantages in order to get an idea which functions most appreciated in a crisis app. These included warnings, named by 46 participants, and a communication function (mentioned 41 times). One feature that 19 participants named was calling for help, especially based on previous experience: *"Since I recently had a traffic accident, I realize more and more that it is more important in this situation to provide help for the victims, and that this, of course, also facilitates the emergency managers' work"* (R286). Slightly less (14) also mentioned they would benefit from instructions for action. New features were not suggested apart from a unified crisis app, more recommendations for action and up-to-date information.

The category of *scenarios* included all cases participants mentioned for the possible usage of such apps. Travelling was particularly frequent, followed by terrorism (24 and 23 each), while several simply stated they would use it for future cases (31 responses). Twelve people indicated they used them professionally or as part of an emergency services. Also, traffic accidents or jams were named a few times. Other scenarios included floods, accidents at chemical factories and natural disasters.

Eventually, especially among responses for question 5.2, we identified *disadvantages*, or reasons not to use crisis apps. These barriers were mostly psychological or circumstantial rather than technical. Most respondents in this category (212) said that they never heard of crisis apps, while 207 stated they would not need it in the past or present: *"I live in a city which I think is quite unremarkable, and most events (terrorists etc.) won't happen here anyway"* (R146). Because of other media (internet, traditional media or on-site information) they could use, 32 respondents did not see crisis apps as necessary. 16 further were not convinced of their overall benefit. The problem mainly lies in a lack of understanding concerning the functionality (14 responses), as there is *"too little information on significance and use*" (R3013). Nine individuals would like to see only one standardized crisis app: *"The biggest problem is that there is no one unified public app, but many different ones. One can never be sure to get all warnings"* (R846).

For six participants, crisis apps even implied a negative effect of over-sensitization: *"I don't want to be paranoid if there is no reason for it"* (R58). 34 individuals did not have a smartphone, mobile internet, storage space or other technical requirements to download and use crisis apps. The limitation to a device means *"we have to keep in mind that the smartphone can probably also be damaged"* (R177).

Concerning the functions, they were described as unreliable and the whole process of app usage as too much effort (twelve and seven statements each). R257 stated: *"In Germany, this kind of information and crisis management is still in its infancy"*. While one participant lamented the lack of recommendations for action, six others indicated they already knew how to behave in a crisis situation or where to seek information, therefore lacking a need for related apps. In five answers, people expressed the concern that crisis apps could even draw gazers to the site of the emergency, thus getting in the way of crisis responses and helpers: *"[I would use crisis apps] simply to be informed about potential natural events or larger damaging events. EXPLICITLY NOT for posting them in social media (!!!!), so that rescue works or rescuers are obstructed by others"* (R628). A concern on privacy and bad functionality was further expressed: *"I am unsure about the trustworthiness of many providers"* (R118).

However, from the problems mentioned, we can derive that such apps should be tailored especially to users' location to give relevant and up-to-date information to them, all the while indicating trustworthy sources to be able to complement or even replace other media channels. Furthermore, the knowledge about crisis apps in general, as well as their purpose, functions and importance, is still too low, where one participant agrees *"this should be better explained and advertised"* (R380). All in all, almost half of participants did not know or never needed a crisis app. By interpreting the responses, we assume that the number is probably much higher including those who did not explicitly state their lack of awareness.

Our study seemed to be a first step towards a higher awareness: *"Due to this survey, I decided to use a crisis app in the future, because I want to always be up to date"* (R15). The prerequisite is a visible advantage above other types of media: *"I would probably use [crisis apps], if it means information is transferred more precisely and faster"* (R798). Still, there will be an issue of acceptance to a certain degree: While 31 participants explicitly stated they would probably use crisis apps in the future, 31 had no interest in this technology at all (not including those who did not need it or had other reasons). Nine participants specifically demonstrated an aversion against apps in general.

Table 7.6 Overview of the code categories in open responses (Q5). Only the five most frequent codes are displayed with the according number of responses in parentheses

Disadvantages (610)	Advantages (317)	Scenarios (124)	Features (121)
Unknown (212)	Current information (129)	Future (31)	Warning (46)
No need (207)	Safety (46)	Travelling (24)	Communication (41)
Technical prerequisites (34)	Interest (39)	Terror (23)	Ask for help (19)
Other media (32)	Fast (26)	Profession (12)	Behaviour instructions (14)
No interest (31)	Increased need (10)	Accident (10)	Capture gazers (1)

7.5   Discussion and Conclusion

While ICT in disasters and emergencies have been used and researched for over 15 years (Reuter & Kaufhold, 2018; Soden & Palen, 2018), specialized applications for mobile technology beyond social media, especially in Central Europe, is an area of research which needs more attention (Al-Akkad & Raffelsberger, 2014; Groneberg et al., 2017). By comparing different crisis apps and interrogating a representative sample of citizens, we could gather important findings on the perception of and user requirements for crisis apps in German society. Our findings not only complement existing studies with a broader target group, but also offer an insight into expectations and needs concerning communication in crises via specific apps as well as practical implications for them to become more use- and helpful in such situations.

7.5.1   Main Results

In the following, we briefly summarize the essential results of our representative survey based on the predefined research questions.

RQ1: What is the awareness and diffusion of mobile crisis apps among the German population? Compared to the earlier studies (Klafft, 2013; Reuter, Kaufhold, Leopold, et al., 2017b) where 16% of participants indicated the use of crisis apps, the awareness did rise: The percentage of participants using or ever having used a crisis app has reached 25%. Furthermore, based on Q7 from our survey, of all crisis app users, 40% considered crisis apps as a very helpful source, which is similar to the perceived helpfulness of radio, television and contacting emergency services (all 41%) and even more useful than social media (36%), other online offers (28%), personal conversations, phone calls (both 23%) and newspapers (22%). This suggests that crisis apps are and will be a worthwhile complement to existing media channels used for crisis response. On the other hand, only 13% are ready to use a crisis app in the future and between 61% and 77% are still refusing to adopt this type of ICT at all. The most frequent reasons for this are that participants simply are not familiar with or not interested in crisis apps due to their safety perception.

RQ2: What are the desired behaviours and features for mobile crisis apps? In our open responses, it became clear that relevant, reliable and up-to-date information plays a large role in crisis app functionalities, as found in prior works (Fischer et al., 2016; Geenen, 2009; Hagar, 2010; Volgger et al., 2006). Over two thirds of participants would like to have only one crisis app and almost half would like it to have everyday functionalities included. A similar number wished for a preinstalled app. Participants appreciated all features, with the warning configurator and recommendations for action being the most popular (over two thirds agreed). At the same time, the trust in warnings was relatively low: Only 35% agreed to this point. Test warnings should occur as seldom as possible, with one third of participants voting for the monthly alternative. Overall, we can state that participants would like to have as little effort as possible while being offered as many functions as possible at the same time.

RQ3: What are characteristics and requirements for the adoption of mobile crisis apps? Demographic aspects do not play a large role in crisis app perception apart from age, in contrast to earlier studies concerned with technology adoption (Choudrie & Dwivedi, 2006; Petersen et al., 2017). However, the more users are already utilizing smartphones and crisis apps, the more they trusted in warnings, wished for functions and accepted effort in using the crisis apps. The main reasons for using crisis apps, according to our respondents, are up-to-date information, reliability as well as increased perceived safety and preparedness. An alternative for communication and fast help was also a criterion. While false information, data security and potential malfunction have been the main barriers for using ICT in disasters (Reuter, Kaufhold, Spielhofer, et al., 2017), lack of knowledge, threat awareness and advantage above other channels are predominant reasons not to use crisis apps in our study. Some participants expressed general scepticism towards this type of ICT such as false alarms, over-sensitization, privacy and gazers.

7.5.2 Implications of Our Findings

Responses to our open question showed that our participants feel relatively safe, which stands in contrast to increasing natural and human-induced disasters in Germany and Central Europe in general (Guha-Sapir et al., 2017). Risk awareness is comparably low in German citizens, but this is not only a national problem (United Nations, 2006). Therefore, our main implication is that crisis apps and their benefits need to be advertised among Central Europeans, for example via widely used traditional media, in order to be diffused, adopted and effectively used in crises. The fact that many indicated to feel safer with a crisis app shows a general need for crisis preparation, both on a mental and infrastructural level. Linking crisis apps to diverse social media and other communication channels could support the spread of information as well as the usage of the app itself. Although in their answers, most participants referred to the response and sometimes to the preparation or pre-crisis phases (Coombs, 2010; Fischer et al., 2016), crisis apps should already be introduced in the mitigation stage to make sure information and communication is guaranteed throughout all stages of crisis. Consequently, they can help avoid larger consequences induced by low threat awareness, the so-called "vulnerability paradox" (Bundesministerium des Inneren, 2009), which means the impact of a disaster is higher due to low material and mental citizen preparedness coming from low threat awareness.

As a prerequisite, crisis apps thus need to be more available, offer useful functions appreciated by users, require minimum effort, include information that is accurate, recent and significant, as well as consider users' privacy and personalization concerns. This would increase trust, which is a crucial factor for adoption of new ICT (Choudrie et al., 2005; Petter et al., 2013) as well as social resilience in case of disasters. At the same time, by reducing the number of required apps and test warnings as wished by our participants, we can address the issues of technical and social overload (Agogo & Hess, 2018; Kaufhold, Rupp, et al., 2020; Maier et al., 2012), which would add to the stress already induced by a crisis. Another factor which should be considered is the effort of adopting a new technology, while it is equally an opportunity for those who do not wish to use social media (Maier et al., 2015).

Functions to which our participants most agreed can be categorized as information and preparation, while communication mostly referred to an all-clear notification (Groneberg et al., 2017). Based on our analysis of existing apps, there is a lack of apps who fulfil these requirements found in our study together with the desired degree of personalization, where specialized apps need to be created for different target groups and cultures. The need for production of own information did not appear often in our sample apart from seeking help. However, communication is an essential part of crisis ICT, in previous work as well as in our open answers; as suggested by Mirbabaie and Zapatka (2017) as well as the UN (2006), users are able and willing to fulfil different roles in crisis prediction and communication across the crisis phases, while crisis apps can offer this opportunity. Crisis communication via ICT is only effective if locals participate accordingly (Jagtman, 2010). In our participants' interest in using the app in the future, also for specific cases such as terrorism, travels or traffic, we can see a tendency in an increased readiness to prepare for potential threats which may be further researched.

7.5.3   Limitations and Outlook

Firstly, there are limitations and potentials for future **quantitative survey design**. As diffusion, use and adoption of crisis apps is still low, most responses have been given in a hypothetical context. Asking our participants whether they had experienced a crisis before could help differentiate between behaviour in the mitigation, preparation, and response phases. As we found especially in our open question, apart from diverse usage contexts, there are different categories of users: for example, members of crisis management organizations and civilians. These stakeholders have different requirements and expectations concerning crisis apps. The open question could also be equipped with a "no answer" option to reduce the number of invalid responses. Instead of limiting the survey to 13 given crisis apps, giving participants the option to add more apps they know and use could paint a more accurate picture of current diffusion and perception, especially since a general usage question produced more positive replies than asking about specific apps.

Since our study has been conducted online, a certain degree of experience with technology is a prerequisite. While our participant sample has been only representative concerning demographic aspects, therefore, results linked to technology usage should be further verified offline and with a broader participant sample. Klafft and Reinhardt (2016) pointed out that in order to be able to address all potential users of crisis apps, language and literacy barriers as well as physical and mental impairments have to be considered. A future prototype of an optimized crisis app should consider these aspects and support all kinds of users by utilizing multimodal, specifically focused information and suggestions. As our sample is restricted to the German population, studies in other countries should expand the results we have gained, especially considering cultural aspects such as risk cultures, where, for example, the responsibility for crisis prevention and management is perceived differently (Cornia et al., 2016). Since Germany is considered a state-oriented risk culture, we estimate that individual-oriented risk cultures, such as the Netherlands, are more motivated to use ICT as a means for individual preparation.

Secondly, we identified potentials for future **qualitative user studies**. For now, our study corresponds to the pre-prototype approach testing user acceptance proposed by Davis and Venkatesh (2004). Similar to their work, we found that perceived ease of use in terms of effort, usefulness and social influence are significant factors for technology acceptance (Davis & Venkatesh, 2004; Venkatesh & Davis, 2000). A user study carried out in an actual or simulated crisis situation could help validate and extend our results, adding to features and characteristics of crisis app required for a disaster.

By conducting a user study based on one or several crisis apps, more insights could further be gained on desired and required functionality and user experience (UX) design, perceived effort and effectivity of as well as satisfaction with this type of ICT. While the desired functionalities were similar to those found in previous studies (Reuter, Kaufhold, Spielhofer, et al., 2017), especially in actual crisis situations, their use, adoption and support of disaster management as well as potential problems should be examined using a case study which addresses local and situational differences (Choudrie & Dwivedi, 2005; Groneberg et al., 2017). Furthermore, as we found that users would like to invest as little effort as possible, a case study would further give the opportunity to directly measure technostress using an established scale (Agogo & Hess, 2015). Since mobile apps can influence behaviour to different levels especially depending on their UX, conformity with expectations, and perceived usefulness (Roidl & Rüsing, 2014), case studies could help research how these dimensions can support user needs and crisis management throughout the different stages. This can also occur via simulations using technologies like VR with embedded mobile apps, which have the potential to prepare and even replace resource-intensive real crisis simulations if further developed and improved (Rother et al., 2015). Here, it would be most interesting to examine to what extent those who answered "Yes, I will use [a crisis app]" are prone to keeping their resolution and which factors influence their adoption of this ICT.

Based on this, further mixed-method research could benefit from the application of acceptance theories, such as the Technology Acceptance Model (TAM) (Davis, 1993) or the Unified Theory of Acceptance and Use of Technology (UTAUT) (Venkatesh et al., 2003) to generate further theoretical insights, as already demonstrated by Fischer et al. (2019).

Thirdly, future research should comprise **technical feasibility studies**. We could compare their added value to the benefit of other ICT, such as social media and websites, for crisis communication. While some crisis apps like Facebook Safety Check are connected to other platforms, the benefits of standalone apps need to be discussed, such as stability during infrastructure breakdowns. The latter often require costly and elaborate recovery plans (Boin & McConnell, 2007), while crisis apps are usually only limited by phone battery life. The latter can be enhanced by power banks, while internet connectivity issues can be overcome by mobile ad-hoc networks (Reuter, Ludwig, Kaufhold, et al., 2017a). Since these networks are often small, local and decentralized, Low Power Wide Area Networks (LoRaWAN) might help to connect local networks, ensuring a widespread distribution of information or warning messages. In future, this could not only be realized by static nodes using emergency generators but also mobile nodes, such as drones, robots or vehicles. The information of all citizens, which was found as insufficiently developed by Nestler (2017), can be reached by an app covering all relevant types of crisis in the users' area, and especially used by a maximum number of locals. As suggested by the United Nations (2006), a global network built on national warning systems will help predict, prepare for and effectively and cooperatively manage emergencies. At the same time, it may increase the perceived benefit and citizens' readiness to use crisis apps.

Part III
Design and Evaluation Findings

Design and Evaluation of Social Media Guidelines for Citizens

8

Social media have been established in many natural disasters or human-induced crises and emergencies. Nowadays, authorities, such as emergency services, and citizens engage with social media in different phases of the emergency management cycle. However, as research in crisis informatics highlights, one remaining issue constitutes the chaotic use of social media by citizens during emergencies, which has the potential to increase the complexity of tasks, uncertainty, and pressure for emergency services. To counter these risks, besides implementing supportive technology, social media guidelines may help putting artefact and theoretical contributions into practical use for authorities and citizens. This paper presents the design and evaluation (with 1,024 participants) of citizens' guidelines for using social media before, during, and after emergencies.

This chapter has been published as the journal article "Avoiding chaotic use of social media before, during, and after emergencies: Design and evaluation of citizens' guidelines" in the Journal of Contingencies and Crisis Management (JCCM) by Marc-André Kaufhold, Alexis Gizikis, Christian Reuter, Matthias Habdank, and Margarita Grinko (Kaufhold, Gizikis, et al., 2019).

8.1 Introduction

Social media are a part of everyday life and are also prevalent during crises, disasters, and emergencies. Reuter & Kaufhold (2018) summarized 15 years of social media use during natural disasters and human-induced crises, identifying different usage, role and perception patterns between administrative and public stakeholders both in the real and virtual realms. New opportunities have emerged for emergency services to alert and warn of crises (Brynielsson et al., 2018) and for

M.-A. Kaufhold, *Information Refinement Technologies for Crisis Informatics*, https://doi.org/10.1007/978-3-658-33341-6_8

social media analytics (Stieglitz, Mirbabaie, Ross, et al., 2018). Palen & Hughes (2018) conclude that social media also promote socio-technical innovations, such as citizen reporting, community-oriented computing, collective intelligence and distributed problem solving, and digital volunteerism. However, one remaining issue is that citizens' activities coordinated via social media have the potential to increase the complexity of tasks, uncertainty, and pressure for emergency services, for instance, if volunteers themselves are endangered (Perng et al., 2012). Given the risks of chaotic social media use (Kaewkitipong et al., 2012), guidelines may help authorities implement social media into their organizational culture (Reuter, Ludwig, Kaufhold, et al., 2016) and encourage citizens to foster good practice and prevent misuse of social media during emergencies (Reuter & Spielhofer, 2017).

While there are plenty of guidelines for using social media in general or during emergencies from the perspective of organizations, few guidelines approach the citizens' perspective and none of them has been evaluated with citizens. Thus, our study aims to contribute with the design and evaluation of citizens' guidelines for using social media in emergencies, while the latter aspect of evaluation is the main focus of this paper. The aim of the guidelines is to approach various issues that occur with social media usage from citizens' perspective during crises and that are not adequately addressed by existing guidelines. Such issues include task complexity and uncertainty, information overload, the need for timely and trustworthy information, and the effective organization of volunteer groups and help for emergency services. We will outline them in detail in the next chapter. Our main research question is: **How do citizens perceive guidelines for social media use in emergencies (RQ1)?** To guide the analysis of our results, we formulated further five sub-questions:

- To what extent do participants agree to the presented guidelines on social media behaviour before, during, and after emergencies (RQ1.1)?
- How consistent is the agreement with guidelines across the categories, i.e. concerning the use of social media before, during, and after emergencies (RQ1.2)?
- To what extent is the attitude towards the guidelines affected by smartphone and social media usage in daily life and in emergencies (RQ1.3)?
- How do demographic factors influence the agreement with guidelines concerning the use of social media in emergencies (RQ1.4)?
- What are caveats and improvement suggestions towards the guidelines on social media in emergencies (RQ 1.5)?

First, related work on the use and challenges of social media in emergencies and an analysis of existing guidelines for the use of social media will be discussed (section 2.8). Based on this, we designed citizens' guidelines for evaluation (section 8.3). To gather quantitative and qualitative feedback on our research questions, we then conducted a representative online survey in Germany in July 2017 (section 8.4). The paper concludes with a short summary of results, discussion, and outlook for future research (section 8.5).

8.2   Related Work: Challenges of Social Media in Emergencies and Existing Guidelines

Based on the potentials of social media in the Emergency Management Cycle (EMC), comprising the phases of mitigation, prevention, response, and recovery (Baird, 2010), we present and discuss existing guidelines for the use of social media. In this paper, we use the term emergency as a state of collective disruption, caused by natural hazards (i.e. earthquakes, floods, hurricanes, and tsunamis) or human-induced disasters (i.e. accidents, attacks, shootings, terrorism, and uprisings) that are of a sudden and unexpected nature. Therefore, they threaten societal values and structures, and require authorities to make decisions during uncertain circumstances (Boin et al., 2005).

8.2.1   Use and Challenges of Social Media in Emergencies

For 15 years, the public has used social media in emergencies (Reuter & Kaufhold, 2018), with 2012 Hurricane Sandy being one of the most prominent cases (Caragea et al., 2014; Homeland Security, 2013; Kogan et al., 2015) but also the 2008 Sichuan earthquake (J. Li & Rao, 2010; Qu et al., 2009) or 2013 European floods (Kaufhold & Reuter, 2016; Reuter & Schröter, 2015). Accordingly, a mixed-method study with authorities and citizens outlines that, in times of crisis, Facebook is perceived as a channel to reach the general public and where people debate and connect with family and friends, whereas Twitter is perceived as an elite channel which primarily serves as an early warning channel (Eriksson & Olsson, 2016). Research outlines a variety of potentials of social media use, i.e. increasing the resilience of citizens (Jurgens & Helsloot, 2017). Panic, looting, and helplessness or dependency on external rescuers in crisis situations are rare, contrary to popular belief (Helsloot & Ruitenberg, 2004). Instead, self-organization and help is a common reaction: Citizens of affected areas assist emergency services by social aftercare for victims and relatives online (Mirbabaie & Zapatka, 2017), search for missing people by sharing and using information on microblogging platforms (Qu et al., 2011), and coordinate and mobilize volunteers through social media (Albris, 2017; Kaufhold & Reuter, 2016).

Despite these potentials, research also outlines a variety of risks associated with the use of social media (Hiltz & Plotnick, 2013; Plotnick & Hiltz, 2016; Reuter, Ludwig, Kaufhold, et al., 2016). A study on the 2011 Norway attacks revealed that citizens' activities coordinated via social media increased the complexity of tasks, uncertainty, and pressure for emergency services, since volunteers endangered themselves by autonomous rescue attempts (Perng et al., 2012). Likewise, Kaewkitipong et al. (2012) report on the risks of chaotic social media use, including "information redundancy, information inconsistency, chaos and rumours among all social communities" (p. 13), that occurred during the 2011 Thailand flooding disaster. A further study on the 2010 Haiti earthquake highlights that uncertainty can be generated by unreliable information and mistakes due to chaotic and disorganized work of volunteers (Valecha et al., 2013). During the 2013 floods in Dresden, Facebook groups such as "Fluthilfe Dresden" (Flood Help Dresden) distributed and spread out citizen volunteers successfully i.e. for filling and piling sandbags, although in some cases more people were mobilized than the response effort could accommodate (Albris, 2017). Thus, the convergence of volunteers led to suboptimal outcomes and authorities feared dilettante behaviour by citizen volunteers, indicating the need for guidance to improve citizens' behaviour both in real and virtual spaces (Reuter et al., 2013). Accordingly, Schmidt et al. (2017) outline the challenges of governing self-organization of citizens, connecting online platforms with on-site response initiatives, fostering inclusiveness, and managing information.

8.2.2   Existing Guidelines for the Use of Social Media

Our search revealed that there are many existing guidelines concerning the use of social media in general and for use in emergency situations (see section 8.3.1). We found no commonly accepted definition for social media guidelines in emergencies, but multiple aspects in terms of their applicability, domains, and goals are highlighted in different definitions. For instance, social media guidelines for emergencies intend to "enable citizens using new mobile and online technologies to actively participate in the response effort" (Belfo et al., 2015) and to "enhance the safety and security of citizens by supporting both citizens, and public authorities, in their use of social media to complement their crisis management efforts"

(Helsloot et al., 2015). Table 8.1 contains guidelines and handbooks from different institutional backgrounds, ranging from civil institutions, cities, and authorities to the military. In addition, organizations, companies and authorities publish guidelines for the use of social media by their employees. These guidelines usually focus on the employee's private use of social media and rules regarding the company or authority, therefore lacking the use perspective of the citizens as public actors. The guidelines mostly point to law, behaviour, data security issues, and non-disclosure obligations. Yet, these handbooks do not concern social media usage with regard to the EMC.

As Table 8.2 shows, there are guidelines for the specific use of social media in emergencies. From our analysed set, only the guidelines of the European projects COSMIC and iSAR+ as well as the Emergency 2.0 Wiki cover the citizens' perspective. The COSMIC guidelines provide recommendations in terms of preparation, seeking aid and information, providing aid, mobilization, as well as recording and sharing (Helsloot et al., 2015). Over 14 pages, the different advices are illustrated by text, examples, pictures, and bullet points. In these guidelines, only the authorities' recommendations (represented by parenthetical crosses in Table 8.2) but not the citizens' guidelines refer to the EMC.

In contrast, the iSAR+ guidelines provide three pages of "recommendations and explanations on how to prepare for a crisis situation, what to consider when the incident is about to happen, ongoing, and when it is over" (Belfo et al., 2015). The guidelines are divided by the phases of preparation, warning, acute incident, and aftermath, giving a short explanation and several bullet points with concrete advice for each. Similarly, the Emergency 2.0 Wiki (2015) provides "tips for the public" on how to behave before, during, and after emergencies, including a general introduction to social media, advice on finding and sharing information during the different phases, communicating with family and friends, or saving battery power. Both, the iSAR+ guidelines and Emergency 2.0 Wiki organize their recommendations based on the EMC, although the mitigation phase is not included. These guidelines mostly use bullet points and pictures for advice and illustration, and they are distributed over different wiki pages.

Table 8.1   List of guidelines for the use of social media in general

Title	Publisher
Verification Handbook—An ultimate guideline on digital age sourcing for emergency coverage (2014)	European Journalism Centre, EU
Social engagement handbook 2.0 (2012)	American Red Cross, US
Social media handbook (2014)	United States Army, US
Ein Leitfaden zum Umgang mit Social Media im DRK (2012)	Deutsches Rotes Kreuz (German Red Cross, DRK), DE
Social-Media-Guideline—Empfehlungen für einen sicheren Umgang mit sozialen Medien (2012)	Berliner Feuerwehr (Berlin Fire Department), DE
Verhalten in sozialen Netzwerken (2011)	Bundesanstalt Technisches Hilfswerk (Federal Agency for Technical Relief), DE
Das soziale ins Netz bringen—die Caritas und soziale Medien (2014)	Deutscher Caritasverband (German Caritas Association), DE
Social media guidelines for IFRC staff (2009)	International Federation of Red Cross and Red Crescent Societies (IFRC)
Social Media—Guidelines for Canadian Red Cross Staff and Volunteers (2013)	Canadian Red Cross, Canada
Rotkreuz-Social-Media-Policy (2010)	Österreichisches Rotes Kreuz (Austrian Red Cross), Austria
ACT Government Social Media Policy Guidelines (2012)	ACT Government, Australia
Social Media in der Hamburgischen Verwaltung—Hinweise, Rahmenbedingungen und Beispiele (2011)	Freie Hansestadt Hamburg (Free Hanseatic City of Hamburg), DE
Social Media Guidelines and Best Practices—Facebook (2012)	Centers for Disease Control and Prevention (CDC), US
Social Media Guidelines and Best Practices—CDC Twitter Profiles (2011)	
The health communicator's social media toolkit (2011)	
CDC's Guide to Writing for Social Media (2012)	

Table 8.2 List of guidelines for the use of social media in emergencies (ES = emergency service guidelines, C = guidelines for citizens, EMC = emergency management cycle)

Title	Publisher	ES	C	EMC
Guidelines for the use of new media in crisis situations (2015)	COSMIC project, EU	x	x	(x)
Warning and informing Scotland using social media in emergencies (2012)	Scottish Government, Scotland	x		
Social media for emergency management—a good practice guide (2014)	Wellington Region Emergency Management Office, New Zealand	x		(x)
Emergency 2.0 Wiki (2015)	Emergency 2.0 Wiki	x	x	x
Social Media in an emergency: Developing a Best Practice Guide Literature Review (2012)	Opus International Consultants Limited, New Zealand	x		(x)
iSAR+ Guidelines (2015)	iSAR+ Project, EU	x	x	x
Using social media for emergency notifications—7 questions for emergency managers to consider	Twenty First Century Communications, Inc., US	x		
Social Media in Emergencies—UNICEF Guidelines for Communication and Public Advocacy (2012)	United Nations Children's Fund (UNICEF)	x		
Smart tips for category 1 responders using social media in emergency management (2012)	Defence Science and Technology Laboratory (DSTL), UK	x		(x)
Crisis communications and social media- A best practice guide to communicating an emergency (2014)	International Air Transport Association (IATA)	x		
Next Steps: Social Media for Emergency Response (2012)	Homeland Security, US	x		
Using Web 2.0 applications and Semantic Technologies to strengthen public resilience to disasters (2013)	Disaster 2.0, EU	x		
Bevölkerungsschutz: Social Media (2014)	Bundesamt für Bevölkerungsschutz und Katastrophenhilfe (Federal Office of Civic Protection and Disaster Assistance, BBK), DE	x		

(continued)

Table 8.2 (continued)

Title	Publisher	ES	C	EMC
The Use of Social Media in Risk and Crisis Communication (2014)	Organization for Economic Co-operation and Development (OECD)	x		
Leitfaden Krisenkommunikation (2014)	Bundesministerium des Innern (Federal Ministry of the Interior, BMI), DE	x		

8.2.3 Research Gap and Objective

Despite the potentials of citizens' use of social media before, during or after emergencies, their activities potentially increase the complexity of tasks, uncertainty, and pressure for emergency services (Perng et al., 2012), for instance, by irrelevant and inconsistent information, information overload and mistakes due to chaotic and disorganized work of volunteers (Kaewkitipong et al., 2012; Valecha et al., 2013). Thus, guidelines may motivate citizens to foster good practice and prevent misuse of social media during emergencies (Reuter & Spielhofer, 2017). Previous research indicates that a variety of general or emergency-specific guidelines exist, but most of them focus on the authorities' or organizations' perspective of social media use. Only three guidelines explicitly contained recommendations for citizens before, during, and after emergencies, varying in length and media richness (Belfo et al., 2015; Emergency 2.0 Wiki, 2015; Helsloot et al., 2015). Considering feedback from practitioners' workshops (section 8.3), these guidelines were perceived as too extensive for reaching a broad audience, and, to our best knowledge, none of them was evaluated with citizens in terms of their practical value. Thus, besides presenting the design of concise citizens' guidelines that are suitable for distribution to large audiences before, during, and after emergencies, the main aim of this paper is the quantitative and qualitative evaluation these guidelines according to our research questions.

8.3 Methodology

This chapter describes the design and evaluation methodology of the citizens' guidelines for using social media before, during, and after emergencies.

8.3.1   Designs of Citizens' Guidelines

The citizens' guidelines were designed to convey general aspects while using social media and information on how to behave before, during, and after an emergency. Firstly, we conducted literature reviews on the potentials and challenges of social media from the perspective of citizens and then reviewed guidelines on the use of social media in general and during emergencies. The guidelines listed in Table 8.1 and Table 8.2 were identified by searching in Google, scanning through related European projects on social media, and considering expert recommendations of an end-user advisory board (EAB) of our European project. The search focused on English guidelines, although the project consortium recommended a few German guidelines that were considered subsequently. Although all found guidelines were considered, only three contained recommendations for citizens in emergencies and thus were relevant for the analysis in this paper.

For the design of our guidelines, no extensive comparative analysis was performed between these three guidelines, but their content was presented as input during an EAB workshop, comprising 18 participants from emergency services (O'Brien et al., 2016). Based on these guidelines, EAB members and practitioners prioritized the most important information that should be communicated to citizens with the intention to keep the guidelines concise. Under consideration of their input, further literature and (revisiting of) guidelines, a first draft of the citizens' guidelines was created. The draft was presented at another EAB workshop, comprising 15 participants emergency services, to gather insights for the design of the final guideline versions (Gizikis, Susaeta, et al., 2017). Since the focus of this paper lies within the evaluation but not the design of the guidelines, please find additional information on a related project deliverable (Gizikis, O'Brien, et al., 2017). The resulting guidelines are presented in Table 8.3.

Table 8.3   Social media guidelines

General Aspects	• Interact with respect and courtesy (Bundesanstalt Technisches Hilfswerk, 2011). • You are responsible for your writing, think of possible consequences (Bundesanstalt Technisches Hilfswerk, 2011). • Protect your privacy and check the privacy settings (Deutsches Rotes Kreuz, 2012). • Respect intellectual property rights, including pictures, graphics, audio and video files (Berliner Feuerwehr, 2012). Add references, make quotations marked (Deutscher Caritasverband, 2014), specify sources and, without approval, do not talk about a third person (Daimler AG, 2012). • Verify your information before posting (Helsloot et al., 2015). • Correct a mistake if you made one (Kaewkitipong et al., 2012).

(continued)

Table 8.3 (continued)

Preparation	Be prepared
	• Know the social media accounts of your local and national ES and follow them. This will help find real-time information during an emergency (Helsloot et al., 2015).
	• Read what to expect from ES in social media. Are they always online? Do they reply to posts in social media (Reuter, Ludwig, Kaufhold, et al., 2016)?
	• Look for apps that ES provide and download them to stay informed during an emergency (Reuter et al., 2017).
	• Follow the information from ES on how to prevent and stay safe during emergencies (Helsloot et al., 2015).
Response	Stay up-to-date
	• Follow official accounts and local organizations to get information updates (Helsloot et al., 2015).
	Social media does not replace 112
	• Remember you can use social media for information updates, but it does not replace emergency calls (Helsloot et al., 2015). If in danger, always call 112 first.
	Be responsible and avoid spreading rumours!
	When you post information about an emergency in social media:
	• Always mention the ES account or include any already used hashtags. When possible report a location and use photos (Helsloot et al., 2015).
	• Tell only facts and don't send information you are not certain about (Helsloot et al., 2015).
	• Share only official and reliable information and avoid spreading rumours! The spreading of false information can threaten the smooth deployment of rescue teams and put you and your relatives at additional risk (Perng et al., 2012).
	• If you spot or shared false information, please correct it (Helsloot et al., 2015).
	• Forward received official messages to your contacts or share them (Helsloot et al., 2015).
	Volunteering initiatives
	• Look for emergent volunteer initiatives in Facebook groups, Google crisis maps or trusted users in Twitter; they may help to increase the impact of your activities (Kaufhold & Reuter, 2016)!
	• If you intend to initiate your emergent volunteer initiative, please check for existing initiatives first and carefully chose the scope of your possible contribution.
Recovery	• Follow official accounts and local organizations to get information updates (Helsloot et al., 2015). Communicate even after a crisis and use social media for the processing of the event.
	• Give feedback to the authorities (Reuter et al., 2016).
	• Restore missing contact and ask for welfare of family and friends (Qu et al., 2011).
	• Help others reconstructing/handling the event (Mirbabaie & Zapatka, 2017).

8.3.2   Evaluation of the Citizens' Guidelines

We conducted a representative online survey (N = 1,024) of the adult German population in July 2017, using the ISO-certified panel provider GapFish (Berlin). They guarantee panel quality, data quality, and security, as well as survey quality through various (segmentation) measurements for each survey within their panel of 180,000 active participants. Our overall survey covered also other topics from our research project. In this work, we are examining five of these (Q22—26), related to social media guidelines. Firstly, participants were asked closed questions on the perceived importance of the guidelines before (Q22), during (Q23), and after (Q24) an emergency (*"How do you rate the following recommendations for usage of social media [on/during/after] an emergency?"*), and general aspects while using social media (Q25: *"How do you rate the following, general aspects for using social media in an emergency?"*). The responses were measured using a five-point Likert scale ranging from "very important" to "not important at all". Secondly, we asked for open-ended criticism or improvement suggestions for the presented guidelines about the use of social media before, during, and after a crisis (Q26).

8.3.2.1 Characteristics of Survey Participants

The sample of survey respondents (N = 1,024) was adapted to the distribution of age, gender, region, education, and income according to the general German population (Bundeszentrale für politische Bildung (bpb), 2016; Statista, 2016; Statistisches Bundesamt, 2016). Preselection of participants was carried out by the external survey panel provider (GapFish) so that we received the already adapted sample. Based on these statistics, our sample consisted of 49.5% female (n = 507) and 50.5% male (n = 517) respondents. They were between 18 and 64 years old, nearly half of them being 45 and older (n = 492, 48%). We recruited participants from every federal state of Germany, where the largest sample came from North Rhine-Westphalia (n = 228, 22%) and Bavaria (n = 165, 16%). Only 1% (n = 9) of participants did not graduate from a school, while 15% (n = 160) held a degree from a university or college. Over two thirds (n = 707, 69%) earned between 1.500€ and 3.500€. Concerning the smartphone usage, almost half of respondents indicated to use it daily (n = 500, 49%). A similar result could be found for daily usage of social media, namely Facebook (n = 475, 46%), instant messaging services (n = 449, 43%), and YouTube (n = 301, 29%). Another third stated an hourly usage of smartphones (n = 441, 43%) and instant messengers (n = 339, 33%), while 19% (n = 196) claimed to actively post in social media daily and an almost equal amount (n = 197) at least once a week. As for behaviour in crisis situations, over half of participants (n = 556, 54%) indicated to have used social media in such an event. Slightly more (n = 607, 59%) expected to see messages by critical infrastructure (CI) operators on social media in the case of an infrastructure failure.

8.3.2.2 Quantitative and Qualitative Analysis

For the quantitative analysis, the survey data was extracted and analysed using IBM SPSS Statistics 23, a software package for analysing quantitative data (IBM,

2014). Microsoft Excel was used for qualitative coding and for the design of the diagrams showing relative amounts of participant responses (see section 8.4). The analysis consisted of three key steps: (1) *Preparing* the data including assignment of missing data values, and combination of categories of demographic background variables. This occurred by excluding participants where an answer to a closed question was missing, which would have made quantitative analysis unreliable, thus reducing the number of participants from N = 1,069 to N = 1,024. (2) *Exploring basic frequencies* for each question. (3) *Using cross-tabulations* with chi-squared tests to explore any significant differences across different types of respondents in relation to gender, age, region, income and education level, as well as smartphone and social media usage. (4) *Determining correlations* between ordinal items using Spearman's Rho.

The qualitative analysis of our open-ended survey question was based on the inductive approach of *grounded theory* (Strauss, 1987). We used *open coding* to derive categories from the more qualitative open-ended responses by carefully reading and aggregating categories. In a first iteration, to achieve a quick overview of the interesting and relevant topics, two coders went through the open-ended responses independently and proposed a set of preliminary categories. The categories were discussed in tandem to agree upon a shared set of category codes. The previously acquired knowledge from the literature review and quantitative analysis was used to increase theoretical sensitivity. In a second iteration, the coders independently assigned each open-ended response to one or multiple categories. Finally, the codes were jointly checked to reach agreement between the coders. Each quotation is referenced with the participant's response identifier.

8.4   Results

This chapter presents the results of the quantitative and qualitative evaluation of citizens' attitudes towards the guidelines on social media use before, during, and after emergencies.

8.4.1   Importance of Individual Recommendations of the Guidelines

To what extent do participants agree to the presented guidelines on social media behaviour before, during, and after emergencies (RQ1.1)? Before an emergency (see Figure 8.1), interestingly, more than two thirds highlighted the importance of knowing the social media accounts of local and national emergency services (ES) (70%) or following their information on how to prevent and stay safe during emergencies (66%). In contrast, only about half think it is essential to look for apps that emergency services provide and download them to stay informed during an emergency (52%). Even fewer participants agree to read what to expect from emergency services in social media (45%).

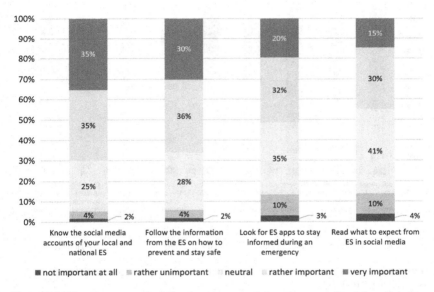

Figure 8.1   How do you rate the following recommendations for usage of social media before an emergency? (Q22)

The guideline which participants consider as most important to follow during an emergency (see Figure 8.2) is the fact that social media do not replace the emergency call and should rather be used for information updates (82%). Seventy-nine percent of the participants think it is essential to share only official and reliable information and avoid spreading rumours. Moreover, three quarters agree it is vital to tell only facts and to not send unreliable information. Seventy-four percent rate it as important to correct false information if they spot or share it, and 67% of the participants want to always stay up-to-date. Also, sixty-six percent find it essential to always mention the emergency service account or include any already used hashtags and forward received official messages to contacts or share them. More than a half of the respondents (55%) assume that checking for existing initiatives if they intend to initiate an emergent volunteer initiative is important. It is least, but still important for the respondents (47%) to look for emergent volunteer initiatives in Facebook groups, Google crisis maps or trusted users on Twitter.

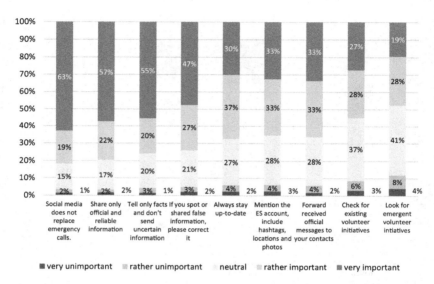

Figure 8.2   How do you rate the following recommendations for usage of social media during an emergency? (Q23)

The question about the importance of different activities after an emergency (see Figure 8.3) received the most negative responses. However, about half of the participants still assume it is essential to give feedback to the authorities (54%). Also, over half of the respondents find it important to follow the official accounts and local organizations to get information updates and to communicate even after a crisis, as well as use social media for the processing of the event (57%). It is even more important for the participants to help others to reconstruct and handle the event, and to restore missing contacts and ask for welfare of family and friends (61% each).

Besides the importance of guidelines during specific phases of an emergency, in terms of general aspects while using social media (see Figure 8.4), about 4

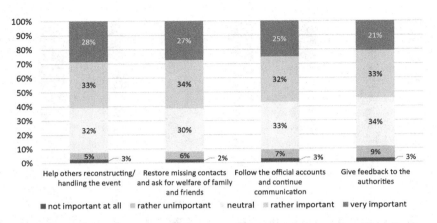

Figure 8.3 How do you rate the following recommendations for usage of social media after an emergency? (Q24)

out of 5 of the participants find it very or rather important to correct a mistake if they made one (83%), and to know it is important that they are responsible for their writing and that they should remember possible consequences (82%). Additionally, they agree it is essential to interact with respect and courtesy (82%) and to verify information before posting (81%). Participants also find it important to protect their privacy and check privacy settings (79%), as well as to respect intellectual property rights (75%).

How consistent is the agreement with guidelines across the categories, i.e. concerning the use of social media before, during, and after emergencies (RQ1.2)? Taking a closer look directly at the questionnaire items concerning guidelines, we revealed a significant correlation (Spearman's Rho) between the responses for all guidelines (p < .0001), the values ranging between r = .131 and r = .785. For all correlations between the responses, see Table 8.6 in the appendix. In the following, we will highlight a few significant findings to explore the research question mentioned above.

If we examine correlations within groups, we observe especially high significances over r = .600. The highest correlation coefficients were found between items of Q25 (General social media guidelines). Those who agreed to follow information provided by ES tended to agree with looking for ES apps to a higher degree (r = .610, p < .0001). There is also particularly high correlation

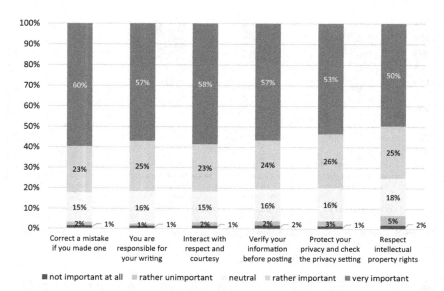

Figure 8.4 How do you rate the following, general aspects for using social media in an emergency? (Q25)

of the guideline "tell only facts" with "share only official and reliable information" (r = .778, p<.0001). Furthermore, attitude towards the two guidelines on helping others and restoring contacts were especially highly correlated (r = .718, p<.0001). Opinions on verification and correction of own posts also showed a significant correlation (r = .785, p<.0001), as do respect, courtesy and responsibility for own writing (r = .740, p<.0001).

To reveal the relationship between the individual categories we addressed in the questions, we also carried out Spearman correlations between the clustered items of each group (see table 8.4). All groups were significantly related and showed a high correlation coefficient, mostly greater than r = .600. Only for Q25 (general guidelines), the relationship to guidelines before and after an emergency as slightly lower than r = .500.

Table 8.4 Correlations between responses for the individual questions

	Q22: Before	Q23: During	Q24: After	Q25: General
Q22: Before		r = .657, p < .001	r = .632, p < .001	r = .425, p < .001
Q23: During	r = .657, p < .001		r = .690, p < .001	r = .634, p < .001
Q24: After	r = .632, p < .001	r = .690, p < .001		r = .481, p < .001
Q25: General	r = .425, p < .001	r = .634, p < .001	r = .481, p < .001	

8.4.2 Influencing Factors on the Attitude towards the Guidelines

Via chi-squared tests, we examined whether demographic factors as well as smartphone and social media usage habits, both in daily life and crisis situations, were related to the responses. The factors have been derived from other questions included in our survey at an earlier point, where we asked for age and gender, education status and income, frequency of smartphone usage, social media usage and publishing posts therein, and familiarity with crisis apps. Similarly, we also considered the degree to which participants expected critical infrastructure operators to use social media in case an infrastructure failure occurs. To provide an overview of these influences, we clustered the response values for each question and carried out chi-squared cross-tabulations for the clustered values. Additionally, we used Spearman correlations to determine the direction of interesting correlations for a more detailed insight into trends. In Table 8.5, all chi-squared test results and correlations are presented. The dependencies between the demographic variables themselves have not been calculated.

To what extent is the attitude towards the guidelines affected by smartphone and social media usage in daily life and in emergencies (RQ1.3)? A higher smartphone usage influenced responses to questions Q23 (r = .083), Q24 (r = .059) and Q25 (r = .019), leaving out guidelines before an emergency. At the same time, the more participants indicate to post in social media, the more they tend to agree to the statements in questions Q22 (r = .212), Q23 (r = .151), and Q24 (r = .169), namely guidelines before, during, and after an emergency. Similarly, for participants' usage of social media in crisis situations and CI operator communication expectations, we found significant effects for all questions:

The more social media are used in a crisis, the more participants agree to the proposed guidelines on it (r = .252; r = .162; r = .186, r = .010). The same is true for expectations from CI operators to use social media (r = .302; r = .282;

r = .284; r = .202). In contrast to this, general social media and crisis app usage do not significantly influence the responses.

How do demographic factors influence the agreement with guidelines concerning the use of social media in emergencies (RQ1.4)? Gender and age only have a significant influence on the items in question Q25 (general aspects for the use of social media). Here, a Spearman correlation indicates that women tended to agree to the general guidelines to a higher extent as compared to men (r = .132). The same applies to older participants as compared to younger ones (r = .149).

We also found that the higher the income, the more participants tend to agree to statements in the question on guidelines before an emergency, Q22 (r = .030). In contrast to this, respondents with lower education show a greater tendency to agree with all statements on guidelines before (r = −.035) and after an emergency (r = −.007), while the opposite was true for general guidelines (r = .039) and those to be adhered to during an emergency (r = .052). However, while chi-squared tests show a significant influence of demographic factor, we could not determine the same for Spearman correlations, which means we cannot make a solid statement on a trend in this case. Region is the only demographic factor which does not have a significant influence on any of the questions.

8.4.3   Citizens' Attitudes and Suggestions concerning the Guidelines

What are caveats and improvement suggestions towards the guidelines on social media in emergencies (RQ1.5)? In Q25, the last item asked the participants about their overall opinion on the guidelines: *"The presented guidelines are helpful".* The participants generally agreed that the guidelines are generally appropriate and helpful, as 72% of participants gave a positive answer. However, the least number of participants agreed with this item (37% compared to more than 50% of all other items in this question). Twenty-four percent were neutral, and only two percent each disagreed. To analyse the reasons for and opinions behind these responses, we further asked for open comments and present the qualitative results in the following.

The answers to our open-ended question Q26 *"Which criticism or improvement suggestions concerning the presented guideline about the use of social media before, during, and after a crisis do you have?"*, responses were classified into ten meta-codes: 1) *fake news*, 2) *gazers*, 3) *no idea*, 4) *no answer*, 5) *no criticism*, 6)

Table 8.5 Results of chi-squared tables for all questions, with correlations where appropriate. Non-significant results are displayed on grey background

	Q22: Before	Q23: During	Q24: After	Q25: General
Gender	$\chi^2(15, 1{,}024)$ = 16.33, p = .361 r = .069	$\chi^2(30, 1{,}024)$ = 33.25, p = .312 r = .061	$\chi^2(15, 1{,}024)$ = 10.35, p = .797 r = .028	$\chi^2(20, 1{,}024)$ = 45.80, p < .001 r = .132
Age	$\chi^2(75, 1{,}024)$ = 65.99, p = .762 r = .002	$\chi^2(150, 1{,}024)$ = 127.10, p = .913 r = .016	$\chi^2(75, 1{,}024)$ = 74.33, p = .500 r = .012	$\chi^2(100, 1{,}024)$ = 126.26, p < .05 r = .149
Education	$\chi^2(75, 1{,}024)$ = 124.28, p < .001 r = −.035	$\chi^2(150, 1{,}024)$ = 300.40, p < .001 r = .052	$\chi^2(75, 1{,}024)$ = 136.64, p < .001 r = −.007	$\chi^2(100, 1{,}024)$ = 150.50, p = .001 r = .039
Income	$\chi^2(45, 1{,}024)$ = 88.26, p < .001 r = .030	$\chi^2(90, 1{,}024)$ = 89.21, p = .504 r = .042	$\chi^2(45, 1{,}024)$ = 50.71, p = .259 r = .036	$\chi^2(60, 1{,}024)$ = 79.69, p < .05 r = .066
Region	$\chi^2(210, 1{,}024)$ = 233.98, p = .123	$\chi^2(420, 1{,}024)$ = 410.58, p = .620	$\chi^2(210, 1{,}024)$ = 176.59, p = .955	$\chi^2(280, 1{,}024)$ = 279.38, p = .449
Overall smartphone usage	$\chi^2(60, 1{,}024)$ = 76.34, p = .076 r = .074	$\chi^2(120, 1{,}024)$ = 162.19, p < .01 r = .083	$\chi^2(60, 1{,}024)$ = 106.78, p < .001 r = .059	$\chi^2(80, 1{,}024)$ = 153.56, p < .001 r = .014
Overall social media usage	$\chi^2(465, 1{,}024)$ = 492.71, p = .181 r = -.030	$\chi^2(930, 1{,}024)$ = 951.63, p = .304 r = .000	$\chi^2(465, 1{,}024)$ = 449.38, p = .690 r = .015	$\chi^2(620, 1{,}024)$ = 528.815, p = .997 r = .000
Posting in social media	$\chi^2(60, 1{,}024)$ = 96.49, p < .005 r = .212	$\chi^2(120, 1{,}024)$ = 164.73, p < .005 r = .151	$\chi^2(60, 1{,}024)$ = 100.98, p < .005 r = .169	$\chi^2(80, 1{,}024)$ = 76.03, p = .605 r = .028
Use of social media in crisis situations	$\chi^2(30, 1{,}024)$ = 95.04, p < .001 r = .252	$\chi^2(60, 1{,}024)$ = 100.76, p < .005 r = .162	$\chi^2(30, 1{,}024)$ = 64.09, p < .001 r = .186	$\chi^2(40, 1{,}024)$ = 65.06, p < .01 r = .010

(continued)

Table 8.5 (continued)

	Q22: Before	Q23: During	Q24: After	Q25: General
Use of crisis apps	$\chi^2(600, 1,024)$ $= 544.48$, p $=$ .949 r $= -.004$	$\chi^2(1200, 1,024)$ $= 1122.61$, p $=$ .945 r $= -.026$	$\chi^2(600, 1,024)$ $= 475.92$, p $=$ 1.000 r $= .027$	$\chi^2(800, 1,024)$ $= 691.863$, p $=$ .998 r $= .000$
Expectations of CI to use social media	$\chi^2(60, 1,024)$ $= 241.73$, p$<.001$ r $= .302$	$\chi^2(120, 1,024)$ $= 296.60$, p$<.001$ r $= .282$	$\chi^2(60, 1,024)$ $= 261.72$, p$<.001$ r $= .284$	$\chi^2(80, 1,024) =$ 195.97, p$<.001$ r $= .202$

presentation, 7) *scepticism,* 8) *extent,* 9) *unnecessary* and 10) *improvement sugge-stion.* The meta-codes are not mutually exclusive, and the responses coded within them may overlap. Firstly, there were two categories of less useful content for analysis. The meta-code *no answer* included answers which were not understan-dable and therefore not usable for the evaluation (n = 155). Respondents who did not know how to answer the question were classified into the meta-code *no idea* (n = 124), for instance: *"I don't know at the moment as these theses are absolutely new and I first have to process them"* (1354).

8.4.3.1 Positive and Thoughtful Attitudes towards the Guidelines

In summary, half of the participants' answers (50%) were classified into the meta-code *no criticism* (n = 515). That means that they, in total, agree with the guidelines and would not make many changes: *"I think the presented guideline is very coherent and complete"* (241) or: *"The guideline is very well elaborated. No further suggestions"* (504). Many respondents are convinced of the guidelines: *"I really like it. It should be short, give clear action hints and should not be too instruc-tive"* (281). Another participant also has no criticism, but he wants to be secured: *"No criticism, it must be guaranteed that the mobile network is not dead in case of an emergency"* (1801). However, some participants have doubts, for instance, on the distribution of smartphones: *"You think that everyone owns a smartphone. What about the elder? People that reject smartphones due to mistrust? Etc. I really like the presented suggestions, but the question is what actually is realized in such situations"* (177). Further participants are sceptical about the application of gui-delines: *"I don't have any criticism, but I don't think that all people will hold on to it"* (262). In this context, it is also interesting that few respondents do not have any improvement suggestions, but they think of people who do not use social media: *"I don't have any criticism, but there are still many people who do not use social media—like me"* (876).

8.4.3.2 Suggestions for Improvement on Content, Extent and Presentation

Furthermore, participants elaborated a variety of constructive *improvement suggestions* (n = 78). Although other categories also include recommendations for improvement and thus responses may overlap, we distinguished this particular meta-code for a better overview of possible contributions to improvement of our guidelines and further research. For instance, participants suggested to add further recommendations: *"Only share correct and trustworthy information, if so, with giving the source. If you detect misinformation, make people aware of it"* (70). Moreover, it is demanded that volunteers coordinate each other and distribute auxiliary goods usefully. One respondent also considers the question of time and has a suggestion: *"I ask myself if I have time to use Twitter or Facebook in case of an emergency. I would like public terminals, which are readily accessible and located in a safe place"* (391). Besides internet and social media, one respondent makes clear that traditional media still should be used for the dissemination of information (1615).

Also, there are some recommendations for the presentation of the guideline (n = 26). For example, it should also be more understandable for children, foreign people, elder people, and disabled persons. One participant recommends add some pictures: *"As a text, no one will really read it through"* (697). In this context, the meta-code *extent* matters as several respondents think the guideline is too extensive (n = 36): *"The guideline should be restricted to basic information and ignore norms that are simply nice or desirable"* (410). One participant asks for a cheat sheet which summarizes the most important aspects of the guideline to have a compact overview. On the other hand, one interviewee even asks for more details: *"For inexperienced users of social media, a more explicit guideline would be interesting, i.e., how I contact official entities via social media or where I have to search on Facebook or Twitter"* (1093).

8.4.3.3 Scepticism about Fake News and the Necessity of Guidelines

As represented by the meta-code *scepticism* (n = 52), several people do not really trust the guideline as they have too many doubts concerning the use, design or privacy. For instance, one participant has doubts about the usage type: *"It should be emphasized VERY clearly that social media do NOT replace the official numbers and channels"* (157). Additionally, privacy is an important factor for few participants: *"Private content of a conversation and chat reports should not be investigated if there's no reason as we wouldn't have any privacy in society"* (507). Furthermore, *gazers* were perceived as disturbing factor in case of an emergency by a few participants (n = 3): *"If you don't want any gazers on the accident site, you shouldn't ask for photos. I generally find it inconvenient to post something about a catastrophe if the police don't ask for it. That quickly leads to the distribution of misinformation as people have misperceptions in such situations"* (512).

Another meta-code which shows the respondents' doubts is the category *fake news* (n = 26). As said before, people worry about the dissemination of misinformation as there is much content and many rumours on the internet and social media: *"I think, in case of an emergency, social media are crap. You simply can create too much fake news, and everyone thinks something happens"* (463). For this case, one participant suggests: *"Actually none. Maybe Facebook, Twitter and co. could develop bots which detect fake news or evaluate user ratings (many dislikes) and automatically delete postings"* (1087). In summary, only nine people rate the guideline as *unnecessary* (n = 9): *"Why do we need it? Are there concrete reasons? Such a guide can spread fear and feed speculations"* (644). They are very sceptical and do not believe that the guideline is very helpful in case of an emergency: *"It generally is totally unnecessary and a waste of time as every normal human-being, who thinks about it a bit, will decide correctly in the concrete situation. And if not, a 'guideline' will not discourage him"* (122).

8.5   Discussion and Conclusion

In this paper, we outlined the risks of chaotic social media use (Kaewkitipong et al., 2012; Perng et al., 2012; Valecha et al., 2013) and presented guidelines for the use of social media in general and in emergencies, revealing that most of them focus on the perspective of emergency services or organizations. Only three guidelines considered the perspective of citizens before, during, and after emergencies (Belfo et al., 2015; Emergency 2.0 Wiki, 2015; Helsloot et al., 2015), and none of them was evaluated in terms of practical value with a larger sample of citizens. That, firstly, led us to the design of guidelines for citizens. Secondly, this paper contributes with the qualitative and quantitative evaluation of the citizens' guidelines based on a representative sample for Germany (N = 1,024) stratified for gender and age, but also ensuring a wide spread of the survey sample in terms of region, education, and income. Considering our research question RQ1: **"How do citizens perceive guidelines for social media use in emergencies?"**, we have divided our results into five separate research questions, which we will summarize in the following.

To what extent do participants agree to the presented guidelines on social media behaviour before, during, and after emergencies (RQ1.1)? Our results indicate that all proposed guidelines for social media in crisis contexts achieve a high acceptance rate exceeding 50%. It is only slightly lower than 50% for anticipating the type of information shared before a crisis, and for helping and volunteering in case an emergency occurs. The present guidelines are also rated as helpful by over two thirds of respondents (72%).

How consistent is the agreement with guidelines across the categories, i.e. concerning the use of social media before, during, and after emergencies (RQ1.2)? Concerning the relationship between responses to the individual guidelines, most participants tended to agree to guidelines related to each other. In total, there is a significant and high correlation between agreements to the specific categories. We can conclude that perception of the guideline importance is consistent across the stages (before, during, and after), while responses to the general category, which show the highest agreement, were not as highly correlated to the others (below $r = .500$).

To what extent is the attitude towards the guidelines affected by smartphone and social media usage in daily life and in emergencies (RQ1.3)? The more often the individuals use their smartphone, post in social media, or tend to use them in crisis situations, the more they agree to our guidelines. Additionally, the agreement is consistent with the expectation to read messages by CI providers in social media during crises. However, overall social media or crisis app usage does not have an effect on the responses. Therefore, guidelines appeal most to active mobile social media users and those who rely on them during emergencies.

How do demographic factors influence the agreement with guidelines concerning the use of social media in emergencies (RQ1.4)? Demographic factors have only a partial influence on the responses, namely education and income, and to a small degree gender and age. We can therefore conclude that they are generally reasonable for most demographic groups, where age or economic situation is not a strongly determining factor, although it might have an influence on overall technology usage.

What are caveats and improvement suggestions towards the guidelines on social media in emergencies (RQ1.5)? Most participants had a positive reaction to the presented guidelines (see RQ1.1), however, some improvements based on participants' suggestions had to be done to make the guidelines more relevant, concise, appealing, and applicable for everyone. A reason for criticism, among others, lies in the mode of presentation, which is often considered too long and text-based, as well as containing information that is not vital in this context.

Our qualitative responses reflect the agreement previously observed in RQ1.1: Half of the participants agree with the guidelines as they are, while many also point out that these were not relevant for people who do not use smartphones or social media, or if there is no time or network for this in case of an emergency. These last two comments are especially similar to results by Sutton, Woods and Vos (2017), who identified the time and effort invested into information search as well as the technological access to reliable information as hurdles to an effective crisis communication via social media or websites. Therefore, it is suggested to expand the scope and apply the guidelines to mass media as well. According to the social-mediated crisis communication model (Jin et al., 2014), the media channel, along with information source and origin of the crisis, is important for recipients' perception of the crisis and the affected organization. Generally, citizens tend to use personal conversations and social media mostly for communication about a crisis, but traditional media is still perceived as most trustworthy and generates most positive attitudes (Austin et al., 2012; B. F. Liu et al., 2013). In terms of our concrete recommendations, the highest importance lies on reliability of sources and correctness when reading and sharing information (Kaewkitipong et al., 2012), as well as using social networks only as supplement to other media like the emergency call. Another concern was fake news with the potential to cause panic and harm, and gazers that could be drawn to the site of emergency.

The results of our study informed the final visualization of the citizens' guidelines (see Figure 8.5 in the appendix). Firstly, the content was reduced: The gathered feedback suggested to even reduce the amount of text, i.e. to omit general norms, and the recommendations for social media use after emergencies received the least consent of citizens. Secondly, some illustrations were used to increase the overall presentation quality of the guideline.

Limitations and future work. In future emergencies, the guidelines may be distributed in social media, i.e. Facebook and Twitter, to observe and analyse their reception by citizens. Although the guidelines were evaluated with a large sample representative for Germany, intercultural differences could lead to different perceptions on the relevance of specific recommendations. For instance, prior research identified different risk cultures across Europe, such as individual-oriented, state-oriented and fatalistic risk cultures, which shape the behaviour and perception of citizens (Dressel, 2015; Dressel & Pfeil, 2017). Despite disasters are common (Höppe, 2015), the number of German citizens who already experienced a crisis is relatively low (Reuter, Kaufhold, Leopold, et al., 2017b). Therefore, this evaluation is based on a hypothetical scenario in contrast to a case study, and the results could be influenced by this fact. A specific case study could examine whether participants who use social media already (partially) apply the strategies addressed in the guidelines before, during, and after an emergency, and how. Apart from an attitude towards the guidelines, we could further analyse to what extent citizens would apply and adhere to them in an actual crisis scenario.

Furthermore, guidelines for the use of social media for emergency services were developed in our project. Despite the feedback of emergency managers during EAB workshops, no evaluation was performed on the final guidelines. However, the specific needs and perceptions of emergency managers in terms of citizens' behaviour and content creation in social media (Flizikowski et al., 2014; Reuter, Ludwig, Kaufhold, et al., 2016) are important precursors to optimize the interplay between authorities and citizens during emergencies, and the design of both guidelines.

8.6 Appendix: Visualization of the Guidelines

Figure 8.5 Visualisation
of the Guidelines

8.7   Appendix: Correlations between Responses for all Guidelines

Table 8.6 Correlation coefficients (Spearman's Rho) between responses for all guidelines (p < .001)

	Q22.1	Q22.2	Q22.3	Q22.4	Q23.1	Q23.2	Q23.3	Q23.4	Q23.5	Q23.6	Q23.7	Q23.8	Q23.9	Q24.1	Q24.2	Q24.3	Q24.4	Q25.1	Q25.2	Q25.3	Q25.4	Q25.5	Q25.6
Q22.1																							
Q22.2	.545																						
Q22.3	.539	.564																					
Q22.4	.593	.442	.610																				
Q23.1	.614	.452	.494	.534																			
Q23.2	.395	.183	.282	.471	.428																		
Q23.3	.440	.425	.431	.462	.532	.469																	
Q23.4	.390	.209	.334	.495	.419	.630	.488																
Q23.5	.397	.201	.309	.480	.418	.608	.483	.778															
Q23.6	.392	.209	.304	.440	.411	.570	.453	.689	.672														
Q23.7	.472	.332	.421	.473	.495	.468	.531	.555	.579	.596													
Q23.8	.394	.486	.426	.370	.422	.237	.503	.289	.294	.342	.510												
Q23.9	.421	.413	.411	.450	.490	.363	.497	.437	.435	.452	.520	.658											
Q24.1	.568	.494	.486	.461	.587	.295	.495	.362	.343	.373	.492	.539	.531										
Q24.2	.412	.379	.401	.400	.422	.268	.399	.316	.310	.390	.452	.434	.435	.533									
Q24.3	.431	.373	.401	.444	.447	.319	.432	.374	.400	.396	.500	.440	.461	.520	.557								
Q24.4	.434	.339	.393	.453	.428	.356	.431	.383	.406	.432	.501	.439	.480	.498	.531	.718							
Q25.1	.388	.169	.300	.433	.412	.510	.360	.558	.538	.512	.439	.229	.383	.355	.312	.387	.383						
Q25.2	.367	.160	.288	.462	.388	.545	.367	.592	.597	.545	.441	.233	.390	.326	.305	.381	.400	.740					
Q25.3	.324	.192	.254	.393	.333	.460	.324	.478	.478	.452	.402	.211	.302	.265	.288	.348	.362	.613	.695				
Q25.4	.329	.185	.268	.376	.311	.423	.370	.448	.496	.432	.392	.289	.371	.306	.275	.372	.382	.528	.605	.625			
Q25.5	.373	.160	.297	.480	.368	.545	.376	.597	.609	.564	.464	.234	.321	.310	.387	.409	.409	.660	.713	.653	.673		
Q25.6	.344	.131	.270	.438	.335	.529	.358	.573	.582	.560	.441	.215	.318	.303	.281	.378	.391	.622	.668	.606	.574	.785	

Design and Evaluation of a Cross Social Media Alerting System

9

The research field of crisis informatics examines, amongst others, the potentials and barriers of social media use during conflicts and crises. Social media allow emergency services to reach the public easily in the context of crisis communication and receive valuable information (e.g. pictures) from social media data. However, the vast amount of data generated during large-scale incidents can lead to issues of information overload and quality. To mitigate these issues, this paper proposes the semi-automatic creation of alerts including keyword, relevance and information quality filters based on cross-platform social media data. We conducted empirical studies and workshops with emergency services across Europe to raise requirements, then iteratively designed and implemented an approach to support emergency services, and performed multiple evaluations, including live demonstrations and field trials, to research the potentials of social media-based alerts. Finally, we present the findings and implications based on semi-structured interviews with emergency services, highlighting the need for usable configurability and white-box algorithm representation.

This chapter has been published as the journal article "Mitigating Information Overload in Social Media during Conflicts and Crises: Design and Evaluation of a Cross-Platform Alerting System" in Behaviour & Information Technology (BIT) by Marc-André Kaufhold, Nicola Rupp, Christian Reuter, and Matthias Habdank (Kaufhold, Rupp, et al., 2020).

9.1   Introduction

As the work of professional bodies, volunteers, and others is increasingly mediated by computer technology, and more specifically by social media, research on

© The Author(s), under exclusive license to Springer Fachmedien Wiesbaden GmbH, part of Springer Nature 2021
M.-A. Kaufhold, *Information Refinement Technologies for Crisis Informatics*,
https://doi.org/10.1007/978-3-658-33341-6_9

conflict and crisis management in HCI has become more common (Hiltz, Diaz, et al., 2011; Palen & Hughes, 2018; Reuter, 2018; Reuter, Hughes, et al., 2018; Reuter, 2019). The emerging research field of *crisis informatics* has revealed interesting and important real-world uses for social media (Soden & Palen, 2018). Coined by Hagar (2007), crisis informatics is 'a multidisciplinary field combining computing and social science knowledge of disasters; its central tenet is that people use personal information and communication technology to respond to disasters in creative ways to cope with uncertainty' (Palen & Anderson, 2016).

During conflicts and crises, it is necessary for emergency services to obtain a comprehensive situational overview for coordination efforts and decision making (M. Imran et al., 2015; Vieweg et al., 2010). In such situations, social media are increasingly used for the exchange of information (Hughes & Palen, 2009) while emergency services encounter issues of information overload and quality (Hughes & Palen, 2014; Mendoza et al., 2010; Plotnick & Hiltz, 2016). Although companies and researchers continuously develop systems to support social media analytics, including the discovery, tracking, preparation and analysis of social data (Stieglitz et al., 2014; Stieglitz, Mirbabaie, Ross, et al., 2018), research indicates that there is still a need for systems that support emergency services by providing manageable amounts of high-quality information (Moi et al., 2015). Furthermore, to overcome the issue of information overload, visual analytics strives for the automatic processing of data (Keim et al., 2008), but user interaction is required to filter and visualise data according to practitioners' requirements (Onorati et al., 2018). Indeed, research suggests that not only a customisation of filtering algorithms is required for an efficient response to specific crisis situations but also that social media analytics tools require a good usability during stressful crisis situations (M. Imran et al., 2015; Stieglitz, Mirbabaie, Fromm, et al., 2018). Based on a communication matrix (Reuter, Hughes, et al., 2018) and social media analytics framework (Stieglitz, Mirbabaie, Ross, et al., 2018), we designed and evaluated a system to support the two information flows of *crisis communication* and *integration of citizen-generated content* featuring social media. Thereby, we seek to answer the following research questions:

- How can social media alerts based on information gathering, mining, and quality filters help to mitigate the issue of information overload (RQ1)?
- How can the trade-off between automation and user interaction be designed to mitigate the issue of information overload (RQ2)?

To reflect the methodology used in our project, the paper is structured as follows: First, based on the analysis of related work and existing technical systems (section 9.2), we conducted interviews with emergency services followed by quantitative empirical studies and workshops with both emergency services and citizens (section 9.3). These results informed the design and development of a prototype including iterative evaluations in different phases (section 9.4). To evaluate the system, we conducted semi-structured interviews after demonstrations, field trials and a workshop exercise (section 9.5). Finally, we analysed (section 9.6) and discussed (section 9.7) the results to draw conclusions on how social media analysis might be improved during emergencies (section 9.8).

This paper contributes findings of how emergency services analyse and use social media with a three years study including empirical pre-studies, the design of a prototype and evaluation in practice. While the empirical pre-studies (Reuter, Ludwig, Kaufhold, et al., 2016; Reuter, Kaufhold, Spielhofer, et al., 2017) and a first round of evaluation with 12 distinct participants based on a preliminary version of the system (Reuter, Amelunxen, et al., 2016) have already been published scientifically, the main contributions of this article are:

- Design of a novel multi-scenario and semi-automatic approach for generating and visualising social media alerts featuring information gathering, mining, and quality filters.
- Evaluation of the designed approach with emergency services using demonstrations, field trials and a workshop exercise to generate empirical insights on the functionality and usability of the approach.

Our results show that social media are more likely to be used in emergencies if alerts, defined as sets of grouped messages sharing a similar context, and information quality (IQ) indicators support the processing of big social data. The paper furthermore highlights the need for HCI research in terms of (1) usable configurability and (2) white-box algorithm representation.

9.2   Conceptual Framing and Related Work

Based on natural and human-induced (large-scale) incidents, such as 2012 Hurricane Sandy (Hughes et al., 2014), the 2013 European Floods (Albris, 2017), or the 2016 Brussels bombings (Stieglitz et al., 2017), a body of research has examined potentials and challenges of social media usage in conflicts and crises by both authorities and citizens (Kaufhold & Reuter, 2019; Reuter & Kaufhold, 2018). On the one hand, social media might enable crowdsourcing of specific tasks (Dittus et al., 2017; Ludwig et al., 2017), communication between authorities and citizens

(Reuter, Ludwig, Kaufhold, et al., 2016; Reuter & Spielhofer, 2017), coordination among citizens and mobilisation of unbound digital or real volunteers (Kaufhold & Reuter, 2016; Reuter et al., 2013; Starbird & Palen, 2011; J. I. White et al., 2014), (sub-)event detection (Pohl et al., 2015; Sakaki et al., 2010) or improved situational awareness (M. Imran et al., 2015; Vieweg et al., 2010).

Considering the challenges, according to a study with US public sector emergency services, the major barriers to social media use are organisational rather than technical (Hiltz et al., 2014). Research suggests that human factors are crucial for effective emergency management, but also technology for conducting respective emergency tasks (Kim et al., 2012). However, once organisational guidelines, policies, and willingness are established (Kaufhold, Gizikis, et al., 2019), technical systems are needed to make sense of the large amount of data. For instance, research has identified barriers and challenges in the authorities' use of social media, such as credibility, liability (Hughes & Palen, 2014), reliability and overload of information (Mendoza et al., 2010), as well as lack of guidance, policy documents, resources, skills and staff within the organisation (Plotnick & Hiltz, 2016).

In this paper, we present an approach for mitigating information overload, which includes the utilisation of a novel information quality algorithm. For the conceptual framing, we refer to the crisis communication matrix by Reuter et al. (2018) and the social media analytics framework by Stieglitz, Mirbabaie, Ross et al. (2018). The crisis communication matrix distinguishes authorities (A) and citizens (C), both as sender and receiver, respectively, to derive four communication flows (Reuter, Hughes, et al., 2018):

- Crisis communication (A2C): Authorities include social media into their crisis communication to disseminate information on how to prevent or behave during emergencies as well as concrete emergency warnings.
- Self-help communities (C2C): Social media enable people, such as affected citizens, real and digital volunteers, to help each other and coordinate emergency response activities among themselves.
- Interorganisational crisis management (A2A): Authorities use social media for the awareness, distribution of information, communication, and networking among themselves.
- Integration of citizen-generated content (C2A): Authorities may enhance situational awareness based on citizen-generated content, such as eyewitness reports, pictures, and videos taken with mobile phones.

Since this paper presents an approach for managing the information overload of social media by authorities, we focus on the communication flow of C2A but also discuss aspects of A2C. Furthermore, the social media analytics framework comprises the steps of discovery, tracking, preparation and analysis of social data (Stieglitz, Mirbabaie, Ross, et al., 2018). In this paper, discovery of social data is driven by the research domain of crisis management and tracking involves a keyword-based use of multiple social media APIs. For pre-processing, heterogeneous data is stored according to a unified exchange format and the analysis comprises content- and metadata-related approaches (cf. section 9.4).

9.2.1 Crisis Communication Perspective: The Authorities' Challenges of Information Overload, Quality and Communication in Emergencies

To leverage social media information as a basis for authorities' decision-making, they are not only required to *integrate citizen-generated content (C2A)*, i.e. monitoring social media, while managing the vast amount of data (Olshannikova et al., 2017). When tens of thousands of social media messages are generated during large-scale emergencies, authorities have to deal with the issue of information overload which is traditionally defined as '[too much] information presented at a rate too fast for a person to process' (Hiltz & Plotnick, 2013). Referring to the information overload problem from the field of visual analytics, Keim et al. (2008) highlight the danger of getting lost in data which may be irrelevant to the current task at hand as well as processed and presented in an inappropriate way. Considering the human capacity of information processing, Miller (1956) suggests 'organizing or grouping the input into familiar units or chunks' (p. 93) to overcome such limitations. Accordingly, functionalities such as filtering and grouping potentially assist in overcoming the issue of information overload (Moi et al., 2015; Plotnick et al., 2015; S. Tucker et al., 2012). This is supported by a survey of 477 U.S. county-level emergency managers which revealed that perceived information overload negatively influences the adaption of social media, while the 'chunking' or grouping of social media messages by specific tools positively influences the intention to use social media during emergencies (Rao et al., 2017).

Besides dealing with information overload, authorities have to select the most accurate information (Shankaranarayanan & Blake, 2017). The spread of misinformation and rumours can be understood as the result of a 'collective sensemaking process whereby people come together and attempt to make sense of imperfect and incomplete information' (Arif et al., 2017; Krafft et al., 2017; Stieglitz et al., 2017). Although research highlights the capabilities of the so-called *self-correcting crowd*, Chauhan and Hughes (2017) suggest emergency services to monitor emerging event-based resources, such as Facebook pages that provided the highest percentage of relevant information, to ensure that the information they provide is accurate. Besides, local news media were observed to provide the timeliest information and highest number of relevant messages around the event. Thus, concepts of *information quality* may support the adaption and evaluation of information (Naumann & Rolker, 2000; Shankaranarayanan & Blake, 2017) and take into account the context-dependent and subjective characteristics of information quality (Ludwig, Reuter, & Pipek, 2015; Reuter, Ludwig, Kaufhold, et al., 2015).

Furthermore, authorities integrate social media into their *crisis communication* (A2C) efforts to share official information with the public on how to avoid accidents or emergencies and how to behave during emergencies (Reuter, Ludwig, Kaufhold, et al., 2016), but also to 'shape social media conversation and mitigate misinformation and false rumour around a crisis event' (Andrews et al., 2016). A study highlights how authorities corrected mistakes caused by the 'emerging risks of the chaotic use of social media' (Chen et al., 2011). Emergency services may establish their trustworthiness by the three dimensions of ability, integrity, and benevolence (Hughes & Chauhan, 2015), e.g. maintaining a public-including expressive communication approach (Denef et al., 2013). Research suggests that citizens share information across multiple platforms during crises (Hughes et al., 2016), indicating that both crisis communication and monitoring is required to encompass cross-platform interactions despite the observed lack of skills and staff by emergency services (Plotnick & Hiltz, 2016).

9.2.2   Social Media Analytics Perspective: Suitability of Existing Systems for the Authorities' Use in Emergencies

As public interfaces (APIs) enable the retrieval and processing of high volume data sets, 'systems, tools and algorithms performing social media analysis have been developed and implemented to automatize monitoring, classification or aggregation tasks' (Pohl, 2013). Here, *social media analytics* is defined as the process of social media data collection, analysis and interpretation in terms of actors, entities and relations (Stieglitz et al., 2014). Accordingly, Stieglitz, Mirbabaie, Ross, et al. (2018) differentiate between the steps of discovery, tracking, preparation, and analysis of social media data:

- Discovery: This first step entails the 'uncovering of latent structures and patterns' (p. 158). Even if—as often is the case in emergency situations—it is clear which topic is relevant, it may still be necessary to identify hashtags or keywords that are used frequently when referring to the emergency.
- Tracking: In this stage, decisions with respect to tracking approaches (keyword-, actor- or URL-related), sources (social platforms), methods (APIs or RSS/HTML parsing), and outputs (structured and unstructured data) are to be made.
- Preparation: This step requires data preprocessing, such as the elimination of stop words, stemming and lemmatisation. With respect to the veracity of data, it is advised to remove low-quality data by 'incorporating a filtering step in the preparation phase' (p. 164), ignore incomplete data or alternatively infer it.
- Analysis: In this step, based on the purpose of analysis (focusing on (1) a structural attribute or being either (2) opinion-/sentiment-related or (3) topic-/trend-related), one may correspondingly choose (1) statistical analysis, social network analysis, (2) sentiment analysis, or (3) content analysis, trend analysis.

Social media data, sometimes referred to as *big social data*, includes the characteristics of *high-volume* (large-scale), *high-velocity* (high speed of data generation), *high-variety* (heterogeneous data with a high degree of complexity due to the underlying social relations) and *highly semantic* (manually created and highly symbolic content with various, often subjective meanings) data (Olshannikova et al., 2017). Furthermore, with respect to the crisis management domain, Castillo (2016) introduces the notion of *big crisis data*, discussing its volume, vagueness, variety, virality, velocity, veracity, validity, visualisation, values, as well as the contribution of volunteers. These characteristics pose challenges for emergency services who need their own concepts of analysis.

Accordingly, specialised systems for social media analytics were developed. Pohl (2013) outlines that there are systems available for different online and offline applications, which consider one or multiple social media platforms for monitoring, are especially developed for crisis management and perform different kinds of analysis. For instance, *Twitinfo* supports the analysis of Twitter feeds by visualising message frequency and popular links, showing geolocated Tweets on a map, and calculating event-relevant tweets and the overall sentiment (Marcus et al., 2011). *Public Sonar* (formerly *Twitcident*) proposes an architecture of (1) incident profiling and filtering as well as (2) faceted search and real-time analytics to explore social media, both including the aggregation and semantic enrichment of social media data (Abel, Hauff, & Stronkman, 2012). Furthermore, the *Semantic Visualization Tool* combines Twitter searches and information categories with configurable visualisation techniques, such as a message list, timeline, tree map, word cloud, bubble chart and animated map, supporting the filtering and visualisation of social networks according to emergency managers' requirements (Onorati et al., 2018). Imran et al. (2014) created the *Artificial Intelligence for Disaster Response (AIDR)* platform, using artificial intelligence for classification of microblog communication in the context of crises, allowing users to search for emergencies located in a specific region and filter with respect to various topics like infrastructure damages or medical needs.

9.2.3 Social Media Analytics Systems for Event Detection and Alert Generation

There is a variety of different tools for event detection or message grouping. To extract situational information in tweet streams, Rudra et al. (Rudra et al., 2015) present a classification-summarisation approach. This is achieved by developing a Support Vector Machine (SVM) classifier using low-level lexical and syntactic features while word coverage in the summarisation process, called COWTS (Content Word-based Tweet Summarization), is optimised by employing an Integer Linear Programming (ILP) framework (Rudra et al., 2015; Rudra, Ganguly, et al., 2018; Sen et al., 2015). Furthermore, using the AIDR framework for classification, Rudra et al. (2018) proposed an approach based on simple algorithms identifying sub-events and creating summaries of a great amount of messages. Nguyen et al. (2015) developed *TSum4act*, offering a summary through constructed event graphs for each topic, which are ranked and offer users a 'summary for recommendation' (p. 4) derived from top-ranked tweets. In detail, it comprises the components of 1) informative tweet identification using a classification algorithm, 2) topic identification using LDA and clustering, and 3) tweet summarisation using event extraction via NER, event graph construction via cosine similarity, ranking via PageRank and filtering via the Simpson equation.

Further works point out the necessity of alert generation. Adam et al. (2012) stress the importance of customisation of alerts, including warning time, physical disabilities, socio-economic factors, location, connectivity and language as well as envisioning 'Full Disaster Lifecycle Alerts'. Their approach *SMART-C* includes an alert app service as well an 'interface to other alerting systems' like IPAWS Open Platform for Emergency Networks, television, web or radio, encapsulating a given alert in a CAP message in the Standard EDXL-DE envelope (Adam et al., 2012). However, the focus of this approach lies in the customised generation of alerts for citizens, but not in the algorithmic generation of alerts for emergency services. Various scholars dedicate themselves to both event detection and alert generation (Avvenuti et al., 2014), yet, often solely referring to the incorporated email notification or early warning system without further elaboration on the parameters of alert generation. Cameron et al. (2012) developed an *Emergency Situation Awareness-Automated Web Text Mining (ESA-AWTM)* system which, in contrast to our factor-inclusive work, uses a burst detection method allowing authorities to distinguish between differently coloured and sized alerts words, both characteristics indicating the size of the burst. Yet, the alert monitor was accompanied by mapping of the tweets' geolocations, offering differentiation in this respect. The spatiotemporal model of earthquake detection by Sakaki et al. (2010) has already implemented this, similar to a lot of other event detection approaches (Simon et al., 2014; Veil, Buehner, & Palenchar, 2011) and reaches users by notification (via email). Used by news agencies and emergency management services, *Dataminr* exemplifies an important alert service, offering real-time information (R. Miller, 2017). Emphasising the need for trust and timely event response, Brynielsson et al. (2018) present the development process of a tool used to analyse social media content which serves as proof of concept and is integrated into the Alert4All environment (Párraga Niebla et al., 2011). The concept comprises an emotional classifier, including the classes "positive", "fear", "anger", and "other", as well as a variety of data filtering operations and interactive charts to visualise emotional content. However, the focus of this concept is not the creation of alerts but the monitoring of public reactions to warning messages. In order to facilitate the ranking of social media alerts, Purohit et al. (2018) propose a 'quantitative model for determining how many and how often should social media updates be generated, while also considering a given bound on the workload for an end user' (p. 212).

9.2.3.1 Comparison of Existing Social Media Analytics Systems

A comparative review of social media analysis tools outlines tool- and data-related barriers, emphasising a lack of capacity to handle large amounts of information and the lack of usability, amongst others (Trilateral Research, 2015). A further market study compares existing systems regarding their management, analytics and visualisation functionalities focusing on their suitability for emergency services (Kaufhold et al., 2017). The study concludes that although some systems feature use cases of the public domain including emergency services, most systems are designed for the specifics of business contexts and none provides a framework for evaluating information quality of social media messages (Shankaranarayanan & Blake, 2017). Furthermore, although these solutions support alarm notifications if specified indicators reach specific thresholds, research suggests to consider the qualitative context of individual messages, such as date, time, location, full text, identified event types or language (Reuter, Amelunxen, et al., 2016); these might be important metadata for the grouping of messages and, consequently, for the mitigation of information overload (G. A. Miller, 1956). An overview of intelligence, management and special systems (Table 9.1) reveals a lack of emphasis on the criterion of information quality regarding existing approaches, architectures, and implemented systems.

We adopted the categorisation and overview from Kaufhold et al. (2017), which distinguishes between intelligence, management and special systems, while expanding and updating each section, respectively. Thus, we replaced old system names, included architectures which were of academic importance but not implemented as systems, and introduced the criterion of event detection. We expanded the table accordingly in order to distinguish between systems intended to fulfil this task and systems not dedicated to detecting specific events (including those showing long-term trends). Thus, we were able to present systems focusing on event detection, thereby offering maps or visualised rankings of topics or events while not integrating an alert. Even though no specific notification based on contextual factors is sent, systems aiming at event detection use GPS data, event-detecting algorithms and allow for filtering with respect to e.g. location, language and issues. Both back- and forward literature review was conducted. Due to the vast body of social media analytics systems and the increasing number of respective papers, we naturally do not offer a complete overview (vom Brocke et al., 2015); yet, we tried to include the ones being used by (economically or socially) relevant actors. At the same time, we focused solely on systems and approaches of social media analytics, thereby excluding e.g. social networks aiming at the collection of information regarding emergencies, relying mainly on volunteers like *OpenCrisis*. We also assumed some systems to represent the work of a (group of) scholar(s), thereby not listing each single variant of their developed system. Conducting a thorough literature review (via Google Scholar, libraries) and online research of services' websites, our overview aims at representing the various strands of approaches (visual analytics, geo-mapping, earthquake-specific, etc.) to reflect dominant work. Review was conducted by searching for work related to e.g. "alert generation", "event detection", "social media analytics".

9.2.4 Research Gap

Multiple studies examine barriers and potentials of social media use by authorities (Plotnick & Hiltz, 2016) and technical solutions supporting the analysis of big social data (Olshannikova et al., 2017). Amongst others, they identified the issues of information overload (Hughes & Palen, 2014; Plotnick & Hiltz, 2016), credibility and reliability (Hughes & Palen, 2014; Mendoza et al., 2010) as critical barriers of organisational social media use. Since the appropriation of social media analytics systems faces barriers in terms of data, tools, organisation and users (Plotnick & Hiltz, 2016; Reuter, Ludwig, Kaufhold, et al., 2016), empirical evaluation studies may offer insights for developing mitigation strategies and improving the quality of supportive technological solutions (Trilateral Research, 2015). Previous research highlights the relevance of designing with users to achieve useful and usable systems, especially in stressful situations, and the requirement of supporting the sense-making and information validation processes of emergency managers (M. Imran et al., 2015; Stieglitz, Mirbabaie, Fromm, et al., 2018). Furthermore, a recent study concludes that most previous research has focused on identifying information that contributes to situational awareness (Zade et al., 2018). Accordingly, the authors introduce and emphasise the concept of actionability, meaning that 'information relevance may vary across responder role, domain, and other factors' (p. 1) and that right information needs to reach the right person at the right time. Current studies on event detection and summarisation focus more on experimental evaluation designs (M.-T. Nguyen et al., 2015; Rudra et al., 2015; Rudra, Goyal, et al., 2018) but less on the user-based evaluation in deployed systems.

Based on the feature gaps of existing systems, especially the absence of comprehensive functionality for information quality assessment (Kaufhold et al., 2017; Pohl, 2013; Trilateral Research, 2015) as well as the need for actionability and usability (M. Imran et al., 2015; Zade et al., 2018), our aim is to contribute with the design, implementation and user-based evaluation of a novel approach and integrated system for overcoming information overload by (1) processing and analysing social media data and transforming the high volume of noisy data into a low volume of rich content useful to emergency personnel (Moi et al., 2015) by grouping messages with regard to their qualitative context (Rao et al., 2017). Furthermore, approach and system will (2) support authorities in the assessment of information quality (Shankaranarayanan & Blake, 2017) and (3) enable communication among authorities and citizens.

Table 9.1 Overview of Intelligence, Management and Special Systems, adapted from Kaufhold et al. (2017), *de facto non-operating systems, **architecture but no implemented system

Systems		Cross-media	Communication	Monitoring	Alert	Event Detection	Collaboration	Influencer	Sentiment	Topic	Quality	Map	Filter	Diagrams
Intelligence Systems	Adobe Social	✓	✗	✗	✗	✗	✓	✓	✓	✓	✗	✓	✓	✓
	Mention	✓	✗	✗	✗	✗	✗	✓	✓	✓	✗	✗	✓	✓
	BrandWatch	✓	✗	✓	✓	✗	✓	✓	✓	✓	✗	✗	✓	✓
	Cogia	✓	✓	✓	✓	✗	✓	✓	✓	✓	✗	✓	✓	✓
	Evolve24	✓	✗	✓	✗	✓	✓	✗	✓	✓	✗	✓	✓	✓
	GeoFeedia*	✓	✗	✓	✗	✓	✓	✓	✓	✓	✗	✓	✓	✓
	Meltwater	✓	✓	✓	✓	✓	✓	✓	✓	✓	✗	✓	✓	✓
	PublicSonar	✓	✗	✓	✓	✓	✗	✓	✓	✓	✗	✓	✓	✓
	Signals	✓	✓	✓	✓	✗	✗	✓	✓	✓	✗	✓	✓	✓
	Socialmention	✓	✗	✗	✗	✗	✗	✗	✓	✓	✗	✗	✓	✓
	Quintly	✓	✓	✓	✗	✓	✓	✗	✓	✗	✗	✗	✓	✓
	Trackur	✓	✗	✓	✗	✗	✗	✓	✗	✓	✗	✗	✓	✗
	TweetTracker	✗	✗	✗	✗	✗	✗	✓	✗	✓	✗	✓	✓	✓
	ubermetrics	✓	✗	✓	✓	✗	✓	✓	✓	✓	✗	✗	✓	✓
	VicoAnalytics	✓	✓	✓	✓	✗	✓	✓	✓	✓	✗	✗	✓	✓
	Dataminr	✓	✗	✗	✓	✓	✗	✗	✗	✓	✗	✓	✓	✗

(continued)

Table 9.1 (continued)

Systems		Cross-media	Communication	Monitoring	Alert	Event Detection	Collaboration	Influencer	Sentiment	Topic	Quality	Map	Filter	Diagrams
Management Systems	Coosto	✓	✓	✓	✗	✗	✗	✓	✓	✓	✗	✗	✓	✓
	Crowdbooster	✓	✓	✗	✗	✗	✓	✓	✗	✗	✗	✗	✓	✓
	Lithium	✓	✓	✗	✗	✗	✓	✓	✓	✓	✗	✗	✓	✓
	HootSuite	✓	✓	✓	✗	✗	✓	✓	✓	✓	✗	✗	✓	✓
	Salesforce	✓	✓	✓	✗	✗	✓	✓	✓	✓	✗	✗	✓	✓
	Simplify360	✓	✓	✓	✗	✗	✓	✓	✓	✓	✗	✗	✓	✓
	Facelift	✓	✓	✗	✓	✗	✓	✓	✓	✓	✗	✗	✓	✓
	SproutSocial	✓	✓	✓	✓	✗	✓	✓	✗	✓	✗	✗	✓	✗
	TweetDeck	✗	✓	✓	✗	✗	✗	✓	✗	✗	✗	✗	✓	✓
	CrowdControlHQ	✓	✓	✓	✓	✗	✓	✓	✓	✗	✗	✓	✓	✓
	Orlo	✓	✓	✓	✓	✗	✓	✓	✓	✓	✗	✗	✓	✓
	MusterPoint	✓	✓	✓	✓	✗	✓	✓	✓	✗	✗	✗	✓	✓
	TalkWalker	✓	✓	✓	✓	✗	✓	✓	✓	✓	✗	✓	✓	✓
Special Systems	AIDR	✗	✗	✗	✗	✓	✗	✗	✗	✗	✗	✗	✓	✗
	CircleCount*	✗	✗	✗	✗	✗	✗	✗	✓	✓	✗	✗	✗	✓
	CrisisTracker*	✗	✗	✗	✗	✓	✗	✓	✗	✓	✗	✓	✓	✓
	SensePlace2	✗	✗	✗	✗	✓	✗	✗	✗	✓	✗	✓	✓	✗
	Tweedr	✗	✗	✗	✗	✓	✗	✗	✗	✓	✗	✗	✗	✗

(continued)

Table 9.1 (continued)

Systems	Cross-media	Communication	Monitoring	Alert	Event Detection	Collaboration	Influencer	Sentiment	Topic	Quality	Map	Filter	Diagrams
TwitInfo	✗	✗	✗	✗	✓	✗	✓	✓	✓	✗	✓		✓
Twitris	✓	✗	✗	✗	✗	✗	✓	✓	✓	✗	✓	✓	✓
Ushahidi	✓	✗	✓	✓	✓	✓	✓	✓	✗	✗	✓	✓	✓
SMART-C**	✓	✓	✗	✓	✓	✗	✗	✗	✓	✗	✗	✗	✗
EARS**	✗	✓	✗	✓	✓	✗	✗	✗	✓	✗	✓	✓	✗
Leadline**	✓	✗	✗	✗	✓	✗	✗	✗	✓	✗	✓	✓	✓
Alert4All*, **	✗	✓	✗	✓	✗	✓	✗	✓	✗	✗	✓	✓	✓
ESA-AWTM	✗	✓	✓	✗	✓	✗	✗	✗	✓	✗	✓	✓	✗
RSOE EDIS	✓	✓	✗	✓	✓	✓	✗	✗	✗	✗	✓	✓	✗
Vox Civitas*	✗	✗	✗	✗	✓	✗	✗	✓	✓	✗	✗	✓	✓
Visual Backchannel**	✗	✗	✓	✗	✗	✗	✗	✗	✗	✗	✗	✓	✓

9.3 Requirements Analysis: Methodology, Pre-Studies and Workshops

One aim of the project was to show the positive impact of gathering, qualifying, mining and routing information from social media on the management of emergencies, i.e. the mitigation of the information overload problem (RQ1, RQ2), which is realised through requirements analysis, and the development and evaluation of artefacts for emergency services and citizens. Thus, Design Science Research (DSR) plays a significant role, which is considered a problem-solving paradigm that 'seeks to create innovations that define ideas, practices, technical capabilities, and products through which the analysis, design, implementation, management, and use of information systems can be effectively and efficiently accomplished' (Hevner et al., 2004). According to Hevner (2007), DSR features three cycles: The central process is the building and evaluation of design artefacts and processes (*design cycle*). That should be grounded in and contribute to the knowledge base (*rigor cycle*). Moreover, field testing is required to raise requirements for the development of technology (*relevance cycle*). Since insisting 'that all design research must be grounded on descriptive theories is unrealistic and even harmful to the field', as Hevner (2007) suggests, we integrated 'several different sources of ideas for the grounding of design science research', such as requirements based on literature findings, existing artefacts and the inquiry of domain experts.

To identify requirements for a supportive approach, we employed a requirements analysis process: (1) scenarios and use cases from real-life operations were chosen to be illustrated and analysed; (2) these were presented in workshops to end users, development teams and experts to discuss different approaches, establish a common understanding and allow interaction with each other; (3) and to involve a broader community, we conducted online survey to collect data on the involved actors' views. All interventions (see Table 9.2) were supported by prior literature reviews to consider the state of the art and to inform the application of appropriate methodologies. The results of several empirical studies have already been published (Reuter, Ludwig, Friberg, et al., 2015; Reuter, Ludwig, Kaufhold, et al., 2016; Reuter, Kaufhold, Spielhofer, et al., 2017; Reuter & Spielhofer, 2017). For illustration, all elicited requirements were aggregated to high-level abstractions (Table 9.3).

Architecture implications were identified by reviewing guidelines, norms, laws (e.g. in terms of ethics, information security and usability), and literature on the technological state of the art (e.g. in terms of availability, stability and scalability). In summary, we decided to develop a web-based standalone solution that is easy

Table 9.2 Empirical pre-studies and workshops

Title and Focus	Year	Quantity
Interviews: Social media in emergencies	2014	11
Workshop I: End-user advisory board (ES)	2014	16
Survey I: Perception of emergency services	2015	761
Workshop II: End-user advisory board (ES)	2015	18
Survey II: Perception of citizens	2016	1,034
Evaluation I: First round of system evaluation	2016	12
Workshop III: End-user advisory board (ES)	2017	15
Survey III: Perception of emergency services	2017	473

to deploy and maintain in multiple system instances. Communication requirements were mainly identified by reviewing case studies on the use of social media during emergencies and thereafter validated in large-scale surveys (Reuter, Ludwig, Kaufhold, et al., 2016; Reuter, Kaufhold, Spielhofer, et al., 2017). Emergency services state that they currently most likely use social media to share information (A2C), but also consider monitoring to enhance situational awareness (C2A). To ensure wide use of social media, a facilitating organisational culture, trained personnel, appropriate knowledge and excellent communication skills are required. On the technical side, it demands an available and reliable internet infrastructure, including easy-to-use software artefacts that support users in dealing with multiple social networks. Our research, for instance, shows that around 45% of citizens use social media during an emergency and 46% expect to get a response to their social media post from emergency services within an hour.

Processing features were largely designed by reviewing the technological state of the art and existing solutions in terms of semantic data models, data gathering, data mining, information quality, and clustering algorithms. The processing components were implemented considering the requirements of tailorability, which were refined iteratively based on feedback gathered from the workshops. Visualisation opportunities were designed and advanced by reviewing the state of the art and elicited user requirements from the workshops (top-down) and by analysing how the gathered data can be visualised in a meaningful manner (bottom-up). Finally, the visualisation was refined upon the feedback of the first round of scenario-based evaluation (Reuter, Amelunxen, et al., 2016), which followed the structure of situated evaluation (Twidale et al., 1994).

Table 9.3 Abstraction of the system requirements

Category	Description
Architecture	Easy-to-use, available, maintainable, privacy-respecting, secure, stable, and scalable web-based standalone solution.
Communication	Reception, publication, response and broadcast of messages with multimedia (audio, photo, video) between authorities and citizens.
Processing	Cross-platform gathering, enrichment, relevancy and quality assessment of social media activities and alert generation.
Tailorability	Filtering of results in terms of geolocation, keywords, relevancy (mining) and information quality.
Visualisation	Display of generated alerts on a list and map view and classification of alerts according to the Common Alerting Protocol.

9.4   Development and Architecture of a Cross-Platform Social Media Based Alerting System

This chapter presents the development and underlying architecture of the web-based Emergency Service Interface (ESI). The system is connected to a mobile application (Kaufhold et al., 2018), which cannot be explained in detail within the scope of this paper but mentioned to define existing interfaces (see 'app alerts' in Figure 9.3). The system supports multiple information flows. First, authorities may use ESI to disseminate messages to multiple social media channels (A2C). Second, emergency services may monitor different social media, whose activities are grouped as social media alerts within the ESI (C2A). While A2C is a simple service forwarding messages to the respective APIs, C2A follows a complex path of processing information, which is described in the next chapter, before it is visualised in ESI.

9.4.1   The Backend: Grouping Messages to Alerts

For the integration of citizen-generated content (C2A), a processing component (PC) manages the interplay of gathering, enrichment, mining, information quality, and alert generation components (Figure 9.1). If the user defines search keywords in the interface (section 9.4.2), these are sent to the processing component. It instructs the gathering component to collect and return the relevant data to the PC, which then serves as an input for the enrichment component. This process is repeated with all components until information is grouped and sent to the interface by the alert generator.

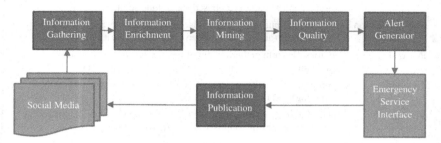

Figure 9.1   The backend with C2A (blue) and A2C (red) information flows

Since emergency services are likely to encounter a variety of different incident types, such as fire, floods, or traffic incidents, which can occur simultaneously, we implemented multi-scenario support. Accordingly, the user can instantiate this process multiple times, whereby we refer to each instance as a separate *pipeline*. The general idea is that each pipeline reflects a different scenario, e.g. a fire or a flood scenario (Table 9.4). A pipeline is then defined by the assignment of search keyword (e.g. "fire, bomb, explosion" or "floods, thunderstorm, water level") and a category (e.g. "Fire" or "Meteorological"). The categories are derived from the Common Alerting Protocol (CAP), which is an exchange format for 'all-hazard emergency alerts and public warnings over all kinds of networks' (OASIS, 2010). For different scenarios, varying characteristics and kinds of messages are relevant. Therefore, based on the defined category, differently trained Naïve Bayes classifiers (e.g. a classifier for fire and a classifier for floods) are used for determining relevant messages within the information mining component. All pipelines are directed to the same interface, but generated alerts are differentiated by respective category icons (section 9.4.2). Since, in contrast to fully automatic approaches, the user has to possibility to adapt the process by changing scenario details, i.e. keyword and category, we refer to this process as a semi-automatic approach.

To provide more detail, the process comprises the following steps: First, the *information gathering (1)* component allows the users to gather social media activities via keywords from Facebook, Google+, Twitter, and YouTube. The keywords are sent to and interpreted by the respective social media provider APIs. As these APIs return results in different exchange formats, we convert and store

Table 9.4 Comparison of two exemplary pipelines and their characteristics based on simplified fire and flood scenarios. The affected component, if any, is indicated in parentheses within the first column

Characteristics	Pipeline I	Pipeline II
Scenario	Analysis of fire-related messages	Analysis of flood-related messages
Keywords (1)	"fire, bomb, explosion"	"floods, thunderstorm, water level"
Category (1)	Fire	Meteorological
Enrichment (2)	Scenario-independent information enrichment	
Relevancy (3)	Use of a classifier trained for fire	Use of a classifier trained for floods
Quality (4)	Scenario-independent information quality assessment	
Alert (5)	Grouping of pipeline I messages	Grouping of pipeline II messages
Routing (6)	Scenario-independent information routing	
Interface (7)	Alerts with fire icon	Alerts with flood icon

all messages according to the Activity Streams 2.0 specification (World Wide Web Consortium, 2016). Since we are interested in multiple metadata that are not provided by different social media APIs but required for the information quality component, the *information enrichment (2)* component computes additional metadata such as entropy, positive or negative sentiment, or the number of characters, punctuation signs, sentences and words.

Since we intend to reduce the large quantity of activities to a manageable amount of high-quality information, the *information mining (IM) (3)* component pre-processes the gathered activities and applies configurable geographic boundary and event type filters. This restricts the data to that generated in the incident specific area and regarding the identified incident only. As the last step, after some phases of pre-processing (basic normalisation, stop word removal, tokenisation, and URL extraction), a relevancy filter based on a trained Naïve Bayes classifier filters out activities whose contents are not related to an emergency. To support both the fire and flood scenario, we trained two respective classifiers based on data gathered from actual incidents, i.e. the 2016 BASF fire and 2013 European floods. The data sets were labelled by single and different labellers. For each data set, a scenario description was created, containing basic information about the incidents that were labelled. This approach aimed to enable the labellers to understand and immerse in the situation from an emergency service operative's perspective. The labellers then were presented with the data and labelled it according to their understanding of the situation on a binary scale (relevant or irrelevant). We recognize the weaknesses of this approach, especially concerning the manual labelling process, using a single labeller, resulting in heavily biased data. Since the focus of the project was on the development of an overall system and not on the optimisation of the classifier, this weakness was accepted. Further research in generally classifying data from emergencies should be done, placing more resources in the classification process, especially the labelling of a broad range of different incidents with sufficient labellers. For the fire classifier, we manually labelled 3785 tweets, whereof 48% were labelled relevant and 52% irrelevant. Furthermore, 2000 tweets were manually labelled for the flood classifier and reached a relevant to irrelevant ratio of 66% to 34%. By comparing the manual labels with the automatically classified messages, the fire scenario classifier reached an accuracy of 73.3% and the flood scenario classifier an accuracy of 76.1%.

Table 9.5 The information quality framework with criteria and indicators

IQ Criteria	IQ Indicators
Believability	Existence of URLs, locality, proximity, existence of media files
Impact	Number of comments, number of shares, involvement, number of likes, number of views
Reputation	Number of followers, number of statuses, verified account, trusted account
Completeness	Existence of URLs, number of characters, number of hashtags, type of information present, time of information present, location of information present
Relevancy	Existence of emergency words, relative frequency of emergency words, amount of contained crawl keywords, number of relevant entities, number of sentences with relevant entities, relative frequency of relevant entities
Timeliness	Closeness, first occurrence of an emergency word, post age
Understandability	Average length of words, readability, existence of media files, information noise, appropriate language

Thereafter, an *information quality (IQ) (4)* component evaluates the remaining social media messages with an IQ framework (Table 9.5) estimating the criteria of believability (including the sub-criteria of impact and reputation), completeness, relevancy, timeliness, and understandability by different indicators (e.g. number of followers). The dependencies between criteria and indicators are modelled as nodes of a Directed Acyclic Graph (DAG). The output of each indicator node lies within [0,1] and criterion nodes collect and aggregate the output of indicator nodes dependent on them. They compute a weighted arithmetic mean and output a value within the interval [0,1]. The weights are attached to the edges of dependent nodes and express the importance of the dependent indicator or criterion. Thus, the higher the importance of a dependency, the bigger its influence of its output for the result of the criterion. Since the output of each node lies between 0 and 1, we compute the overall IQ in a single value. Although, in principle, it is possible that end-users manually set the weights according to their own preferences, for our evaluations, we trained the weights of the IQ graph using the Backpropagation Algorithm (BPA), which consists of two phases (Werbos, 1994). Three experts used an IQ assignment tool to create a training set of 2.500 posts with assigned IQ values. In the first phase, the training data was propagated through the neuronal network and the IQ values were calculated automatically. The results were reported back, compared to the results the human evaluators provided, and the difference between automatically and manually rated IQ values was calculated. If both values differed, the second phase of the BPA was executed. In the second phase, the weights were adjusted to better fit the manually rated IQ value. These steps were repeated until the performance reached a satisfactory threshold.

Finally, the *alert generator (5)* component groups messages to provide meaningful and manageable bits of information for emergency services. We refer to an alert as a set of classified messages sharing a similar context, which is defined by event type, keywords, language, location, platform, quality, relevancy, and time. The relevant contextual filters are described in Table 9.6. Due to the project's time constraints, we were not able to integrate the configurability of all contextual factors in the frontend. Thus, some of them were preconfigured by experts in the backend depending on the requirements of the respective exercise or field trial. After the contextual filters are applied, in a last step, a geographical database is used to group geo-located messages at the nearest geographical named entity of the database. Using the *information routing (6)* component, these alerts are sent to the *user interface (7)*.

9.4.2   The Front-end: Visualisation of Alerts

The web interface is split up into the four different pages *(1) Dashboard*, *(2) Social Media (SM) activity*, *(3) App activity* and *(4) Settings*. The dashboard is the default page featuring the display of alerts within the *map view* and *list of alerts* to provide a quick orientation (Figure 9.2). Following the CAP standard, alerts are categorised as either 'Fire', 'Meteorological', 'Transport' or 'Other',

Table 9.6 The contextual filters influencing the generation of alerts. Some factors are configurable in the backend (BE) and some in the frontend (FE)

Factor	Description	BE	FE
Event Type	The algorithm only groups messages of the same event type (options: Fire, Meteorological, Transport, or Other).		x
Keywords	The algorithm includes keywords that match the defined Boolean search query (options: use of Boolean operators, such as AND, OR, NOT, etc.).		x
Language	The algorithm excludes messages that are not within the range of specified languages (options: allow a single, multiple and all languages).	x	
Location	The algorithm excludes messages that are disseminated outside of a specified bounding box (default: no bounding box, but for the field trials bounding boxes were created).	x	
Platform	The algorithm includes the configured social media platforms (options: one or multiple platforms of Facebook, Google+, Instagram, YouTube, Twitter).	x	
Quality	The algorithm excludes messages based on information quality thresholds (options: include low, medium and high-quality messages).		x
Relevancy	The algorithm excludes irrelevant messages (Naïve Bayes classifier) and retweets (options: include also irrelevant or only relevant messages).		x
Time	The algorithm excludes messages that are older than a specified number of hours (default: two hours).	x	

the latter comprising all other CAP categories. We chose to only integrate the most relevant subset of categories for our scenarios (see section 9.5.3) to keep the interface clear and simple. Each category has its own symbol, which is used on the map, list and alert counters above. Furthermore, using text input fields, the user can define a distinct set of complex Boolean keywords for each category, such as 'Dortmund (fire, bomb)' for the fire category. In this example, the results are filtered by social media messages that contain the term 'Dortmund' as well as 'fire' and/or 'bomb'. At the top-right corner, the user may enter the settings page to define the sets of keywords (information gathering).

For each alert in the list view, the overall IQ score is indicated by the indicators 'high', 'medium' and 'low'. Since preliminary results suggested to keep the UI simple (Reuter, Amelunxen, et al., 2016), we chose this approach instead of visualising the whole IQ graph of criteria for each alert. The user can filter the list of alerts by the relevance filter (IM) and an IQ threshold. If the user clicks on an alert, a window containing the list of individual social media messages are listed including their IQ value (Figure 9.3). Three icons at the top-left corner represent the information sharing functionality allowing the user to login to private social media accounts and to share information on Facebook and Twitter (A2C). Lastly, while the *SM activity* page lists individual, non-grouped social media activities using the same layout, the *App activity* page displays app alerts and allows emergency services to reply to them individually.

Figure 9.2   The Emergency Service Interface (ESI): dashboard view

9.5   Methodology of the Systems' Evaluation

This chapter deals with the second evaluation of the EmerGent ICT system using different surveys. The need for a second evaluation became apparent due to several reasons: Firstly, based on the first evaluation and new requirements, the ESI had been redesigned completely and the quality of the new interface was to

Figure 9.3   ESI: details of an alert

be evaluated. Secondly, in the first version of ESI, most features were not fully implemented; therefore, the value of app alerts, potential alerts, keyword performance and information quality could not be tested thoroughly. Finally, due to the unfinished state of the first ESI version, we could only perform a constructed scenario-based evaluation. Thus, the evaluation based on a functional prototype and in real-world scenarios e.g. via field trials promised richer feedback from the users.

To evaluate and achieve productive feedback on the system's components, we decided to conduct semi-structured interviews. The interview guideline was derived from the first evaluation (Reuter, Amelunxen, et al., 2016) but the questions were refined to get better feedback on individual functionalities. In terms of personal details, the guideline asks for the type of organisation, main role, command level, work years, age, gender, and country. Additionally, organisational details such as current and future role of social media as well as organisational barriers were asked. The survey consisted of six guiding questions, which were open-ended unless indicated differently by footnotes:

- Q1: What is your first impression?

- Q2: How would you evaluate the functions according to their importance in your job?
- Q3: How do you evaluate 'social media alerts'?
- Q4: How do you evaluate 'information quality'?
- Q5: What functionality of the application do you find most useful or has potential?
- Q6: Is there any additional functionality you would like the application to have?

For Q2, participants had to indicate the importance of the C2A and A2C functions on a 4-point scale of max (4), high (3), moderate (2) and min (1). Furthermore, For Q3 and Q4, Participants had to indicate the benefit on a 4-point scale of high (3), medium (2), low (1) and none (0) and were asked for further open-ended feedback. Overall, 21 interviews with emergency services were conducted during the second evaluation (Table 9.7). Interviews were audio-recorded and transcribed for further analysis. In our subsequent analysis, we employed open coding (Strauss & Corbin, 1998), i.e. gathering data into approximate categories to reflect the issues raised by respondents based on repeated readings of the data, organising them into similar statements. As most of the analysis was conducted in German, selected quotes were translated into English by the authors.

Table 9.7 Second evaluation: personal details of participants

Category	Data
Roles	crew (10), head (1), incident commander (8), other (6), press (5), PSAP operator (1), PSAP supervisor (5), section leader (3)
Level	gold (3), silver (9), bronze (8), none (1)
Age	20–29 (2), 30–39 (10), 40–49 (6), 50–59 (3)
Gender	male (18), female (3)
Country	Germany (13), Poland (7), Slovenia (1)
Type	Field trial (10), live and paper-based demonstration (9), workshop exercise (2)

To allow for different degrees of emergency services involvement, considering their potentially limited time resources, we conducted different types of evaluation, which are presented in the following chapters.

9.5.1   Live & Paper-based Demonstrations (2017)

During a live system demonstration, based upon a short introduction of its functionalities, the participants could interact with the system, which was preconfigured with fire and flood scenario keywords (see Figure 9.3), before and during the guideline-based inquiry. In a paper-based demonstration, the interviewer introduced prepared screenshots of the system and explained its functionality as a foundation for the inquiry. One survey was conducted in 2017 in Warsaw, Poland, by the Scientific and Research Centre for Fire Protection—National Research Institute (CNBOP-PIB), and further two interviews with members of volunteer fire departments (FD) in Germany.

9.5.2   Workshop Exercise (2017)

The integrated system was tested at a convention in Salzburg, Austria. During the live exercise, a video of an incident (a fire) was shown, and the audience was asked to participate by using their Facebook and Twitter accounts. The audience, representing the role of active citizens, used the video for taking pictures of the incident scene and sending them along other valuable information to a simulated command and control (C&C) room (Figure 9.4). The video started with a general introduction, but then visualised a fire which grew bigger and was followed by explosions as well as response efforts from emergency response teams. Furthermore, social media demo accounts were used to flood the system with prepared messages simulating a constant flow of false or irrelevant information, which are common in real-world scenarios. The C&C room, amongst others, was manned with an incident commander (gold level firefighter from FD Dortmund) and a social media manager (bronze level firefighter from FD Ljubljana) who used the ESI to get relevant incident information from social media and to broadcast relevant information to ESI users on Facebook and Twitter.

9.5.3   Field Trials (2017)

For longer-lasting testing periods in real-world scenarios, three field trials were conducted. The field trial of FD Dortmund was in March/April 2017 (four interviews) and of FD Hamburg in April/May 2017 (three interviews), both lasting four weeks. Another one was conducted in July 2017 with FD Hamburg during the G20 event (three interviews). During these trials, the system was used by

Figure 9.4   ESI in a simulated C&C room during the workshop

different functions of the organisation alongside their regular duties: The public relations (PR) department, the head of the dispatchers and the department of strategic planning. Dortmund decided to use the system with keywords for the topics 'fire', 'rescue' and 'severe weather' which remained the same over the whole field trial. Hamburg preferred the topics 'fire/terror', 'CBRN', 'malfunction of subway/bus' and warnings about 'contamination'. During the first Hamburg field trial, the keywords were regularly discussed and adapted with the help of technical experts and using a *Telegram* messenger channel. For instance, after regular revisions, Hamburg used the following final keyword set for the 'fire' scenario, as translated from German: '(Hamburg, HH, hvv, TMC) (fire, flames, incident location, unsecured, chemical accident, police operation, incident, breaking news, bomb threat, blaze, bomb, injured, smell, gas leak, poisoning, weather alert, RTW, personal damage, rescue mission, protest, conflict)'. The second field trial in Hamburg was not planned from the beginning, but due to the fire department's positive reception of the system, they encouraged the conduction of a field trial during the G20 event where extensive demonstrations and riots from left-wing groups were announced and expected.

9.6   Results of the Systems' Evaluation

The main results of the first evaluation, which have already been published (Reuter, Amelunxen, et al., 2016), include the potential of the system to identify risks and to filter using own criteria. It was mentioned that precise information (alerts) is needed, especially in mass events. Approaches that allow both individual settings and automatic processing of data can help here. As for the threats, false information is available and negative consequences might occur. To summarise one key requirement: '*Keep it simple on the UI and complex in the backend*'. In the following chapters, we focus on the results of the second evaluation. Furthermore, we use the identifiers I1-I21 to reference statements by participants.

9.6.1   General Attitudes and Impressions (Q1)

9.6.1.1 Positive Reception of the System

Overall, 13 of 21 participants expressed a positive attitude towards the system. While two considered it to be useful (I5, I21) and to provide important information (I2) in a general manner, others explicated more specific benefits: It can be used to support decision-making (I8) and the reporting of incidents (I6), inform the population (I1, I15), reduce reaction time to emergencies and thus improve overall safety (I1). Six participants highlight the simplicity of ESI, although one participant found the map handling to be difficult (I15) and one indicated that training is required, e.g. in terms of selecting suitable keywords to get the most out of the tool (I14). The workshop exercise was perceived as a proof of concept with limitations: '*The demonstration during the final workshop was a good proof of concept, although it was not directly integrated into the whole [control room] system*' (I9). Some participants were sceptical about the interface's aesthetic look (I12, I13).

9.6.1.2 Negative Reception of the System

However, besides mentioning the system being supportive but not a telephone replacement (I7), two participants have mediocre (I12, I13) and one participant a strong negative or sceptical attitude towards the system, listing several issues (I10): Some important events in Dortmund were not properly detected, the system often provided results from the same sources, information was often hours or days old and rather from '*press offices than from normal users*'. Another participant did not perceive the system to be not useful in the current state unless some minor changes were applied (I9).

9.6.1.3 Importance of Keyword Management

The way of dealing with social media keywords is likely to be the reason for some important events in Dortmund not being detected. While we pre-designed the keywords in Dortmund—the preparation of scenarios was perceived as important (I15)—to be used during the whole field trial, we developed and regularly changed the keywords during the Hamburg field trials. In Hamburg, I14 was responsible for keyword management and adapted the keywords on demand if new relevant hashtags or topics emerged. These were not only extracted from social media but also from other internal and external information sources at hand. Thus, five participants involved in field trials highlighted that the regular adaption of keywords is essential and required, with participants from Dortmund reporting negative (I10) and those from Hamburg reporting positive results (I14-16). While in Hamburg, specific keywords were also used to increase the probability of location-specific results, one participant mentioned that a geographical restriction in terms of the towns' administrative area was required (I16).

9.6.2   Social Media Alerts (Q3)

9.6.2.1 Situational Overview and Specific Information

Evaluating 'social media alerts', eleven participants indicated a huge, six a medium, three a small and one no benefit of the functionality (Table 9.8). The respondent indicating no benefit said that '*when there is not yet any direct threat, just a potential one, people will not take it seriously*' (I1). Besides the risk of false information and spread of panic (I6), other feedback was more positive. Social media alerts were considered useful (I17) and assessed as '*one of the most relevant aspects of the system*' (I8). Five participants indicated that it was a good opportunity to get a general situational overview of local events and developments (I10, I12, I15, I19), which was especially useful for press offices (I16), but also to get more specific information sometimes, e.g. '*to recognise emerging situations [and] where something is brewing*' (I10). For instance, Hamburg was able to prepare for a train with thousands of protesters during G20 which they detected via social media alerts (I18). Although, according to two participants, most times the social media alerts did not deliver faster information than other media and control room systems, the information retrieval was perceived to be fast, and there were some occurrences, e.g. about road conditions and road closures which were delivered faster via social media alerts (I15, I16).

Table 9.8   Indicated benefits of social media alerts (Q3) and information quality (Q4)

Benefit	Huge (3)	Moderate (2)	Small (1)	None (0)	Ø
SM alerts	13	6	0	2	2.43
IQ	8	8	4	0	2.20

9.6.2.2   Access to Unfiltered Data

As already indicated, one key aspect regarding the performance of the system was the careful selection and adaption of relevant keywords: '*The keywords and the algorithms were tuned properly. Now the messages are really goal-driven*' (I17). Although the filtering was perceived as good, one participant wished to access non-filtered data as well (I17). During G20, Hamburg also used Twitter and TweetDeck to search for individual and popular keywords, which was sometimes faster with respect to achieving specific information (I18, I19). However, both interviewees would appreciate a combination of all functions within an integrated tool such as ESI, e.g. to support documentation of activities (I18).

9.6.2.3   Issues of Grouping and Geolocation

Moreover, the grouping by geolocation was not perceived as a sufficient means of defining an actual alert (I12): 'However, the grouping by geolocation is not enough, because it is not available in every message. Equally important are the content (text analysing, e.g. 'smiles', capitals), keywords, and psychological aspects: How many exclamation marks are used? Are there any emotions reflected in the message and if yes, which ones? What is the letter case?' (I9). Moreover, it was not clear how and if the geotagging worked properly: 'Only geotagged information should be on the map' (I14). Since geolocation data is either extracted directly from metadata (accurate GPS position), indirectly from metadata by using the attached bounding boxes of towns, or indirectly by analysing and extracting it from the actual message content (both inaccurate indications of positions), it should be indicated which method of location determination was used, allowing the user to assess how accurate the displayed location of the alert is (I15).

9.6.3   Information Quality (Q4)

9.6.3.1   Reliability and perceived Preference for Official Accounts

Eight participants indicated a huge, further eight a moderate and four a small benefit from the IQ component (Table 9.8). On the positive side, on high settings, the component was perceived as a useful filter (I12), if the algorithm was trained correctly (I20), for the most crucial alerts (I2), worked reliably and allowed a focus on important results (I15). Thus, only a small amount of misclassification was observed (I15, I16). Three participants observed that, by tendency, authorities' and media messages were assigned a higher quality than citizens' messages which was viewed sceptically since potential eyewitness reports were rated lower than media reports from hours or days ago (I9, I10, I11). Moreover, two participants assumed that too many messages were filtered out (I21) and thus, a performance feedback (of the different layers of filters) was required (I16).

9.6.3.2 Issues of Transparency and Tailorability
Notably, seven participants criticised the lack of transparency concerning how the algorithm operates (I1, I9, I17): *'For users, it is unclear what happens when the filters are turned on'* (I14). Moreover, they were sceptical about fixed quality criteria, e.g. the number of followers was not perceived as a crucial factor for IQ (I10). Thus, more delicate and visible criteria (I10) and the possibility to parametrise underlying quality criteria were wished for since *'the determination of quality criteria [is done internally] in the organisation'* (I8, I11). On the one hand, the user's knowledge, e.g. about the credibility of specific authors, was recognised as an important resource (I8) that could contribute to an algorithm or a system that learns from user input (I10). On the other hand, one participant was sceptical whether the actual user was capable of parametrising quality criteria, suggesting that the job should be performed by the system's administrator (I9). Nevertheless, it was seen as an important option to define 'trusted users' whose quality level should be estimated as high: *'Expert groups, trusted people, THW relatives, potential app users, etc.—would be a high-quality group of users'* (I10, I18).

9.6.3.3 Impossible to avoid Misinformation
Besides the technical aspects, other participants stated that it was generally difficult to choose relevant information (I4), hard to determine true or false information (I5) and impossible to avoid 'fake' information (I6). Since a huge benefit was expected in cases of components working properly, but a lot of scepticism and potentials for improvement were mentioned, further research is required on this topic.

9.6.4   Importance and Usefulness of Functionality (Q2+Q5)

9.6.4.1 Importance of Functionality

Each functionality, representing a communication flow, was assigned high or maximum importance by at least two-thirds of the participants (Q2, Table 9.9). On average, A2C received a slightly higher value (3.19) than C2A (3.05). Although the average values are quite similar, there seems to be a small preference for A2C over C2A communication, which matches a qualitative snowball study indicating that emergency services are more likely to share information than to monitor or receive messages from social media (Reuter, Ludwig, Kaufhold, et al., 2016).

Table 9.9 Indicated importance of functionality (very important to not important at all)

Importance	max (4)	high (3)	low (2)	min (1)	Ø
C2A indirect	7	7	6	0	3.05
A2C indirect	9	8	3	1	3.19

9.6.4.2 Usefulness of Functionality

Most participants' answers could be assigned to a specific information flow. The C2A flow, represented by 'social media alerts', was the most recognised (by five participants; I9, I10, I15, I16, I19). One emphasised the importance of IQ to get only the most important messages during large-scale emergencies (I12). Three participants valued the C2A flow on a more general level: '*It's an additional way of contacting emergency services and in situations when lives are endangered, all ways are welcome and increase the possibility of helping a victim*' (I5). It has the potential of an information advantage: '*Before the control room or personnel receives the information, it is on the ground. Information can be received which otherwise would have to be manually searched for*' (I14). Thus, in everyday life, '*the information acquisition is most useful*' (I17).

On the other hand, two participants highlight the relevance of A2C communication (e.g. sending a message or broadcast from ESI): '*Being in the 'hot zone', I am receiving a proper message directly on my smartphone, at least I will consider that it is serious and I will follow the instructions*' (I1). Thus, '*it may be helpful to tell people what to do in an emergency*' and '*they may feel comforted as they know that they are not alone*' (I6). Finally, three participants valued the way how information is presented on the dashboard '*to evaluate how the situation is*' *at that moment'* (I14): *'The graphical representation was great'* (I16). *'The most useful feature is to get information presented this way in everyday life'* (I17, I18). Two participants highlighted the map to be the most useful function to get a first situational overview (I12) for the disposition of forces (I11).

9.6.5   Additional Functionality (Q6)

9.6.5.1 Improving Keyword Management and Visualisation
With only three people indicating that the application already includes the most important functions (I2, I4, I6), the participants offered broad feedback on additional functionality for ESI. Since participants emphasised the relevance of adequate keywords, keyword highlighting was desired to develop an idea regarding keywords producing certain results (I9). Also, an enhanced keyword management was wished for: *'All dispatchers would use the same keywords; they should grow in number and be intact forever. It should be possible to add more keywords on UI level'* (I9). Since the keywords, IM and (potentially) IQ components filter the number of incoming messages, it was seen as important to get performance feedback *'on the dashboard because it is important to know how many messages were mined and went through the system'* (I8, I15).

9.6.5.2 Enhancing the Alert Management
Further functionality was desired in terms of 'app alerts' and 'social media alerts'. Four participants mentioned a sound notification if new alerts came in, with the option to turn it on or off, (I10, I11, I16) to *'support the information advantage because the system can't be watched the whole time'* (I14) or configurable mail or push notifications, e.g. based on keywords (I13, I18). To support further investigation of incoming alerts, two participants suggested to display the username, provide a link to the source platform of the message and add a symbol to indicate from which platform the message was received (I14, I17) since it is important to assess the quality of information (I10). Moreover, several management operations for the list of alerts were named: to mark alerts as read or done (I9), set custom priorities (I9) and custom categories for alerts (I19), modify the grouping of alerts manually (I11), manage favourites, pin important alerts or attach notes to alerts (I10).

9.6.5.3 Accessing and Visualising Historical Data
Additionally, an alert archive with a search and filter functionality was demanded by five participants to allow post-processing of social media posts after an

incident. Two participants wished for a Telegram integration to forward all alerts to a private Telegram channel for use as an archive (I14) or to send messages to colleagues (I16). Besides a list-based archive, one interviewee emphasised the need for a chart-based analysis and filtering of past or current alerts, e.g. to show the volume of alerts during a specific timeframe (I11, I13, I19).

9.6.5.4 Adding Collaborative Features

Since in the current state, the interface shows the same view to all users and (private) information management features were suggested, the topics of collaborative work (I10, I18) and role management (I17) were also discussed, e.g. to provide different views for different functions such as press office and situation service (I12, I16). In contrast, another participant spoke against the need for additional collaboration: '*As this is not an incident command system, it does not need additional collaboration features. If a colleague replies to an alert on another computer, he would know that because the message would be marked.*' (I9).

9.6.5.5 Enhancement of Map Functionality

The map view also received critical reception. First, considering the space it takes, the map was barely used, and the list of alerts regarded as more important: '*We have our own maps on which we plot things. For that, we wouldn't use ESI*' (I15). More accurate location information was requested, e.g. with the option to show the individual positions of the messages grouped in an alert or to indicate their distribution with a polygon (I11, I12, I15). The map, moreover, should only present alerts from the emergency services' authoritative area, e.g. the bounds of Hamburg (I16). To improve the utility of the map, one participant wished for the integration of live stream (Facebook, Periscope), radio or webcam layers (I14). Another recommended connecting pictures and videos to geolocations and displaying them on the map as an additional layer (I8). Furthermore, he suggested implementing a multimedia view where only data such as pictures and videos are displayed. During large-scale events with plenty of alerts, such as G20, where up to 160 alerts were recognised by the system (I18), a solution is required if multiple markers overlap in a certain area (I11). Moreover, a better distinction of app alerts and social media alerts was requested (I19).

9.7   Discussion

We designed and evaluated a social media alerting system for emergency services to mitigate the potential information overload in social media during large-scale conflicts and crises. Overall, most participants emphasised a positive attitude towards the system, including the statement that we delivered a good proof of concept. Several benefits for decision-making, reporting of incidents or informing the population were stressed, but participants also valued the simplicity of ESI and contributed potentials of enhancement (Table 9.10). In the following chapters, the main findings are presented and contextualised into existing literature, while recommendations of future research are discussed.

Table 9.10 Outline of requested features in terms of display, alerts, filters, and map

Class	Feature
Display	Custom information management and role-based views.
	Separate multimedia view (e.g. pictures, videos).
	Accessing, searching and filtering historical data.
	Chart-based filtering and visualisation of data.
Alerts	Improve the algorithmic message grouping into alerts.
	Further ways of notification (e.g. e-mail, push, sound).
	Management operations (e.g. read, done, notes, pinning, priority).
	Communication threads (e.g. response relations).
Filters	Improve and simplify the management of keywords.
	Allow keyword highlighting within the message texts.
	Show the performance of keyword, IM, and IQ filters.
	Allow tailoring of IQ graph to user or organisational preferences.
	Illustrate computation of IQ values on demand.
	Support the management of trusted and blocked users.
	Consider IM, IQ algorithms learning from user input.
Map	Allow the restriction of alerts by authoritative area.
	Indicate the precision of geolocation (e.g. GPS or city level).
	Show individual messages and comprising polygon for each alert.
	Allow display of further layers (e.g. radio, streams, webcams).
	Distinction of alert types and overlapping alerts.

9.7.1   Information Quality and White-Box Algorithm Representation: Supporting the Subjectivity, Tailorability and Transparency of Filtering

The combination of social media, mobile and wireless technologies have significantly reduced the time lag between the capture and dissemination of data, and the analysis of big social data is likely to impact decision-making in the future (M. Imran et al., 2015; Shankaranarayanan & Blake, 2017). Besides timeliness, information quality is defined by a variety of dimensions that only become visible in practice which is why we evaluated an information quality framework for social media with practitioners in this paper.

On high settings, the IQ component was perceived as a filter for the most crucial alerts, worked reliable and allowed a focus on important results. However, the performance of IQ should be compared to other ML algorithms, for instance, which learn from user input continuously. Furthermore, fake news, online rumours (Arif et al., 2017; Starbird et al., 2016) and the propagation of social bots (Ferrara et al., 2016) increasingly affect the landscape of big social data and thus should be examined in the light of IQ. Notably, many participants mentioned that more transparency on how the overall IQ score is estimated by the system would increase the comprehensibility. Furthermore, participants demanded more delicate and visible criteria that could be parametrised by the organisation, as already indicated by Reuter et al. (2015), and stressed the importance of qualified or trusted users. In accordance with literature (Hilligoss & Rieh, 2008; Ludwig, Reuter, & Pipek, 2015), it was emphasised that quality has a subjective component. Thus, future versions of ESI should present more detailed IQ scores on demand and allow manually weighting IQ indicators to evaluate the appropriation, assessment, and performance of the IQ component.

The findings highlight the importance of an accurate representation of the system's state and its sub-processes as well as the adaptability of systems (McKinney, 2011). The current 'black box' of algorithms does not allow the users to understand and 'fix the system so that its behaviour becomes more useful to their needs' (Burnett et al., 2017, p. 235). In accordance with the desired feedback on keyword and mining performance (see section 9.7.2), a 'white-box' representation of algorithms—indicators and filters, which make the procedures transparent for the user—seems worth examining in future research to support the assessment of gathering, mining and quality performance as well as their adaption to situational demands. Since research in the education domain highlights the potential of white-box approaches for increasing the users' acceptance of algorithms (Delibaši et al., 2013; Romero et al., 2013), it seems a promising area for HCI to research the requirements, challenges and potentials of white-box algorithms and their visualisation across different types of algorithms, domains and users.

9.7.2 Information Overload and Usable Configurability: Improving the Algorithmic Performance and Configurability of Social Media Alerts by Users

The increasing use of social media and thus the creation of big social data during emergencies raises the risk of information overload (Mendoza et al., 2010; Olshannikova et al., 2017). Since emergency services encounter a scarcity of personnel and time resources (Plotnick & Hiltz, 2016), technological solutions might assist in the filtering of relevant data (M. Imran et al., 2015; Moi et al., 2015). Although there are existing architectures and systems that enable the filtering of big social data, e.g. Public Sonar (Abel, Hauff, & Stronkman, 2012), only few of them integrated a social media alert generation feature, which is why we introduced the concept of social media alerts.

Social media alerts were perceived as a good opportunity to get a general situational overview of local events and developments in social media, but also to get specific information, e.g. to prepare for or predict emergencies. However, amongst others, participants wished for an improved social media alert grouping, e.g. a more sophisticated grouping algorithm, more user metadata and more detailed location information of included messages. Thus, the implementation of more advanced classification (Habdank et al., 2017), clustering and role-based summarisation algorithms (M.-T. Nguyen et al., 2015; Rudra et al., 2015; Rudra, Goyal, et al., 2018), incorporating similarity measures, for event or sub-event detection are likely to increase the algorithmic performance of our approach (M. Imran et al., 2015; Pohl et al., 2015). Comparing the field trials in Dortmund (using pre-defined keywords) and Hamburg (regularly adapting pre-defined keywords), it became apparent that the definition and maintenance of suitable keywords is one key success factor for the system. Thus, an enhanced and more usable keyword management and a performance feedback, e.g. regarding the performance of keywords (in social media) and different filters, would improve the overall handling of the system and allow to adapt more quickly to changing situations. Moreover, several alert management functions were demanded such as: Mark alerts as read, prioritise alerts manually, pin alerts or manage favourites and provide an archive of past alerts, e.g. for the post-processing of deployments.

HCI should further research the issue of 'usable configurability' which demands, on the one hand, easy-to-use and integrated systems and, on the other hand, a configurability of (complex) components regarding the users' and organisations' use cases to achieve, in this case, the goal of a low volume of rich and useful content for emergency services. Based on the 'white-box' representation of algorithms (section 9.7.1), concepts of end-user development, which comprise 'methods, techniques, and tools that allow users of software systems, who are acting as non-professional software developers, at some point to create, modify or extend a software artefact' (Lieberman et al., 2006) and usability engineering (Nielsen, 1993) may be applied to achieve usable configurability.

9.8   Conclusion

In this paper, we analysed emergency services' potentials and barriers of using social media during emergencies as well as existing social media analytics systems identifying a need to support emergency services regarding the assessment and the prevention of information overload (section 9.2). Based on empirical pre-studies, workshops and requirements analyses (section 9.3), we presented the development of the system (ESI) which supports the monitoring of social media via alerts, enables interactions between authorities and citizens, and supports the assessment of IQ (section 9.4). Using semi-structured interviews in different settings such as exercises, live demonstrations, and field trials (section 9.5), we conducted two iterations of evaluation whose results are presented (section 9.6) and discussed (section 9.7) in this paper in order to answer the following research questions:

How can social media alerts based on information gathering, mining, and quality filters help to mitigate the issue of information overload (RQ1)? With the Emergency Service Interface (ESI), we developed a novel approach for generating social media alerts, which transforms the high volume of big social data into a low volume of rich content that is useful to emergency personnel and aims to mitigate the issue of information overload. In comparison to existing social media analytics systems (Kaufhold et al., 2017; Pohl, 2013; Trilateral Research, 2015), ESI utilises an alert generation feature that considers the qualitative context of individual social media messages and integrates a filter layer based upon an information quality framework. During the evaluations, the approach was valued especially during large-scale incidents since it facilitates the adjustment of social media alerts by keyword (information gathering), relevance (information mining) and quality (information gathering) filters. The results suggest that a 'white-box' representation of algorithms would help emergency managers to better understand their computational behaviour, allowing to improve the users' utilisation of these filters and thus the mitigation of information overload. **How can the trade-off between automation and user interaction be designed to mitigate the issue of information overload (RQ2)?** Besides user input in terms of setting or changing keywords as well as activating or deactivating the relevancy and quality filters, after an initial developer- and expert-based configuration, the backend algorithms work automatically. While the evaluation outlines the need for improving the algorithmic performance, such as a more sophisticated grouping algorithm, end-users required the configuration of algorithms according to personal or organisational preferences and requirements, i.e. to adapt the weight of different information quality criteria and indicators. Thus, end-users required a 'usable configurability' combining easy-to-use and integrated systems with a sufficient configurability of complex algorithms or components, that could be further improved by the application of end-user development concepts (Paternò & Wulf, 2017) which we aim to realise with our application. Furthermore, real-time feedback and historic information is required for the end-user to assess the performance of the filter configurations and facilitate the gradual improvement of social media alert generation with regard to the dynamic and partially unforeseeable character of conflicts and crises.

After implementing a revised version of the system with proper alignment to related concepts of the knowledge base, an additional round of evaluation could contribute to these research areas. However, some limitations of the study must be considered. Firstly, the evaluation was mainly conducted with fire services, limiting the applicability of results to other types of organisations. After implementing the gathered user feedback, further evaluations could examine requirements and specifics of danger prediction and prevention by the police using social media via ESI. Secondly, while focusing on communication flows between authorities and citizens (A2C, C2A), inter- and intra-organisational crisis management (A2A) and self-help communities (C2C) were not in the direct scope of this evaluation. Although there are concepts of inter-organisational crisis management (Convertino et al., 2011; Ley et al., 2014; Reuter, Ludwig, & Pipek, 2014; C. White et al., 2009), research should examine opportunities of social media collaboration including instant messengers like Telegram or WhatsApp. Moreover, while self-help communities at times act autonomously, authorities' and citizens' mutual awareness and cooperation, e.g. via Virtual Operations Support Teams (VOST) (St. Denis et al., 2014), could be mediated via ICT such as ESI.

Design and Evaluation of an Active Relevance Classification System

10

The research field of crisis informatics examines, amongst others, the potentials and barriers of social media use during disasters and emergencies. Social media allow emergency services to receive valuable information (e.g., eyewitness reports, pictures, or videos) from social media. However, the vast amount of data generated during large-scale incidents can lead to issue of information overload. Research indicates that supervised machine learning techniques are suitable for identifying relevant messages and filter out irrelevant messages, thus mitigating information overload. However, they require a considerable amount of labelled data, clear criteria for relevance classification, a usable interface to facilitate the labelling process and a mechanism to rapidly deploy retrained classifiers. To overcome these issues, we present (1) a system for social media monitoring, analysis and relevance classification, (2) abstract and precise criteria for relevance classification in social media during disasters and emergencies, (3) the evaluation of a well-performing Random Forest algorithm for relevance classification incorporating metadata from social media into a batch learning approach (e.g., 91.28%/89.19% accuracy, 98.3%/89.6% precision and 80.4%/87.5% recall with a fast training time with feature subset selection on the European floods/BASF SE incident datasets), as well as (4) an approach and preliminary evaluation for relevance classification including active, incremental and online learning to reduce the amount of required labelled data and to correct misclassifications of the algorithm by feedback classification. Using the latter approach, we achieved a well-performing classifier based on the European floods dataset by only requiring a quarter of labelled data compared to the traditional batch learning approach. Despite a lesser effect on the BASF SE incident dataset, still a substantial improvement could be determined.

© The Author(s), under exclusive license to Springer Fachmedien Wiesbaden GmbH, part of Springer Nature 2021
M.-A. Kaufhold, *Information Refinement Technologies for Crisis Informatics*,
https://doi.org/10.1007/978-3-658-33341-6_10

215

This chapter has been published as the journal article "Rapid Relevance Classification of Social Media Posts in Disasters and Emergencies: A System and Evaluation Featuring Active, Incremental and Online Learning" in Information Processing & Management (IPM) by Marc-André Kaufhold, Markus Bayer, and Christian Reuter (Kaufhold, Bayer, et al., 2020).

10.1   Introduction

As the work of professional bodies, volunteers, and others is increasingly mediated by computer technology, and more specifically by social media, research on crisis management has become more common (Hiltz, Diaz, et al., 2011; Palen & Hughes, 2018; Reuter, Hughes, et al., 2018). The emerging research field of *crisis informatics* has revealed interesting and important real-world uses for social media (Soden & Palen, 2018). Coined by Hagar (2007), crisis informatics is 'a multidisciplinary field combining computing and social science knowledge of disasters; its central tenet is that people use personal information and communication technology to respond to disasters in creative ways to cope with uncertainty' (Palen & Anderson, 2016).

During disasters and emergencies, it is necessary for emergency services to obtain a comprehensive situational overview for coordination efforts and decision making (M. Imran et al., 2015; Vieweg et al., 2010). In such situations, social media are increasingly used for the exchange of information (Hughes & Palen, 2009) while emergency services encounter the issue of information overload, amongst others (Hughes & Palen, 2014; Mendoza et al., 2010; Plotnick & Hiltz, 2016). Research indicates that supervised machine learning techniques are suitable for identifying relevant messages and filter out irrelevant messages, thus mitigating information overload (Habdank et al., 2017). Besides the potential of improving the performance of algorithms for relevance classification, supervised machine learning techniques require a considerable amount of labelled data, which constitutes an issue due to the time-sensitive nature of disasters and emergencies (M. Imran et al., 2018). Furthermore, clear criteria for relevance classification are required, a usable interface to facilitate the labelling process (Stieglitz, Mirbabaie, Fromm, et al., 2018) and a mechanism to rapidly deploy retrained classifiers.

Based on a communication matrix (Reuter, Hughes, et al., 2018) and social media analytics framework (Stieglitz, Mirbabaie, Ross, et al., 2018), we designed and evaluated a system featuring active and online learning to support the information flow of *integration of citizen-generated content* featuring social media. Thereby, we seek to answer the following research questions:

- What are suitable criteria for relevance classification and labelling in disasters and emergencies (RQ1)?
- How can existing supervised machine learning techniques for relevance classification be improved for use in real disaster and emergency environments (RQ2)?
- How can the amount of labelled data required for relevance classification be reduced by active incremental learning and transparent visualization of the classifier's quality (RQ3)?
- How can the dynamic retraining of relevance classifiers be supported by user feedback performance-wise using batch learning with feature subset selection (RQ4)?

To reflect the methodology used in our project, the paper is structured as follows: First, based on the analysis of related work on relevance classification in social media during disasters and emergencies (section 10.2), we present the architecture of a system for social media analysis and relevance classification (section 10.3). Thereafter, we present an approach and evaluation of relevance classification via batch learning (section 10.4) as well as an approach and preliminary evaluation of relevance classification via active and online learning (section 10.5). Finally, we discuss the results, outlining practical implications and theoretical contributions, to draw conclusions on how relevance classification in social media might be further improved during disasters and emergencies (section 10.6).

This paper contributes to the research areas of machine learning and social media analytics with (1) a system for social media monitoring, analysis, and relevance classification, (2) the definition of criteria (content, keywords, language, length, links, location of author and post, media, named entities, retweets, time) for relevance classification in social media during disasters and emergencies, (3) the evaluation of a well-performing classifier for relevance classification using batch learning, (4) an approach for relevance classification including active and online learning, (5) an approach for reclassifying wrongly classified messages using batch learning with feature subset selection, as well as (6) the preliminary evaluation and outlook for relevance classification using active, incremental and online learning.

10.2   Literature Review

In order to get insights into **RQ1 ("What are suitable criteria for relevance classification and labelling in disasters and emergencies?")**, we conducted a literature review whose method (section 10.2.1), results (section 10.2.2–2.4) and research gap (section 10.2.5) are described in the following.

10.2.1 Method

Considering the search scope framework of vom Brocke et al. (2015), this literature review followed a sequential search *process (I)*. The literature review motivates the use of social media in disasters and relevance classification to mitigate information overload (section 10.2.2), it discusses abstract and precise criteria for relevance classification (section 10.2.3), outlines artificial intelligence and social media analytics approaches for relevance classification (section 10.2.4) and formulates a research gap (section 10.2.5). In terms of *sources (II)*, the bibliographic databases of IEEE Xplore and Google Scholar were searched to achieve a representative *coverage (III)* of the topics by using the *technique (IV)* of keyword search.

However, since we want to establish a more detailed understanding on definitions and criteria of relevance in social media during disasters and emergencies, in section 10.2.2, we strived for a comprehensive coverage. This part of the literature review is structured into two phases to firstly identify abstract and interpretative relevance criteria, whereby we understand relevance as a component of information quality, and secondly precise and factual relevance criteria. Since keyword searches such as *"(relevance OR relevant OR importance OR important) AND (social media OR social network OR twitter) AND (crisis OR disaster OR emergency)"* did not yield a relevant number of useful results, we also conducted backward, based on works such as Habdank et al. (2017), and forward searches. The review outlined that most publications do not discuss their definition or understanding of relevance comprehensively, only discussing some criteria and indicators selectively.

10.2.2   Social Media in Disasters and Relevance Classification as Means to Mitigate Information Overload

Motivated by citizen behaviour during both natural and human-induced (large-scale) incidents, such as 2012 Hurricane Sandy (Hughes et al., 2014), the 2013 European Floods (Albris, 2017), or the 2016 Brussels bombings (Stieglitz et al., 2017), crisis informatics has examined potentials and challenges of social media usage in disasters and emergencies by both authorities and citizens (M. Imran et al., 2015; Reuter & Kaufhold, 2018; Soden & Palen, 2018). Amongst others, social media might enable crowdsourcing of specific tasks (Dittus et al., 2017; Ludwig et al., 2017), communication between authorities and citizens (Reuter, Ludwig, Kaufhold, et al., 2016; Reuter & Spielhofer, 2017), coordination among citizens and mobilization of unbound digital or real volunteers (Kaufhold & Reuter, 2016; Starbird & Palen, 2011; J. I. White et al., 2014), (sub-)event detection (Pohl et al., 2015; Sakaki et al., 2010) or improved situational awareness (M. Imran et al., 2015; Vieweg et al., 2010).

However, considering the emergency services' use of social media, according to a study with the US public sector, the major barriers to social media use are organizational rather than technical (Hiltz et al., 2014). Research suggests that human factors are crucial for effective emergency management, but also technology for conducting respective emergency tasks (Kim et al., 2012). However, once organizational willingness is established, technical systems are needed to make sense of the large amount of data. For instance, research has identified barriers and challenges in the authorities' use of social media, such as credibility, liability (Hughes & Palen, 2014), reliability and overload of information (Mendoza et al., 2010) as well as lack of guidance, policy documents, resources, skills and staff within the organization (Plotnick & Hiltz, 2016).

In this paper, we present an approach for relevance classification, which potentially mitigates the issue of information overload by filtering out irrelevant messages. For the conceptual framing, we refer to the crisis communication matrix by Reuter et al. (2018) and the social media analytics framework by Stieglitz et al. (2018). The crisis communication matrix distinguishes authorities (A) and citizens (C), both as sender and receiver, respectively, to derive four communication flows of crisis communication (A2C), self-help communities (C2C), (inter-)organizational crisis management (A2A) and integration of citizen-generated content (C2A) (Reuter, Hughes, et al., 2018).

To leverage social media information such as eyewitness reports, pictures, and videos taken with mobile phones as a basis for authorities' decision-making, they are not only required to *integrate citizen-generated content (C2A)*, i.e., monitoring social media, while managing the vast amount of data (Olshannikova et al., 2017). When tens of thousands of social media messages are generated during large-scale emergencies, authorities have to deal with the issue of information overload which is traditionally defined as '[too much] information presented at a rate too fast for a person to process' (Hiltz & Plotnick, 2013, p. 823). Referring to the information overload problem from the field of visual analytics, Keim et al. (2008) highlight the danger of getting lost in data which may be firstly irrelevant to the current task at hand and secondly processed and presented in an inappropriate way. Considering the human capacity of information processing, Miller (1956) suggests 'organizing or grouping the input into familiar units or chunks' (p. 93) to overcome such limitations. Accordingly, functionalities such as filtering and grouping potentially assist in overcoming the issue of information overload (Moi et al., 2015; Plotnick et al., 2015; S. Tucker et al., 2012). This is supported by a survey of 477 U.S. county-level emergency managers, which revealed that perceived information overload negatively influences the adaptation of social media, while the 'chunking' or grouping of social media messages by specific tools positively influences the intention to use social media during emergencies (Rao et al., 2017). Since this paper focuses on relevance classification for noise reduction via supervised machine learning, we will introduce abstract and precise relevance criteria (section 10.2.3) and existing social media analytics approaches for relevance classification (section 10.2.4) before outlining a research gap (section 10.2.5).

10.2.3 Abstract and Precise Criteria for Relevance Classification

Firstly, Saracevic defines relevance as an intuitive understanding: "Intuitively, we understand quite well what relevance means. It is a primitive "y' know" concept, as is information for which we hardly need a definition" (Saracevic, 1975, p. 324). With regard to information science, he postulates that "relevance is considered as a measure of the effectiveness of the contact between a source and a destination in a communication process", including the aspects of subject knowledge and literature, other linguistic or symbolic representations, source and destination (of files), information systems, environment, realities, functions and values (Saracevic, 1975). Accordingly, Schamber and Eisenberg introduce a "user-centric" approach to relevance, emphasizing subjective realities in assessing relevance (Schamber & Eisenberg, 1988), which is extended in a later publication: "As based on the Sense-Making approach, the locus of relevance is within individuals' perceptions of information and information environment—not in information as represented in a document or some other concrete form. [...] Relevance, then, is a *dynamic* concept that depends on users' individual judgments of the quality of the relationship between information and information need at a certain point in time" (Schamber et al., 1990, p. 770).

In their publication about quality of information, Rohweder et al. (2011) define that a piece of information is relevant if it contains information *necessary* for the user. Necessity in this context means that the respective information facilitates reaching a certain achievement. Other than relevance, the authors name the terms *immediacy* and *appropriate scope* as quality of information. During the literature review, these terms have shown to be frequently seized upon. Furthermore, Shankaranarayanan et al. (2012) point out that for defining *relevance* the respective context of use has to be considered. Saracevic adds that "In information science, we consider relevance as a relation between information or information objects [...] on the one hand and contexts, which include cognitive and affective states and situations (information need, intent, topic, problem, task; [...]) on the other hand, based on some property reflecting a desired manifestation of relevance (topicality, utility, cognitive match; [...])" (Saracevic, 2007, p. 1918). Accordingly, Agarwal and Yiliyasi (2010) define *contextual relevance* regarding quality of information in social media. Data from social media might be relevant for certain actors while being irrelevant for others. They describe *relevance* as a degree which determines how useful data are for a certain task. Specifically referring to crisis situations, Jensen (2012) defines *relevance* as *useful* and moreover *adequate* and *valuable* information. This is also in line with Eisenberg, how adds that "two common choices for definitions seem to be *topicality* and *usefulness*" (Eisenberg, 1988, p. 387).

Borlund differentiates between multidimensional, dynamic and situational relevance (Borlund, 2003, p. 913). Regarding crisis situations, Sriram et al. (2010) describe a classifier of Twitter posts relating to news, events, opinions, offers and private messages. They define events as *"something that happens at a given place and time"* (Sriram et al., 2010). Verma et al. (2011) develop a classifier which recognizes relevant Twitter posts regarding situational awareness in crisis situations. They describe a Tweet as relevant if it contains tactical and usable information. Other than in the thesis at hand, the authors limit themselves to textual features. Their result is that posts, which show situational awareness, tend to exhibit objective, impersonal and formal features. Objective posts are based on factual information, do not express an opinion and do not include emotional language. Impersonality is characterized by an emotional distance between user and event. Formal tweets are such which are grammatically coherent and express complete thoughts (Verma et al., 2011). Vieweg (2012a) describes in her dissertation the usage behaviour of Twitter users in crisis situations. She uses the aforementioned classifier developed by Verma et al. (2011) to minimize the dataset to a manageable amount of posts. Vieweg describes how Tweets that contribute to situational awareness in mass events can be recognized. The posts are divided into three categories:

1. "O: Off-topic"
2. "R: On-topic, relevant to situational awareness"
3. "N: On-topic, irrelevant to situational awareness"

O means that none of the information contribute to the event as such. The *N*-division relates to posts that refer to the event but do not add to the relevance. As an example, she mentions persons who call for donations or express their sympathy. All posts which "*contain information that provides tactical, actionable information that can aid people in making decisions, advise others on how to obtain specific information from various sources, or offer immediate postimpact help to those affected by the mass emergency*" (Vieweg, 2012a, p. 164) are allocated to the *R*-Classification. In the dissertation, the author does not offer any specific criteria to explain how the relevance filters were labelled. Instead she refers to other classification of the data. In the first part of their publication, Imran et al. (2013a) also filter the relevance regarding situational awareness in crisis situations. First, they refer to the definition of relevance Vieweg (2012a) employs for the *R*-Classification but also offer their own definition of relevance. They differentiate Twitter posts as follows:

1. "Personal Only"
2. "Informative (Direct or Indirect)"
3. "Other"

A post is personal ("Personal Only") if it only relates to the author himself and close friends or family and does not disclose any useful information to persons unbeknownst to the author. Informative posts ("Informative") include information which might be useful for persons who do not know the author. Informative posts are divided into direct, if the post was written by an eyewitness, and indirect, if the post was written by a person based on information from news, radio or television. All Tweets not written in English are categorized as "Other" (M. Imran et al., 2013b, 2013a). Table 10.1 summarizes the afore-presented relevance criteria.

Table 10.1   Abstract and interpretative relevance criteria

Criteria	Description	Sources
subjective	depending on the individuals' perception of information	(Schamber et al., 1990; Schamber & Eisenberg, 1988)
necessary	information that facilitate reaching a certain achievement	(Rohweder et al., 2011)
useful, adequate, valuable	–	(Jensen, 2012)
context-related	depending from the actual use	(Agarwal & Yiliyasi, 2010; Shankaranarayanan et al., 2012)
tactical	–	(Verma et al., 2011; Vieweg, 2012b)
Usable	–	(Verma et al., 2011)
objective	objective, do not express an opinion and do not include emotional language	(Verma et al., 2011)
impersonal	emotional distance between user and event	(M. Imran et al., 2013b; Verma et al., 2011)
formal	complete and grammatically coherent	(Verma et al., 2011)
contributing to decision making	–	(M. Imran et al., 2013a; Vieweg, 2012b)
assisting and advisory	–	(M. Imran et al., 2013a; Vieweg, 2012b)
informative	not only relating to close friends or family of the author	(M. Imran et al., 2013a, 2013b)

Besides abstract and interpretative criteria, precise criteria to classify relevance in crisis situations were also identified. Abel et al. (2012) present "Twitcident", a filter system for crisis situation, and describe that *keyword-based filters* are one of the two steps towards the recognition of relevant posts in their system. Within the scope of their research project "Alert4All", Johansson et al. (2012) work on the importance of terminology regarding events. Furthermore, Vieweg (2012a) provides a list of terms that are not allowed in the tweets regarding a specific dataset. She removes all posts which include terms regarding donations, sympathies or anger.

Vieweg et al. (2010) point out in their crisis-specific Twitter analysis that the *geographic position* of the Twitter user and references to a certain location can be an indication for relevant posts, especially in terms of situational awareness. Further publications emphasize the relevance of these criteria for authorities (Ludwig, Reuter, & Pipek, 2015; Reuter, Ludwig, Ritzkatis, et al., 2015; Sriram et al., 2010; Verma et al., 2011; Vieweg, 2012a). De Albuquerque et al. (2015) analyse the importance of the geographic position on the basis of the flooding of the river Elbe in 2013. They point out that tweets which are close to the event (up to 10 kilometres) have a higher likelihood to be involved in the flooding.

Rohweder et al. (2011) identify the chronological correlation as a criterion for the quality of information. Moreover, Ludwig et al. (2015) emphasize the importance of *currency* and temporal proximity as criterion for relevance. Sriram et al. (2010) and Palen et al. (2010) also mention timely information regarding the event as an indication for relevance. Similarly, for example Vieweg (2012a), almost any analysis of user-generated data in crisis situations limits the dataset on the basis of time without explicitly mentioning this as a filter for relevance.

Starbird and Palen (2010) show that posts which were *retweeted* have a stronger connection to the crisis event. Furthermore, it became evident that mainly persons who were close to the event used the retweet function. The authors therefore conclude that a focus on retweets might help to minimize uncertainties in the dataset of crisis situations. Reuter et al. (2013) also observe the behaviour of Twitter users in crisis situations. They find that 22.32% of all retweets during an event were sent by only 51 users and therefore argue that the information is highly relevant. The papers of Uysal and Croft (2011) as well as Mendoza and Poblete (2010) describe informally that the retweet behaviour of users points to relevance and are therefore in accordance with the aforementioned publications.

Reuter et al. (2013) observe, besides the relevance of retweets, that *links* to other webpages are a particularity. Accordingly, 39% of the collected posts during the "Super Outbreak" of 2011 contain links. Uysal and Croft (2011) suggest another approach to filtering relevant posts on Twitter but their publication is abstracted from crisis situations. Through machine learning they show that the presence of an URL improves the exactness of the classification. In addition, Habdank et al. (2017) write that links in the form of pictures or videos are an important indication for relevance. Pictures, URLs and videos offer the opportunity to extend postings trough external content (Ludwig, Reuter, & Pipek, 2015).

The evaluation of the machine relevance classification done by Imran et al. (2013a) underlines that the length of a tweet contributes to its relevance. According to Rohweder et al. (2011), the criterion of a reasonable length has to be considered. Furthermore, Sriram et al. (2010) and Abel et al. (2012) eliminate in their preselection of relevant posts those which are too short. The threshold is three words in the case of Sriram et al. (2010) whereas Abel et al. (2012) eliminate all posts with less than 100 characters. In addition to that, Sriram et al. (2010) and Imran et al. (2013a) only consider English tweets as relevant. Abel et al. (2012) specify *language* as criterion for relevance; accordingly, the tool "Twitcident" is able to translate the posts. In Table 10.2 these characteristics are compiled with a general description as well as their sources.

Table 10.2 Precise and factual relevance criteria

Criteria	Description	Sources
keywords	inclusion and removal of posts with specific terms	(Abel, Hauff, & Stronkman, 2012; Johansson et al., 2012; Vieweg, 2012b)
geographic location and referenced positions	actual location of a Twitter user or included positions in the post's text	(de Albuquerque et al., 2015; Ludwig, Reuter, & Pipek, 2015; Reuter, Ludwig, Ritzkatis, et al., 2015; Sriram et al., 2010; Starbird & Palen, 2010; Verma et al., 2011; Vieweg, 2012b; Vieweg et al., 2010)
currency	temporal correlation between Twitter post and event	(Ludwig, Reuter, & Pipek, 2015; Palen et al., 2010; Rohweder et al., 2011; Sriram et al., 2010)
retweet behaviour	exact repetition of a Twitter post by other authors	(Mendoza et al., 2010; Reuter et al., 2013; Starbird & Palen, 2010; Uysal & Croft, 2011)
linking	links in the form of URL, pictures or videos	(Habdank et al., 2017; Reuter et al., 2013; Uysal & Croft, 2011)
length of tweets	–	(Abel, Hauff, Houben, et al., 2012; M. Imran et al., 2013a; Rohweder et al., 2011; Sriram et al., 2010)
language	–	(Abel, Hauff, Houben, et al., 2012; M. Imran et al., 2013a; Sriram et al., 2010)

10.2.4 Machine Learning and Social Media Analytics for Relevance Classification

As public interfaces (APIs) enable the retrieval and processing of high volume datasets, 'systems, tools and algorithms performing social media analysis have been developed and implemented to automatize monitoring, classification or aggregation tasks' (Pohl, 2013). Here, *social media analytics* is defined as the process of social media data collection, analysis and interpretation in terms of actors, entities and relations (Stieglitz et al., 2014). Accordingly, Stieglitz et al. (2018) differentiate between the steps of discovery, tracking, preparation and analysis of social media data. Social media data, sometimes referred to as *big social data*, includes the characteristics of *high-volume* (large-scale), *high-velocity* (high speed of data generation), *high-variety* (heterogeneous data with a high degree of complexity due to the underlying social relations) and *highly semantic* (manually created and highly symbolic content with various, often subjective meanings) data (Olshannikova et al., 2017). These characteristics pose challenges for emergency services who need their own analytical concepts.

In this paper, we focus on machine learning as a subset of artificial intelligence, which can be applied as a technique (e.g., automatically classifying messages as relevant or irrelevant) to solve problems (e.g., mitigating information overload of emergency managers by reducing the volume of incoming irrelevant data) in the domain of social media analytics. Many researchers already examined the classification of posts in disaster scenarios; Table 10.3 illustrates a comprehensive selection of research papers. In most cases, the motivation of supervised relevance classification algorithms lies in reducing the large volume of noisy data, e.g., to facilitate analysis by emergency services (Habdank et al., 2017; M. Imran et al., 2013a). Relevance classification is a binary problem, where a post is either marked as 'relevant' or 'not relevant'. For instance, in a study of Habdank et al. (2017), the Random Forest classifier outperformed a Support Vector and Neuronal Network classifier, achieving an accuracy of 88%, precision of 86%, and recall of 89%. However, in the work of Li, Caragea, Caragea, and Herndon (2017) and Caragea, Silvescu, and Tapia (2016) it was shown that Convolution Neural Networks seem to be well suited for classification of disaster posts with the potential to outperform Random Forests.

However, the issue of existing approaches is that well-performing classifiers require a considerable amount of labelled data, which is often not available at the beginning of a disaster or emergency (M. Imran et al., 2017; H. Li et al., 2015). Thus, in using *domain adaptation*, multiple approaches incorporate labelled data from past disasters in a new situation so that few or no new labels have to

Table 10.3 Overview of works examining textual classification problems in disasters or emergencies. If a comparison was conducted, the best performing method is marked bold; abbreviations: Artificial/Convolutional Neuronal Networks (ANN/CNN), Decision Trees (DT), Jaccard Similarity (JS), K-Nearest Neighbours (KNN), Logistic Regression (LR), Maximum Entropy (ME), Naïve Bayes (NB), Neuronal Networks (NN), Relevance-Based Language Models (RBLM), Random Forest (RF), Support Vector Machines (SVM)

Authors	Classification Problems	Used Methods
(Habdank et al., 2017)	relevance (binary)	NB, DT, RF, SVM, NN
(M. Imran et al., 2013a)	informative/relevant posts (multi-class), direct (eyewitnesses) and indirect posts (binary), different information types (multi-class)	NB
(Verma et al., 2011)	situational awareness/relevance (binary)	NB, ME
(Purohit, Castillo, et al., 2014)	help request or offer (multi-class), types of help (multi-class), resources for help (multi-class)	NB, RF
(Caragea et al., 2011)	different information types (multi-label)	SVM
(Markham & Muddiman, 2016)	relevance	NB
(M. Imran, Mitra, & Castillo, 2016)	different information types (multi-class)	SVM, RF, NB
(Ashktorab et al., 2014)	damage reports (binary)	KNN, DT, NB, LR
(Abel, Hauff, & Stronkman, 2012; Abel, Hauff, Houben, et al., 2012)	relevance (binary)	RBLM, JS
(H. Li et al., 2015)	relevance (binary), help offer (binary), emotions for victims (binary)	NB
(M. Imran et al., 2017; M. Imran, Mitra, & Srivastava, 2016)	different information types (multi-class)	RF
(H. Li et al., 2017)	relevance (binary)	NB, RF, SVM, LR
(Caragea et al., 2016)	informative posts (binary)	CNN, ANN, SVM
(D. T. Nguyen et al., 2016)	informative/relevant posts (binary), different information types (multi-class)	LR, RF, SVM, CNN

be created (M. Imran, Mitra, & Srivastava, 2016; M. Imran et al., 2017; H. Li et al., 2017, 2015). These strategies worked well, although different factors such as language may reduce the usefulness of labelled data from past events (M. Imran, Mitra, & Srivastava, 2016).

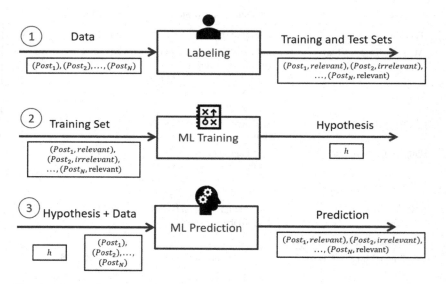

Figure 10.1   Supervised machine learning steps

Most approaches apply *offline learning*, where the phases of labelling data, training a hypothesis and prediction (Figure 10.1) are performed disjointedly (Khouzam, 2009). A version of offline learning is called batch learning, where a batch of data is labelled and then provided for the learning algorithm (Sebastiani, 2002). Due to direct availability of the whole dataset, the algorithm is able to analyse the data in greater detail (Kulessa, 2015). However, research outlines the potentials of *online learning*, which "helps models dynamically adapt to new change and patterns in the data" (M. Imran et al., 2018, p. 510). In case of online learning, data is not provided in a single batch but as a sequential stream of data, allowing for updating the model continuously as new (labelled) data becomes available. This has the potential to help with overcoming the sparsity of labelled data at the beginning of an disaster and increasing the classifier's accuracy over time (M. Imran et al., 2017; H. Li et al., 2015). Furthermore, *incremental learning* describes the capability of an algorithm to be enhanced sequentially (Khouzam, 2009). The benefits of incremental algorithms are that they can be trained rapidly and the training data does not have to be loaded into the central memory completely (Ma et al., 2009). However, since they require the capability of processing an indefinite stream of data rapidly, it is possible that the classification quality decreases (Kulessa, 2015).

Finally, *active learning* may help to reduce the amount of labelled data required for a well-performing classifier. The basic idea is that "a machine learning algorithm can achieve greater accuracy with fewer training labels if it is allowed to choose the data from which it learns" (Settles, 2010). Using active learning instead of labelling random data, the classifier proposes data whose labelling is most likely to increase the classifier's accuracy (Fürnkranz, 2018). Thus, active learning strives for increasing the classifier's accuracy with least effort in terms of the amount of labelled data (Y. Yang & Loog, 2017).

10.2.5 Research Gap

During disasters and emergencies, social media allow for the creation of large but potentially noisy volumes of citizen-generated content, often referred to as big social data (Moi et al., 2015; Olshannikova et al., 2017). Emergency services may extract actionable or situational information from eyewitness reports, photos or videos (Zade et al., 2018), yet due to limited personal or organizational resources (Reuter, Ludwig, Kaufhold, et al., 2016) the issue of information overload constitutes a severe problem (Plotnick & Hiltz, 2016). Here, technology might assist in overcoming information overload, both in terms of algorithmic quality as well as usable and tailorable user interfaces (Plotnick & Hiltz, 2018; Stieglitz, Mirbabaie, Fromm, et al., 2018). From an algorithmic point of view, research examined the potentials of alert generation (Adam et al., 2012; Avvenuti et al., 2014; Cameron et al., 2012; Párraga Niebla et al., 2011; Purohit et al., 2018; Reuter, Amelunxen, et al., 2016), event summarization (M.-T. Nguyen et al., 2015; Rudra et al., 2015; Rudra, Goyal, et al., 2018), precise search keyword selection (Abel, Hauff, & Stronkman, 2012; Johansson et al., 2012; Vieweg, 2012b), and relevance classification (Abel, Hauff, & Stronkman, 2012; Abel, Hauff, Houben, et al., 2012; Habdank et al., 2017; H. Li et al., 2017, 2015; Markham & Muddiman, 2016; D. T. Nguyen et al., 2016; Verma et al., 2011) for mitigating information overload.

Although research developed a variety of relevance classification algorithms, these face practical issues in real-world disaster and emergency scenarios: at the beginning of such events, there is a lack of labelled data which is required in considerable quantity for supervised machine learning algorithms to perform well (M. Imran et al., 2017; H. Li et al., 2015). Research suggests that active and online learning potentially mitigate that issue: Active learning is capable of reducing the required amount of labelled data (Settles, 2010), thus reducing the effort for emergency services, while online learning facilitates the continuous improvement of the classifier's accuracy, including the rapid adaptation to changes and new patterns in the data (M. Imran et al., 2018). To harness these potentials, we implemented a real-time evaluation mechanism using online and active learning (section 10.5.1.1) and a feedback classification mechanism which uses a batch learner with feature subset selection for fast classifier retraining (section 10.5.1.2).

From a user interface point of view, the literature revealed criteria whose consideration and display could assist the labelling and classification of relevant messages. These include currency (Ludwig, Reuter, & Pipek, 2015; Palen et al., 2010; Rohweder et al., 2011; Sriram et al., 2010), geolocation (de Albuquerque et al., 2015; Ludwig, Reuter, & Pipek, 2015; Reuter, Ludwig, Ritzkatis, et al., 2015; Sriram et al., 2010; Starbird & Palen, 2010; Verma et al., 2011; Vieweg, 2012b; Vieweg et al., 2010), keywords (Abel, Hauff, & Stronkman, 2012; Johansson et al., 2012; Vieweg, 2012b) and language (Abel, Hauff, Houben, et al., 2012; M. Imran et al., 2013a; Sriram et al., 2010), amongst others. Also, usable interfaces are required to exploit the potentials of active and online learning. Thus, we integrated our algorithms into a user interface (section 10.3.4).

10.3 Technological Basis: Social Data Management and Analysis

To facilitate relevance classification, we integrated active, incremental and online learning into our social media platform whose design method (section 10.3.1) and architecture (section 10.3.2), comprising the Social Media API (SMA) as a backend (section 10.3.3) and the Social Media Observatory (SMO) as frontend (section 10.3.4), are described in the following as relevant prerequisites for the understanding of our evaluations in section 10.4 and 10.5.

10.3.1 Method

The architecture underwent several iterations of development before integrating machine learning components. The development of the SMA is not documented in a research publication since it was primarily driven by the requirements of the

SMO. However, we provide a detailed description of its design, also to convey the necessary steps of pre-processing required for the implementation of our classifiers. The SMO underwent three iterations of development based on *user-centred evaluations* which are documented in a research paper (Reuter, Ludwig, Kotthaus, et al., 2016). Based on the outlined design requirements, a fourth iteration was conducted to (1) implement the design requirements, (2) upgrade it with state-of-the-art libraries and technologies as well as to (3) improve its hedonic and pragmatic quality. Besides its basic functionality, we describe the integration of active real-time data labelling and feedback classification, which are evaluated in section 10.5, into the graphical user interface.

10.3.2 Overall Architecture

The overall architecture (Figure 10.2) comprises a frontend called *Social Media Observatory* (SMO) and a backend called *Social Media API* (SMA). The SMO is a web application based on *Vue.js* as the overall framework, *Bootstrap* for responsive design and *Chart.js* for data visualization. All actions of the SMO, such as searching for posts in social media, disseminating posts to social media or managing users, are forwarded to the SMA. The SMA is realized as a service following the paradigm of a web-based and service-oriented architecture (SOA). It is a Java Tomcat application using the Jersey Framework for RESTful Web services and the MongoDB database via Hibernate Object/Grid Mapper (OGM) for document-oriented data management. The implementation allows for using a local or remote instance of MongoDB. Several libraries facilitate the integration of social media source APIs such as Facebook Graph API or Twitter Search API. To overcome the diversity of data access and structures, all gathered social media entities are processed and stored according to the Activity Streams 2.0 Core Syntax in JavaScript Object Notation (JSON). In the following chapters, we will give a brief overview of the Social Media Service (section 10.3.3) and Social Media Observatory (section 10.3.4).

10.3.3 Backend: The Social Media Service

The *Social Media API* (SMA) allows for the gathering, processing, storing, and re-querying of social media data. Based on underlying social media, the SMA contains different services that are used by several client applications. Although it was developed as enabling technology for crisis management applications, its

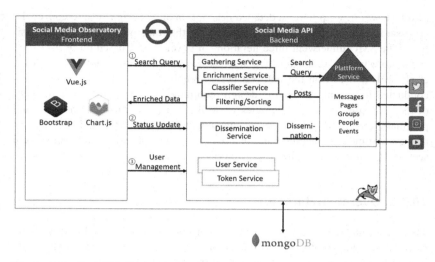

Figure 10.2   Social Media Architecture comprising Social Media Observatory and Social Media Service

implementation allows for supporting a variety of use cases in different fields of application, e.g., to examine the impact of a product image within the field of market research. To enable access to social big data and allow for subsequent analysis, our first step was to specify a service for gathering and processing social media content. With *gathering,* we refer to the ability to uniquely or continuously collect social media activities (e.g., messages, photos, videos) from different sources (Facebook, Google+, Twitter, and YouTube) in a unified manner using multiple searches or filter criteria. *Processing* means that the SMA is able to access, disseminate, enrich, manipulate and store social media activities.

10.3.3.1 Endpoint Overview and Pre-Processing

SMA comprises five main services, each providing a multitude of service functions: The *Gathering Service* contains endpoints for gathering and loading social media activities. The main components are the Search Service, enabling one-time search requests, and Crawl Service, which continuously queries new social media activities across a specified timeframe. Using the *Enrichment Service,* gathered social media activities are enriched with further computed and valuable metadata. For the initialization of a crawl job, a POST call is sent to the *crawlService* endpoint, which contains a payload matching the *application/json* Content-Type header. In its basic configuration, a keyword, at least one source and an interval value to determine the timeframe between each gathering request are required; further configuration parameters may be specified to filter the scope of the crawl job (Table 10.4). To query gathered results of a certain crawl job, a GET call is sent to the *crawlService/{crawljobId}* endpoint with *{crawljobId}* being a concrete instance of an identifier. The identifier may be retrieved from the response of the crawl job's initialization or from the list of the *crawlService/allJobs* endpoint.

Table 10.4 Required and optional query parameters

Parameters	Type	Description
keyword	String	Required. The search keywords.
sources	String	Required. A list (Facebook, Google+, Instagram, Twitter, YouTube).
since/ until	Long	Search Service. Lower/upper bound of the searched timeframe (Unix time).
start/ end	String	Crawl Service. Starting/termination point of the crawl job (Unix time).
latitude/ longitude	Double	Latitude/longitude for geo search (decimal degree).

The *Classifier Service* realizes our supervised machine learning components. Before the classification can take place (see section 10.4 and 10.5), several steps of pre-processing are required. Firstly, the removal of specific characters is realized with simple Java operations. The CISTEM algorithm is used to stem German words (Weißweiler & Fraser, 2017) and the lemmatization is realized via the LanguageTool library (LanguageTool, 2019). During the training phase and for the translation of words into machine-readable characters from all training tweets, a modified Bag-of-Words (BoW) is created. In Java, this is realized as a hash map, where every word as a key is mapped to a number. With regard to efficiency and Inverse Document Frequency (IDF), a word in the training BoW is mapped to the number of documents that comprise the respective word. For each word in the BoW, a feature for each post is acquired. Thus, for each tweet the algorithm iterates through the BoW and generates a vectorized number for each of these words. There, classifier supports two approaches to compute these numbers. Firstly, the *normalized term frequency* (TF), where for each word its frequency in a tweet is divided through the maximum frequency of any word in the tweet. Secondly, this normalized term frequency can be extended to TF.IDF by multiplying the TF with IDF. To determine whether a geolocation is available in a post, the German version of the CoreNLP toolkit (Manning et al., 2014) is used.

Furthermore, the algorithm evaluated in section 10.4.2 requires the computation of geographical and temporal distances between any post and the event's, e.g., a disaster or emergency, location and time. The geographical distance is based on the latitude and longitude of the authors and posts location: *distance in kilometres* $= sqrt(dx^2 + dy^2)$, *whereby* $dx = 71.5 * (longitude - longitude_2)$ *and* $dy = 111.3 * (latitude_1 - latitude_2)$. Since the geolocation of an author is often not represented in latitude and longitude but rather in a textual description, the geocoding REST interface of "here" is used (here, 2019). The temporal distance is computed by the *Duration.between()* function of the *Java.Time* library. Finally, the Naïve Bayes and Random Forest classifiers are implemented using the Java library Weka (Hall et al., 2009). All processed data is stored in *Attribute-Relation File Format (ARFF)* files, which can be processed by Weka learning algorithms. Such a file comprises a list of all used features as well as their internal representation (see Table 10.6).

The *Dissemination Service* allows for the publishing of messages in social media. Finally, the *Source Service* constitutes the interface to individual social media sources and allows for the integration of social media in a standardized manner.

10.3.3.2 Data Specification Using Activity Streams 2.0

The Activity Streams 2.0 Core Syntax (AS2) defines that "an activity is a semantic description of potential or completed actions" (World Wide Web Consortium, 2016), which has at least a verb (the type of activity, e.g., like, post, share), an actor (e.g., the creator) and an object (e.g., an image or message object). There are already many verbs and object types defined within a specification, for instance, a place object may contain the attributes latitude, longitude, and altitude (Table 10.5). Although the specification allows modelling the activities of liking, sharing and so on, there are no attributes designated to carry information like "20 users liked this post". While the specification may be extended with own verbs and object types such as our "enrichedData" object, foreign implementations possibly have not enough knowledge to process them in an intended way. Activity objects must be encapsulated in a collection object before returning them as a JSON object.

10.3.3.3 Data Storage Using Hibernate OGM/MongoDB

For storing and retrieving the collected data in and from a MongoDB database instance, we deploy the Java framework Hibernate OGM. We selected MongoDB as a document-oriented NoSQL solution due to its good performance in reading,

Table 10.5   Activity Streams with EnrichedData object

```
{
  "actor": {
    "content": "48, 2 Kinder, Sarkasmus, private Meinung",
    "displayName": "anonymised",
    "id": "twitter:844424271",
    "type": "person",
    "url": " https://goo.gl/QqV2q6"
  },
  "object": {
    "content": "RT @bzberlin: #Debüt mit 1:0 gegen @SERCWildWings https://t.co/UNlq698PlJ",
    "enrichedData": {
      "absFearFactor": 0,
      "absHappinessFactor": 0,
      "embeddedUrls": ["https://t.co/UNlq698PlJ"],
      "language": "de",
      "tags": ["Debüt"],
      "media": [{
        "mediaType": "image/jpeg",
        "type": "photo",
        "url": "https://goo.gl/QqV2q6"
      }],
      "mentions": ["bzberlin", "SERCWildWings"],
      "numOfCharacters": 133,
      "numOfSentences": 7,
      "numOfWords": 11,
      "numRetweets": 3,
    },
    "id": "twitter:823724465664883940",
    "location": {
      "displayName": "Neunkirchen, Deutschland",
      "latitude": 50.78506988,
      "longitude": 8.00512706,
      "type": "place"
    },
    "startTime": "2017-02-01T10:30:47.000+01:00",
    "type": "post",
    "url": " https://goo.gl/QqV2q6"
  }
}
```

writing and deleting operations on large datasets and, compared to SQL solutions, flexible document schemas and the option of *sharding*, a method for distributing data across multiple instances or machines (Y. Li & Manoharan, 2013). Furthermore, with the aid of the OGM tools (object-grid mapper), we could operate without direct database commands because they are encapsulated in the framework, e.g., as save, update or delete functions. The generation of the database scheme and the storage of the corresponding instances of the objects is done automatically. Only the annotations of the appropriate classes and their attributes are needed for Hibernate to transform the Java classes into database query commands. Moreover, we use a compound unique index based on the activity's source and ID to prevent duplicate activities on database level.

10.3.4 Frontend: The Social Media Observatory

The *Social Media Observatory* (SMO) is a user interface which utilizes the SMA for social media monitoring, analysis, and relevance classification (Figure 10.3). More specifically, it facilitates the creation of continuous search jobs (Crawl-job tab) and single-time searches (Search tab), management of users (Admin tab), creation of classifiers (Classifier tab) as well as dissemination of messages (Dissemination tab).

10.3.4.1 Data Gathering and Visualization

After entering the SMO, the dashboard view (Figure 10.3) visualizes the number of posts crawled, crawljobs created, classifiers learned, users created and comprises an overview of functionality.

After entering the Crawljob tab, the user has an overview of his own crawl-jobs, (Figure 10.4), each displaying the search keyword, user-based description, author of the job, location, sources, results and creation time, and is able to create new crawljobs (Figure 10.5) based on keywords, description, selection of social networks, the definition of geographic boundaries (optional) and a time period (optional). The user can visit the results as a list view (Figure 10.8), each entity containing the time of creation, username, content (text, photos, videos), and metadata (language, location, retweets). Furthermore, the Search tab enables the user to perform single searches based on keywords and selected social media. The Admin tab allows administrators to manage existing users or create new users with a name, password and role (e.g., admin or user).

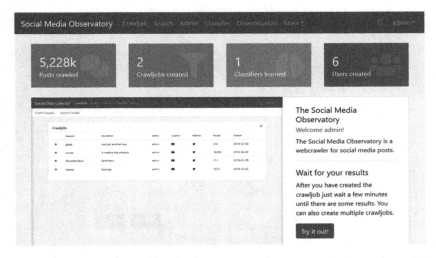

Figure 10.3 Dashboard overview for administrators

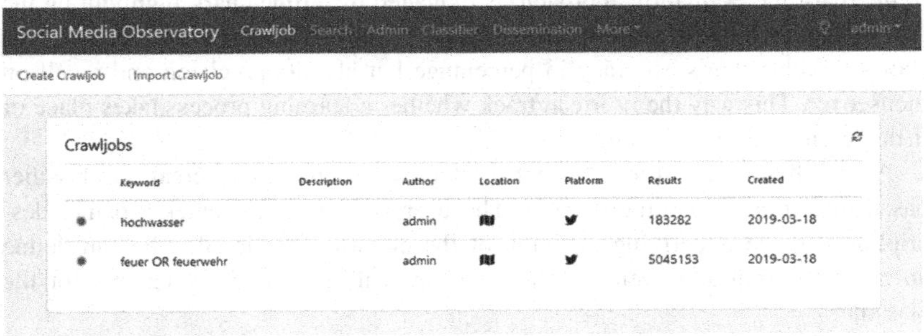

Figure 10.4 Overview of crawljobs

10.3.4.2 Data Labelling and Filtering

For each crawljob, the user may label postings according to their relevance (Figure 10.6). In the beginning a single post with its content, multimedia files and metadata is loaded into the view. After the user labelled it as 'relevant' (green button) or 'not relevant' (red button) the next post is loaded. After each fiftieth post the estimated accuracy of the classifier (see section 10.5.1.1 for the specification

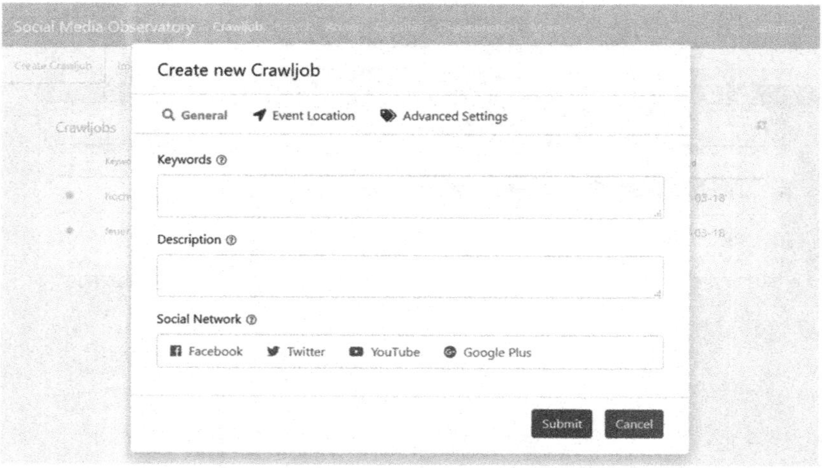

Figure 10.5   Creation of a crawljob

of the real-time evaluation approach) is indicated by a time chart, highlighting the development of the classifier's accuracy over time as well as a pie chart, which shows the classifier's accuracy in percentage but also its precision and recall on mouseover. This way the user can track whether a learning process takes place or if the learning curve converges.

After labelling a sufficient number of posts the user may create a classifier based on a crawljob (Figure 10.7). The user is required to enter a name, description and the underlying crawljob of the classifier. Optionally, he can define an event location and event date as geographical and temporal features for the classifier.

By entering the post list of a crawljob the user is able to apply the trained classifier (Figure 10.8). In the default configuration 'not relevant' posts are greyed out. However, if the user disagrees with the classification, he can mark a 'relevant' post as 'not relevant' (by a tick in the top-right corner of each posting) and vice versa. As soon as the user leaves the list view or navigates to the next page of results the classifier is retrained accordingly (see section 10.5.1.2 for the specification of the feedback classification approach).

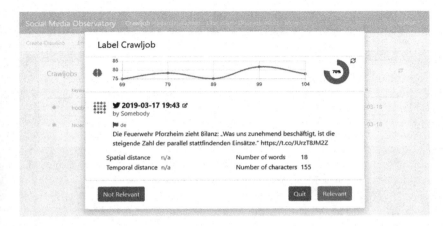

Figure 10.6   Relevance labelling for a crawljob

Figure 10.7   Create a relevance classifier based on labelled crawljob posts

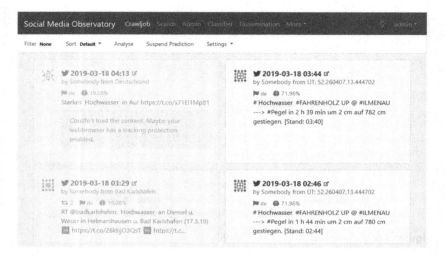

Figure 10.8   Filtered List of crawljob posts

10.4   Evaluation I: Relevance Classification via Batch Learning

To achieve insights into **RQ2 ("How can existing supervised machine learning techniques for relevance classification be improved for use in real disaster and emergency environments?"**), we conducted our first evaluation whose approach (section 10.4.1) and conduction (section 10.4.2) are described in the following.

10.4.1   Approach

Information is relevant when it is useful, valuable, appropriate and necessary (Jensen, 2012; Rohweder et al., 2011). To do justice to these characteristics, one must consider the exact context (Agarwal & Yiliyasi, 2010). In the following and based on the definitions and criteria from section 10.2.3, the contextual relevance related to crises or emergency situations serves as the foundation for labelling and the classification features of the machine learning algorithm. As a result, the term relevance as defined in this paper refers to C2A content, which contributes to information acquisition of and decision support for emergency services. Content of this kind is often up-to-date, refers to the location of the incident, contains facts and references, is retweeted several times and may contain hyperlinks. It is to be stressed that relevance, as described in this paper, must be separated from an all-encompassing information quality. To obtain a high density of high-quality information, customizable filters are of greater use after the classification of relevance. For example, Reuter et al. (2014) present a customizable rating service in which end-users can adjust weightings of several characteristics regarding their information quality. The classification of relevance can thus be a first step towards maintaining information quality.

As the labelling is done by humans, an interpretative and analytical approach is possible (in contrast to the classification feature selection). Therefore, the criteria from Table 10.1 and Table 10.2 can be used for defining relevance in the labelling process (section 10.4.1.1). The labellers are provided with an abstract definition and precise criteria regarding relevance in disasters. By contrast, an algorithmic relevance classification (section 10.4.1.2) can only contain exact and measurable criteria, which can be found in Table 10.2. Ideally, the algorithm, based on the training data of the labelling process and the elaborated classification features, should predict a post as relevant if it suffices the requirements of relevance as described in this paper.

10.4.1.1 Relevance Features for Labelling

The specification and explanation of relevance is important for two aspects of the labelling process for training data. On the one hand, this is the only way to ensure a comprehensive insight into the classifier developed here. On the other hand, several persons who have reached a consensus regarding this clarification can participate in the creation of training data.

In their analysis of tweet behaviour in disasters, Verma et al. (2011) found that objective, impersonal and formal information characteristics tend to be included in their definition of relevant content. However, such a classification would disregard a large number of important tweets. For example, a tweet can be written in a personal or in an informal style and contain an opinion but still provide important information for emergency services. A witness of an incident might not have the time to write a formal post but might be able to share important information due to the short distance and high timeliness. Cheong and Lee (2011) also point out that the civilian sentiment during terror attacks can be of great use for decision-makers and authorities. The concept of relevance as defined in this paper, in contrast, complies with the description of Vieweg (2012a). The labellers label posts as relevant when information can contribute to decision-making, contain advice or when important sources are referenced. A further approach of determining relevance is the distinction between informative, personal and other tweets by Imran et al. (2013a). According to them, a tweet is considered to be informative when it is of interest to people beyond the author's immediate circle. Hence, their labelling process sorts out all tweets which are of interest to the author and his/her immediate circle of family/friends only and do not provide any useful information to people who do not know the author.

Imran et al. (2013a) further describe that all tweets not written in English are sorted out. However, language is not perceived to be a restriction for the concept of relevance in this paper. For example, Wilson et al. (2017) analyse online rumouring behaviours in crises based on languages. They point out that the main language of the affected population, independent of English, is important in the context of the event. Finally, they state that language is an aspect that requires further consideration when conducting studies of social media. Imran et al. (2013a) further describe that an exact distinction between informative and personal content was not always clear to the labellers. In this paper, a tweet is always considered to be relevant in case of doubt. For the emergency services to use the data in a beneficial way, an all-encompassing and strong reduction of the number of posts is helpful but can lead to a loss of relevant content. As in the scenario of Markham and Muddiman (2016), a certain amount of posts incorrectly classified as relevant is acceptable. Furthermore, tweets are tested to see whether or not they actually contribute to situational awareness in crises. As outlined by Vieweg (2012a), all tweets containing condolences or calls for donation only are classified as irrelevant. Content is further classified as relevant if the information it contains is of use or important for the Virtual Operation Support Teams (VOST). The role of virtual communities, as described by Reuter et al. (2013) as well as Kaufhold and Reuter (2016), should also be taken into account. Helpers and reporters directly or indirectly provide important and up-to date information on the incident. Retweeters and repeaters distribute the most important information (which was generated by other users), thus filtering important information that should not be overlooked. Such content is often associated with facts. Labellers ensure that all types of facts related to a crisis are labelled as relevant. False facts can also provide important information for authorities. They can identify the false information and correct it using A2C communication (Reuter, Hughes, et al., 2018). Furthermore, labellers label posts in which help is requested or offered as relevant.

For each tweet, the assessors are provided with an overview of various characteristics of it, which do not have to be extracted from the Twitter message itself (section 10.3.4.2, Figure 10.6). These characteristics may be helpful in identifying relevance but should not be seen as exclusive criteria. From the definition of an event by Sriram et al. (2010), the criteria geographical location, referenced location and timeliness of a post emerge. All three criteria are important for filtering the relevance for VOSTs. While the referenced location in a tweet indicates a connection of the author to the incident, the local proximity can even point to witnesses, helpers and people in need of help at the incident's location. Currency is important because, for example, messages clearly sent before the incident can be irrelevant. Therefore, the assessors are given the location of a tweet and the calculated local proximity to the incident.

However, as only 10% of all Twitter users worldwide allow the locating of a tweet, as described by Ludwig et al. (2015), the assessors are additionally advised to analyse the tweet for referenced locations. Furthermore, the example dataset from section 10.4.2.1 shows that many users indicate the federal state they live in in their user profile. As this can also indicate local proximity to the post, this information is also shown. In addition, the assessor is provided with the date and time of the tweet as well as with the temporal distance of the post from the incident. It is especially important to determine whether a tweet was written before or after the incident, as in some emergency situations no information is available shortly before or until the emergency services arrive.

As shown in Table 10.2, the number of retweets of each post is made available to the assessors. The role of retweeters, more specifically the observation that retweet behaviour may indicate relevance, as described by Reuter et al. (2013), is supported by examinations of Starbird and Palen (2010), Uysal and Croft (2011) and others. However, the following tweet from the 2017 Las Vegas shooting illustrates that the assessors should not treat this criterion as an exclusive indication for relevance: *"[..] As I #prayforlasvegas I pray for us all. Find each other out there....* https://www.instagram.com/p/BZwx8oVle7s/" (Perry, 2017).

Katy Perry published this post on October 3, 2017, to communicate her condolences to the victims and their families of the Las Vegas shooting. So far, the tweet has received about 2.800 retweets and can thus be considered as a very often shared post of this incident. However, in terms of content, this tweet is not relevant for the emergency services. The high number of retweets can be explained by the enormous fan base of Katy Perry. In addition to the number of retweets, the number of likes/favourites is also submitted to the assessors. To like a tweet does, however, not necessarily mean that the person likes the content of it.

As described by Gorrell and Bontcheva (2016), Twitter users also use a like/favourite, for example, to say "thank you" or to create a reminder for themselves. Assessors also see extended content in the form of pictures, videos and URLs. This extension allows Twitter users to bypass the 280-character limitation on Twitter and is therefore likely to be of importance for the emergency services, as described in section 10.2.3. For example, pictures and videos can provide important information about burning buildings. URLs can refer to blogs or other posts in social media which might as well provide necessary information.

Finally, the number of words and letters is shown to the assessors, whereby the lower bound to label a tweet is selected by the assessors themselves. Abel et al. (2012) chose a very high lower limit of at least 100 characters per relevant post. For example, the following tweet contains only 88 characters, but according to Verma et al. (2011), it is a relevant post related to the Red River floods in 2009: *"Country residents outside of Fargo are surrounded by flood waters. Some R being rescued"* (Wise Bitch, 2009). The minimum requirement of Sriram et al. (2010) of three words per Twitter message seems to be a well-chosen parameter regarding the classification of relevance in crises.

10.4.1.2 Relevance Features for Machine Learning

For machine learning, abstract and interpretative relevance criteria are not practicable. Precise and measurable criteria must be defined for the algorithm to be able to conduct a classification of relevance. The selected classification features form the resulting classifiers and define the concept of relevance for machine learning. To form a basis of this concept, the criteria from Table 10.2 are discussed. Also, additional features are considered which do not necessarily have to indicate relevance, as the selected algorithm in section 10.4.2 is evaluated for various classification feature set combinations.

Keywords or terms are essential for identifying relevance in crises. Search-filtering takes place before the actual relevance classification so that the amount of user-generated content is reduced to a manageable subset. The further textual content of a tweet is particularly important for identifying relevance for humans as well as for machines. As in the publication of Vieweg (2012b), a sorting out of tweets containing certain terms, for example, relating to condolences, could be considered. However, it was decided for this paper that the tweet's content or the sequence of words which the algorithm can weight as positive or negative is sufficient as a feature.

The classification results are also used to test whether the results of the classifier can be improved by text-based *Named Entity Recognition* (NER). Named Entities are phrases that contain the names of persons, organizations or places (Tjong Kim Sang & De Meulder, 2003). Especially the identification of the location can be of great importance. Even though the algorithm is given the distance of the tweet to the incident, tweets rarely contain location information. Here, NER can help to identify further locations and thus support the computation of geographical distances of the author and posts in relation to the incident's location. Furthermore, the collection of Twitter posts takes place within a time frame. However, as collected messages may lie in the past and the time frame may not always match, the difference between the creation of the tweet and the time of the incident is another criterion which has to be taken into account by the classifier. In the classification, the retweet behaviour can also indicate the relevance of a tweet. Yet, since the procedure described in this paper is based on many other characteristics, the tweet example of Katy Perrey would probably still be excluded. Moreover, links in the form of URLs, videos or picture can be important for the emergency services. This information is versatile and exceeds the limitation of 280 characters per tweet. An analysis of the content of pictures, videos or Internet pages is, however, not included in the classifier. Furthermore, the length of a tweet is of great importance for the machines and should not be underestimated. Too short tweets (based on the number of letters) are classified as less relevant by the algorithm. Also, as mentioned by Abel et al. (2012), there is the assumption that the underlying language of a tweet may be relevant. In view of the globalized world, a number of important issues could be overlooked. The algorithm is given the original language as an independent classification feature, but the actual text is translated if it was not written in German. Table 10.6 shows the previously mentioned classification features of the algorithm in a compact design.

10.4.2 Evaluation

In the evaluation, various classification features of a classifier detecting a tweet's relevance in disasters and emergencies are tested. Thus, different pre-processing steps (section 10.3.3.1) and subsets of all classification features and methods have to be considered. In the process, approaches from previous work are combined to achieve a high-performance solution. For example, Habdank et al. (2017) and Verma et al. (2011) evaluate various algorithms with purely textual classification features, while Imran et al. (2013a) test one algorithm with several different classification features. In the evaluation procedure, two different algorithms with several classification features are tested for effectiveness and—in contrast to previous work—efficiency. The test is based on an event that is used as a proxy for further scenarios. The evaluations are performed on a computer with Windows 10, Intel i7-7500 with two physical and two logical cores at 2.90 GHz and 12 GB Ram in Java.

Table 10.6 Features, description and internal representation in the classifier (*means optional, [] means that it is no parameter feature)

Feature	Description	Internal Representation
[Keyword filtering]	Pre-filtering by keywords of the incident	Social Media API Crawljob—Collections
[Temporal filtering]	Pre-filtering with regard to the time of the dataset	Social Media API Crawljob—Collections
Content	Word sequence based on unigrams	TF and TF-IDF values based on a Bag of Words of the training data—Double (per word)
NER	Extraction of persons, names, and locations from text	Availability of a geolocation—Boolean
Post distance*	Geographical distance between the post and the event	Kilometre—Double (-1.0, if no data available)
Author distance*	Geographical distance between the author and the event	Kilometre—Double (-1.0, if no data available)
Temporal distance	Temporal difference between the post and the event	Minutes—Double
Retweet count	Number of repetitions of a tweet by other authors	#Retweets—Integer
URLs	The availability of links	Availability—Boolean
Media	The availability of images or videos	Availability—Boolean
Length of post	–	#Characters—Integer
Language	–	Country code—Nominal

10.4.2.1 Method

As classifiers require training data in the form of labelled or coded text (from which they learn how to distinguish between different types of discourse), all tweets are pre-processed and stored in an ARFF file as a training dataset at the beginning of each evaluation step. Based on this file, a classifier is trained. Depending on what is to be tested, each evaluation step inherently adapts the set of characteristics, parameters of the algorithm or the algorithm itself.

Datasets. This study is based on two datasets, which were qualitatively assessed and labelled as relevant or irrelevant by a single expert per dataset. The first underlying dataset for the test is based on the 2013 European floods but focusing on German data in this case (Reuter, Ludwig, Kaufhold, et al., 2015). As reported by the German Federal Agency for Civic Education (Bundeszentrale für politische Bildung (bpb), 2013), there was a lot of commitment from volunteers. Most notably, social media were used to disseminate information and offer help. According to this information, many appeals for donations were made via the Internet, which were labelled as irrelevant.

A total of 3923 Twitter messages sent over a period from 30 May to 28 June 2013 were collected and classified. This corresponds approximately to the amount of data analysed by Habdank et al. (2017) and Imran et al. (2013a). There are 1626 relevant and 2297 irrelevant tweets. In this case, all texts were written in German. However, content not written in German was translated. As can be seen from the dataset, a large number of Twitter users provided status updates on the current water level in addition to offers of and requests for help.

For another evaluation and interpretation of the different phases, we use a dataset from the incident of BASF SE in Ludwigshafen on October 17th, 2016. A facility of BASF caught fire because of working on a pipeline route. The fire lasted 10 hours and was extended by two explosions (Habdank et al., 2017). Five people lost their lives and 28 people were seriously injured.

This dataset was already used for relevance classification by Habdank et al. (2017). They only relied on textual information of the posts. It contains 3790 posts with 1816 being relevant to the accident. Unfortunately, about 5% of the posts only have textual information and no metadata linked to them. There are even more posts that are missing only some of the metadata.

Cross-Validation. As in Verma et al (2011) and Habdank et al. (2017), a 10-fold cross-validation was used for analysis. Whereas in a regular validation the data is partitioned into a training set (to train the model) and a test set (to evaluate it), in a 10-fold cross-validation the data is randomly portioned into ten equal subsets (Habdank et al., 2017). Of the ten subsets, a single subset is retained as the validation data for testing the model and the remaining 10-1 subsets are used as training data. The cross-validation process is then repeated ten times, with each of the ten subsets used exactly once as the validation data (Hastie et al., 2009, pp. 241–249). This prevents certain patterns from occurring which could falsify the classification quality.

Criteria. All evaluation combinations are evaluated using the criteria a*ccuracy, recall, precision* and *time (in seconds)*. Accuracy indicates the percentage of tweets correctly classified as relevant or irrelevant by the classifier. This means that all tweets correctly predicted as relevant (true positives (TP)) and all tweets correctly predicted as irrelevant (true negative (TN)) are divided by all tweets available (Habdank et al., 2017).

Recall is the ratio of tweets predicted as relevant to all tweets classified as relevant (Powers, 2011). The number of all tweets classified as relevant can be expressed by the number of TP plus the number of tweets incorrectly predicted as irrelevant (false negatives (FN)).

Precision, in contrast, describes the ratio of all TP to all tweets predicted as relevant. The number of all tweets predicted as relevant can be expressed by the number of TP plus the number of tweets falsely predicted as relevant (false positives (FP)) (Powers, 2011). This measure is important to minimize the number of FP. In the context of crisis informatics, many false positives mean that many irrelevant tweets were reported to the emergency services. Since this would result in a greater time consumption, the recall unit should be weighted higher in the following examination. The higher the recall, the fewer tweets were falsely predicted as irrelevant (Habdank et al., 2017). Minimizing the FN is particularly important as otherwise valuable information could be lost; in the context of crisis informatics, this could cost lives (Habdank et al., 2017).

The time consumption of classification is particularly important for the use in a productive system. Emergency services should be able to conduct a relevance classification as quickly as possible without being hindered by a long, internal pre-processing or classification process. In the following, time data refer to a complete analysis run including pre-processing (Processing-Duration (PD)), training (Training-Duration (TD)) and 10-fold cross-validation (Validation-Duration (VD)). Thus, time data do not indicate the training or classification in particular. However, conclusions can be drawn for its use in a productive system.

Steps. Given the numerous possibilities to achieve a high quality and temporal efficiency of the algorithm, a systematic approach is necessary. In the text processing phase, the text processing with individual options is tested. The feature set phase consists of a classification feature analysis in which a classifier is tested on various feature combinations. The parameter optimization phase is constitutive for the setting of the learning algorithm. In the algorithm phase, two classification algorithms are compared. Finally, in the fifth phase, a filter approach is used to minimize the vast amount of classification features to the most important subset of it.

- **Text Processing Phase:** In the first phase, text processing options are tested with a Random Forest. The vectorizations Term Frequency (TF) and Term Frequency–Inverse Document Frequency (TF.IDF) are compared, stemming is compared to the more complex lemmatization and the appropriateness of Named Entity Recognition (NER) is discussed. As the NER process takes place in the text processing phase, the results are discussed in this phase and not in the classification feature set phase.

- **Classification Feature Set Phase:** In this phase, different classification feature combinations are tested. The best algorithm of the first phase is used as a reference for the text-based Random Forest. Based on this, a combinatorial adding of the classification features number of retweets, length, geographical distance (author distance and tweet distance), temporal distance and the existence of media and URLs follows. In the last test run, the non-textual characteristics are tested to see whether they indicate relevance without the underlying tweet's text.

- **Parameter Optimization Phase:** The best Random Forest of the last phase is optimized regarding various parameters. For example, it is possible to determine the depth or the number of decision trees. Furthermore, the so-called "threshold-moving" is particularly suitable to compensate for an imbalance of recall and precision. The percentage threshold from which a tweet is classified as relevant or irrelevant is changed (Zhou & Liu, 2006).

- **Algorithm Phase:** The best result of the last two phases is used to compare a Naïve Bayes algorithm with Random Forest. The decision for these two algorithms is justified by the fact that they stand out in multiple reference works (Habdank et al., 2017; M. Imran et al., 2013a; Markham & Muddiman, 2016; Verma et al., 2011). We most importantly focus on Random Forests because already other authors, like Habdank et al. (2017), showed that these are well performing. Even if Convolution Neural Networks are also gaining popularity in this research area, we are not considering them for our production system, since they are hard to train and adapt for new datasets.

- **Feature Subset Selection Phase:** The last phase serves as an outlook on possible reductions of the large and complex classification feature set. The Bag of Words (BoW) is generated based on all words of the training dataset. Most of the several thousand words are not substantial for the actual classification and can thus be eliminated. This could potentially result in fewer adaptations in the learning process and a faster classification process. A so-called filter approach is used here: a heuristic ranking algorithm is used to find the best subset of classification features (John et al., 1994). An analysis of the resulting classification features can yield information on the intelligence of the algorithm.

10.4.2.2 Results of the 2013 European Floods Dataset

It has to be noted that the following results are only substantial for this paper's dataset and the corresponding scenarios. The use of other scenarios and other tweets, e.g., tweets that are not exclusively German, could lead to different results. Comparing the results with related work, the Random Forest of Habdank et al.

(2017) achieved the very good results within a similar setting in this research field with an accuracy of 88.7%. Using a Naïve Bayes classifier, Spielhofer et al. (2016) achieved an accuracy of 77.2%. However, it is worth mentioning that all algorithms were evaluated using different datasets. For example, the inclusion of many retweets in the test dataset could greatly improve the results due to the analogousness of tweets. Information about the occurrence of retweets was not provided by any author of the related work.

Text Processing Phase. All steps of this phase are evaluated with a Random Forest and exclusively textual classification features. In each run, the words are processed according to the pre-processing explained in section 10.3.3.1. Characters, like "\n", "\r", double whitespaces and characters that could not be encoded properly, are removed and the tweets are tokenized. The content is not translated since the dataset contains German tweets only. The tests showed that the stop word list of Porter (2018) was most suitable. Therefore, it is used in all following procedures. If no other information is given, the classifiers are tested with simple stemming and TF.

Comparison between TF and TF.IDF. The two vectorizations TF and TF.IDF are implemented according to the specifications in section 10.3.3.1. Table 10.7 shows the classification quality with the percentage results for accuracy, precision and recall as well as the time consumption in seconds. With the TF.IDF unit, the classifier achieves a classification quality minimally higher than with the TF unit with an Accuracy of 90.84%. However, this difference is considered as not substantial. The use of the more complex vectorization takes almost two minutes longer than using the simple TF method. Both approaches would thus prove to be practicable during real-world application. As the TF has proven to be faster and easier to implement, the following test steps are only given in this unit.

Table 10.7 Classification quality of TF and TF.IDF vectorizations based on a textual Random Forest

	Accuracy [%]	Precision [%]	Recall [%]	Time [s]
TF	90.8	91.3	81.1	**850.271**
TF.IDF	**90.84**	**91.4**	**81.2**	957.981

Comparison between stemming and lemmatization. The classification quality for both word stem creations is also not substantially different, as Table 10.8

shows. However, as the time for stemming of approximately 14 minutes is substantially better than the time for lemmatization of approximately 32 minutes, stemming is preferred in the following tests.

Table 10.8 Classification quality and time of the word stem creations stemming and lemmatization on the basis of a textual Random Forest

	Accuracy [%]	Precision [%]	Recall [%]	Time [s]
Stemming	90.8	91.3	81.1	**850.271**
Lemmatization	**90.87**	**91.4**	**81.2**	1936.385

Appropriateness of NER. The inclusion of the NER classification feature produces the best classifier so far considering the quality, as shown in Table 10.9. However, a 0.3% improvement of the recall does not justify a time of approximately 132 minutes. In all three test procedures, the standard implementation with simple TF, stemming and without NER processing proves to be the most useful for the use in a productive system. The classification quality is not substantially worse. For example, the accuracy does not differ by more than 0.15% in all methods. Thus, an additional expenditure in time is not compensated.

Table 10.9 Classification quality and time with and without the NER classification feature based on a textual Random Forest

	Accuracy [%]	Precision [%]	Recall [%]	Time [s]
Without NER	90.8	91.3	81.1	**850.271**
With NER	**90.95**	**91.4**	**81.5**	7952.865

Feature Set Phase. The various feature set combinations are created based on the results of the text processing phase. Thus, if word characteristics occur in the combination of the set of characteristics, character removal, tokenization, stemming, and TF are used.

Table 10.10 shows a section of the evaluation with various classification feature combinations specified in the first column. The highlighted values represent the best result of the respective column. As in the previous test procedures, the classification quality is shown with the results of accuracy, precision and recall; the time in seconds represents the effectiveness and efficiency in a productive system. As can be seen, all classification features from Table 10.10 have been

Table 10.10 Classifications with different feature set combinations. A Random Forest with the respective classification features is used

Classification Features Used	Accuracy	Precision	Recall	Time (s)
Words	90.8	91.3	81.1	850.271
Words + Number of Retweets	90.82	91.4	81.1	851.14
Words + Length	90.89	91.4	81.3	862.69
Words + Number or Retweets + Length	90.85	91.4	81.1	841.78
Words + Temporal Distance	90.93	91.4	81.3	901.22
Words + Geographical Distance (Author Distance and Tweet Distance)	91.03	91.6	81.3	1021.663
Words + Geographical Distance (Author Distance and Tweet Distance) + Temporal Distance	91.21	91.8	81.4	1078.092
Words + Distance (Author Distance and Tweet Distance) + Temporal Distance + Length	91.23	91.8	81.5	1110.276
Words + URLs	90.9	91.4	81.4	850.22
Words + Media	91	91.5	81.5	860.12
All Classification Features	91	91.6	81.1	1071.79
No Words + All Other Classification Features	84.35	84.4	75.1	281.14

combined with the classification feature words at least one time, except for NER and language (for reasons mentioned above). Further combinations have been listed as an example or if they could improve the algorithm. Slight fluctuations in time can be explained by internal factors of the computer, e.g., a more efficient use of the cache.

Adding classification features to the existing textual classification features slightly improves the quality of the algorithm in every case. Adding classification features further, however, can have a negative impact on the classification quality, as can be seen in rows 3 and 4 with the example of the number of retweets. This is probably due to an overfitting resulting from too much information being available to the algorithm. In this dataset, the Random Forest has reached the best classification quality when the classification features words, geographical distance, temporal distance and length are combined. The accuracy of 91.123% exceeds the accuracy of the purely textual classifier by 0.323%. This means that by adding these features, 17 tweets previously misclassified could now be classified correctly. However, an additional time of about four minutes is required.

The best performance in time can be achieved without textual classification features (see Table 10.10, last row). Although this procedure results in the worst classification quality with an accuracy of 84.35%, it confirms the assumption that the non-textual classification features of Table 10.10 point to relevance without the underlying tweet's text. Nonetheless, the result is still more than acceptable as algorithms of related work achieved even worse results.

The following phases are based on the Random Forest with the classification features words, geographical distance, temporal distance and length as this combination has proven to achieve the best classification quality.

Parameter Optimization Phase. The first runs of the standard implementation of the Random Forest in Weka show discrepancies in the recall and precision value (Table 10.11). This imbalance could be inherent in the dataset but is improved by changing the threshold from which a result is classified as relevant or irrelevant (Zhou & Liu, 2006). We used a threshold selector meta classifier in combination with the RF classifier in Weka. Other default settings, such as the number or depth of decision trees, were tested within a random search. The default values proved to be good for the classification problem at hand.

Table 10.11 Modification of the threshold value (threshold-moving) of the Random Forest with the classification features words, geographical distance, temporal distance and length

	Accuracy [%]	Precision [%]	Recall [%]
Threshold: 0.5	91.23	**96.9**	81.5
Threshold: 0.3	**91.64**	94	**85.2**

In the standard implementation of a Random Forest in Weka, the threshold value is set to 0.5. As can be seen in Table 10.11, a threshold value of 0.3 has proven to be useful, as recall and precision converge and accuracy increases to 91.64%. While the recall value increases as well, the precision value decreases. According to Habdank et al. (2017), this change is to be assessed as positive. Even though a decreased precision value means that the emergency services are shown more irrelevant tweets, which may result in an additional time consumption, a decreased recall value implies that more relevant tweets are incorrectly classified as irrelevant, which could have severe consequences in crises.

Algorithm Phase. Since the combination of the classification features words, geographical distance, temporal distance and length have provided the best results, it is used in the following comparison of the Naïve Bayes classifiers and the Random Forest. As in the last phase, the Naïve Bayes classifier was tested for various parameters. A threshold-moving is not necessary because the recall and precision value are very balanced. As in the Random Forest classification, other parameters prove to be useful in the default setting. For the Random Forest classifier, the threshold of 0.3 was transferred from the last phase.

Table 10.12 Classification quality and time of the Naïve Bayes and Random Forest classifier based on the classification features words, geographical distance, temporal distance and length

	Accuracy [%]	Precision [%]	Recall [%]	Time [s]
Random Forest	**91.64**	**94**	**85.2**	1120.22
Naïve Bayes	85.19	81.8	85.2	**263.749**

Table 10.12 illustrates that the Random Forest performs better than the Bayes classifier in every aspect of the classification quality. The important recall value of the Random Forest is at 85.2% and the recall value of the Naïve Bayes implementation is at 82.7%. Thus, the Naïve Bayes approach incorrectly classifies 282 tweets as irrelevant, whereas the Random Forest approach produces only 240 FN. However, the pre-processing, training and 10-fold cross-validation of the Naïve Bayes classification with about 4 minutes is much faster than the Random Forest with 18 minutes.

For completeness and to make an all-encompassing statement, both algorithms have to be tested on all classification feature combinations and all pre-processing steps. As the results are very clear and in accordance with those of Habdank et al. (2017), this dataset assumes that the Naïve Bayes classifier is to be used for faster classification and the Random Forest for a higher classification quality of relevance.

Feature Subset Selection Phase. The Bag-of-Words (BoW) of this dataset comprises 10149 words, with all of them included as classification features in the classification. In the following, a filter approach is applied to the Random Forest algorithm in an attempt to reduce the time required. The filter procedure is applied directly to the dataset independently of the classifier using information gain as evaluator with a threshold of 0.005.

In this way, the feature set was reduced substantially. Without the filter approach, 10149 words in combination with geographical distance, length and temporal distance were previously used for classification. Using the feature set search, the number can be reduced to 148. These 148 classification features are already decisive for the classification of relevance. Moreover, the resulting classifier still achieves a high classification quality compared to the best algorithm of the previous phase and, as expected, the required time is lower (Table 10.13). With about 3 minutes and 15 seconds this algorithm is also faster than the Naïve Bayes approach of the algorithm phase.

When analysing these 148 classification features, it can be seen that the algorithm selected all four features gained from the feature set phase for the relevance classification. Furthermore, the set contains words that provide direct conclusions about the term relevance as defined in section 10.2.3. This includes words such as "water level" (German: "Pegel"), "alarm level" (German: "Alarmstufe") and "cm" (short for "centimetre") which point to information and facts of the tweets. The terms "help" (German: "Hilfe") and "helpers" (German: "Helfer") speak in favour of the term relevance as defined by Vieweg (2012b). Terms such as "donation" (German: "Spende") are in the end probably weighted negatively by the algorithm as they indicate irrelevance.

Table 10.13 Classification quality and time with and without a feature set elimination procedure based on the classification features words, geographical distance, temporal distance and length

	Accuracy [%]	Precision [%]	Recall [%]	Time [s]
10153 Features	**91.64**	94	**85.2**	1120.22
148 Features	91.28	**98.2**	80.4	**204.326**

10.4.2.3 Results of the 2016 BASF SE Incident Dataset

For a second evaluation of the before stated phases, we use the BASF incident dataset. The best combination of features is given by taking the tweet length and the temporal distance into account. With these features the random forest classifier reached a very good accuracy of 90.3%. The shortened results can be seen in Table 10.14. Adding the distance features, as in the flooding scenario, did not improve the result. We expect that this could be due to the incomplete dataset. Only about 50% of the tweets have an author location specification, whereas in the flooding dataset we were able to calculate the author distance of about 85%. This sparsity implies that the entropy of these features is naturally higher, causing the decision trees in the end to use other features for building the trees.

Threshold-moving did not change the result in a noticeable way, since recall and precision are already good balanced.

In the feature reduction phase, we were able to successfully reduce the size of the features without diminishing much of the quality of the classifier. The filter approach in combination with a threshold of 0.005 reduced the initial 9164 features to remaining 173 also containing temporal difference and length. When inspecting these values, it can be seen that information gain ranked the time difference as the highest feature. Furthermore, there are several words that are expressive for relevance in this fire and explosion situation. "Explosion" (German: "Explosion"), "toxic alarm" (German: "gift-alarm") and "safety note" (German: "Sicherheitshinweis") are words that are probably involved in tweets containing facts and advisory information.

Table 10.14 Classification quality and time of two different additional features sets and a feature reduction on the BASF incident dataset

Classification Features Used	Accuracy [%]	Precision [%]	Recall [%]	Time [s]
Words	90.08	90.1	89.2	1109.04
Words + Temporal Distance + Length	90.32	89.8	90.0	1224.10
Words + Temporal Distance + Length [Reduced]	89.13	89.6	87.5	143.00

10.4.2.4 Summary

In all phases of the evaluation, the classification quality for this dataset was continuously improved and the required time reduced. Each of the steps covered with a different sub-area. Hence, the resulting findings are different. The *text processing phase* showed that the TF vectorization, the stemming and the exclusion of the NER classification feature are preferable to the complex procedure TF.IDF, lemmatization and the use of the NER classification feature because they offer higher efficiency. The additional required time does not justify the small increase in classification quality. In the *feature set phase*, the use of the classification features word, geographical distance, temporal distance and length proved to be the best combination regarding classification quality in the flooding scenario. For both datasets, the exclusion of textual classification features and the use of all other classification features results in an acceptable classification quality and a considerably better (lower) time consumption.

In the case of the flooding scenario, the *parameter optimization phase* showed that the use of the Random Forest with a threshold value of 0.3 instead of 0.5 improves the classification quality by increasing accuracy and recall. According to the results of the *algorithm phase*, Random Forests achieve a better classification quality than Naïve Bayes classifiers. However, Naïve Bayes classifiers are preferable when aiming for a lower time consumption. In the *feature subset selection phase*, the time consumption is considerably reduced when using only the most important classification features. The classification quality decreases minimally. Examining the remaining classification features, it can be seen that these reflect the terminology of relevance as defined in this paper.

10.5   Evaluation II: Relevance Classification via Active and Online Learning

Finally, for insights into **RQ3 ("How can the amount of labelled data required for relevance classification be reduced by active incremental learning and transparent visualization of the classifier's quality?")** and **RQ4 ("How can the dynamic retraining of relevance classifiers be supported by user feedback performance-wise using batch learning with feature subset selection?")**, we conducted our second evaluation whose approach (section 10.5.1) and conduction (section 10.5.2) are described in the following.

10.5.1 Approach

As already indicated by a variety of publications, active learning units can substantially reduce the amount of labelled data required to reach a certain accuracy threshold in different application domains of machine learning (Bernard et al., 2018; M. Imran et al., 2017; Settles, 2010). Since time is limited in disaster scenarios, active learning could be of use for creating suitable classifiers. For instance, Imran et al. (M. Imran et al., 2017) demonstrate an active learning unit which supports the labelling of events by suggestions based on past events. Motivated by this potential, this chapter conceptualizes and evaluates an active learning unit which directly works with the dataset of an event, requiring no previously labelled data or events.

The active learning is realized via *uncertainty sampling*. According to Lewis and Catlett (2014), it is especially reasonable to label the posts where the classifier has the lowest confidence. The learning unit occurs in a *pool-based sampling* scenario using the *least confidence* measure. Settles (2010) describes pool-based sampling as an environment in which the algorithm checks the set or subset of the non-labelled data with regard to its information content and returns the most reasonable datum. By applying the least confidence measure to a binary classification problem, the instance is returned where the classifier's prediction confidence is nearest to 50% (Settles, 2010).

Table 10.15 Time for the retraining of the classifier with Random Forest Naïve Bayes, Hoeffding Tree, and k-Nearest Neighbour (IBk)

	RF	NB	HT	IBk
Time [s]	129.294	0.017	0.19	3

For the presented approach it is obvious that classifiers, which are able to compute prediction probabilities, are already required at the labelling stage. For the execution of the active learning unit it is important that these classifiers can be rapidly adapted by newly labelled posts to improve their quality and to allow an application in real-time. Lewis and Catlett note that "uncertainty sampling requires the construction of large numbers (perhaps thousands) of classifiers which are applied to very large numbers of example" (Lewis & Catlett, 2014).

The aim is to develop an online learning environment where sequentially labelled data incrementally enhance the classifier. The RF classifier of section 10.4 is not suitable for this task since it is a batch learner which requires a completely new training run for each additional datum. Incremental learning, then again, allows the sequential integration of new posts without looking at the previously labelled data again. To clarify this aspect, we compared the RF batch classifier with different incremental methods regarding retraining time. Weka supports multiple incremental learning approaches and Hoeffding Tree (HT) (Ren et al., 2014), Incremental Naïve Bayes (iNB) (Hulten et al., 2001) and k-Nearest Neighbour (IBk) (Aha et al., 1991) are candidates for our scenario. In case of IBk, we additional use a k-d-tree structure to improve the search for the $k = 50$ neighbours (Moore, 1991). We use the dataset of section 10.4.2.1 for comparison, dividing it into 3902 posts for initial training and 20 posts for retraining. As seen in Table 10.15, iNB and HT require less than a second for retraining. IBk with $k = 50$ requires three seconds for the task and still seems suitable for the use in SMO. However, the RF classifier requires more than two minutes for the same task.

In addition, we have found that incremental methods for our purposes produce worse classification grades than RF (section 10.5.2). However, the maximization of classification accuracy is not required during the labelling process using an active learning unit as, thereafter, a different algorithm can be used for the creation of the classifier based on the actively created dataset. As Lewis and Catlett (2014) outline, for uncertainty sampling it is suitable to use simple or 'cheap' classifiers for the active selection of data to be labelled and to create complex or 'expensive' classifiers based on this. Despite the heterogeneity of the type of simple or complex classifiers, a reduction of the required labelled data is possible.

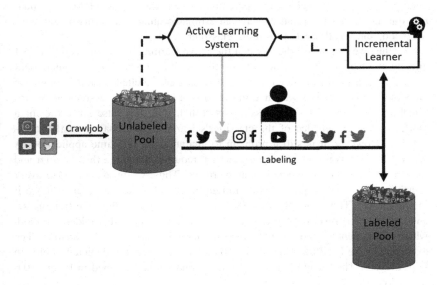

Figure 10.9   Incremental active learning in the labelling process of SMO

In the labelling process of SMO, we use IBk classifiers with k-d-trees that are extended incrementally with each new labelled post. The process is visualized in Figure 10.9. After each third to fifth labelling, the classifier is used to compute an active learning request (marked with a *-sign and in yellow in comparison to the black non-active labelled posts). In this case, the classifier predicts the probability with regard to the relevant class of 50 to 100 random unlabelled posts (unlabelled pool) und returns the post that is most ambiguous, i.e., nearest to the 0.5 value (50%). By only using each fifth post for the active learning unit, bigger amounts of the pool can be included and excessive influence of the first increments can be prevented. The indicated values are well-suited for real-time use but are dependent on the performance of the underlying machine. After the labelling process, RF classifiers are trained to achieve the best classification accuracy based on the labelled posts (labelled pool).

10.5.1.1 Real-time Evaluation using Active and Incremental Learning

In the labelling process, labellers and researchers ask themselves the question of how much data needs to be labelled to achieve a classifier with an acceptable quality. To get one step closer to a potential answer, we propose real-time evaluation during the labelling that displays an estimated quality, in terms of accuracy, precision, and recall, for the resulting classifier.

For this, we can utilize the fact that the active learning unit from section 10.5.1 already conducts online learning during the labelling process. Due to incremental learning, a new classifier is available after each labelled post. Although the classification quality is lower than in the case of using a RF classifier, an approximation regarding the final classifier is sufficient in this case since the main aim is to convey an idea of the classifier's learning curve based on the labelled data. However, it is still a problem to guarantee its real-time application. A cross-validation is not recommended since it requires retraining the classifier and does not utilize the incremental characteristics. Thus, we propose to skip every fourth or fifth labelled post for the training of the classifier in order to move it into a test set. Thus, roughly 75% (80%) of the posts constitute the training set and 25% (20%) constitute the test set. This strategy reflects the holdout method, which is frequently used as an evaluation method besides cross validation. The extension to the labelling process is marked with a *-sign and depicted in blue in Figure 10.10. The integration of real-time evaluation is described in Figure 10.6 of section 10.3.4.2.

10.5.1.2 Feedback Classification using Feature Subset Selection

Active learning indicates the potential of labelling posts where the classifier is unconfident. Another (even better) option would be to select and correct the posts which were misclassified by the classifier. This cannot be achieved proactively, but a reactive method, comparable to spam detection, would be necessary to reduce false positive and false negatives after final classifier creation. Based on the system of a European research project, Markham and Muddiman describe feedback classification as a potential extension, since continuous feedback "leads to a system that adapts and improves over time" (Markham & Muddiman, 2016, p. 3).

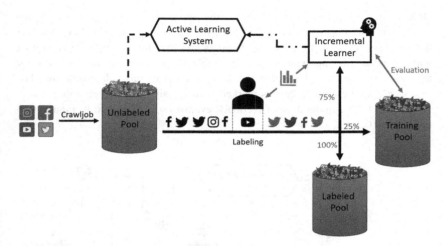

Figure 10.10 Real-time evaluation during the labelling process in SMO

The correction of false predictions has a considerable impact on the classifier's quality and allows for the adaptation of the classifier along the event's process.

Table 10.16 Comparison of time for the recreation of the RF classifier with and without Feature Subset Selection (FSS) using the 2013 European floods dataset

	RF without FSS	**RF with FSS**
Time [s]	129.294	4.704

For the practical implementation, again, the issue emerged that batch RF classifiers, which performed better than incremental classifiers, require too much time for retraining since they are not extensible with new data. We address this issue by the use of supervised Feature Subset Selection (FSS), as most of the thousands of features are not decisive for the actual classification and can be eliminated (see section 10.4.2.2 for the feature subset selection phase of the first evaluation). As a consequence, the training is performed much faster and less overfitting occurs during learning. As a side note, this approach is not suitable for real-time evaluation and active learning (section 10.5.1.1), because supervised FSS requires a considerable amount of labelled data. However, for this problem we apply a filter approach. By using a search algorithm the best subset of features is searched

(John et al., 1994). To reduce the time of this approach, we perform a bottom-up-search with the upper bound of 200 features. The time evaluation revealed that the RF classifier of section 10.4.2.2 requires less than five seconds for retraining with 20 posts after application of FSS (Table 10.16). This is still too slow for the labelling process but suitable for the correction of false classifications. The feedback classification process is marked with a *-sign and depicted in green in Figure 10.11. The integration of feedback evaluation is described in Figure 10.8 of section 10.3.4.2.

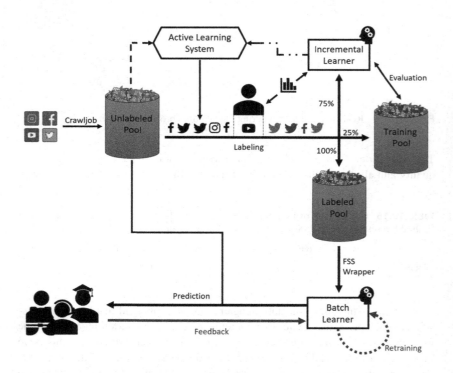

Figure 10.11 The classification process comprising active learning, real-time evaluation and feedback classification

To summarize the overall approach, Figure 10.11 also comprises the labelling process with the active learning and real-time evaluation methods. The active learning unit allows to use a small pool of labelled data to create a classifier of good quality. The real-time evaluation based on online learning supports the labeller to get an idea at which time a sufficient amount of posts was labelled. Subsequently, a RF classifier using FSS can be trained. Furthermore, the false predictions of this classifier can be corrected by the user, e.g., the emergency manager, reactively; based on the feedback, a new classifier is trained in the background. All units are designed for multithreading and are thread safe so that delays constitute no problems, allowing all units to work consistent and as intended.

10.5.2   Preliminary Evaluation

10.5.2.1   Method

In this chapter, we simulate a real application of the active learning unit in combination with the real-time evaluation component using online learning. We use the collected data from the 2013 European floods and the BASF SE incident in 2016. We assume a scenario where we have no labelled data available but intend to build a relevance classifier. The first phase of the evaluation describes the comparison of three incremental classification methods on the flooding scenario. The second phase compares the learning success with and without active learning on both datasets. In summary, several datasets, each comprising 1000 posts, were labelled within the SMO. The evaluations are performed on a computer with Windows 10, Intel i7–7500 with two physical and two logical cores at 2.90 GHz and 12 GB Ram in Java.

10.5.2.2   Results

The first phase reveals which incremental classifier achieves good results, so that the SMO can provide accurate estimations on the classifier's quality and suggest useful posts for active learning. For that purpose, the SMO labelling process was conducted three times for Incremental Naïve Bayes (iNB), Hoeffding Trees (HT) and a k-Nearest Neighbour algorithm (IBk) based on the 2013 European floods dataset and using the active learner. For iNB and HT, standard parameters from Weka were used. For IBk, we used a k-d-tree as representation model to facilitate a fast classification. Furthermore, 50 neighbours were selected as parameter for k.

   The area under the ROC curve (AUC) of the real-time evaluation was noted down every 50 labels. The results are displayed in Figure 10.12. The evaluation revealed that iNB and HT are not suitable for the present classification problem. After 200 labelled posts, both algorithms score consistently low AUC values

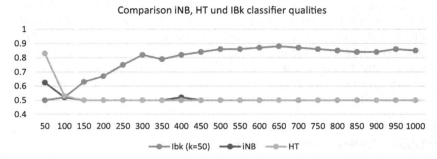

Figure 10.12 Comparison of the incremental learning methods iNB, HT, and IBk using AUC values from the real-time evaluation in SMO

around 0.5, thus showing a random classification behaviour. Besides low suitability for real-time evaluation, these algorithms are not able to suggest useful posts for active learning. In contrast, IBk is suitable for the present classification problem, showing a good learning curve. After labelling 1000 posts, an accuracy of 84.5% was achieved, which missed the accuracy of the resulting RF at that point (89.3%) by only about 5%.

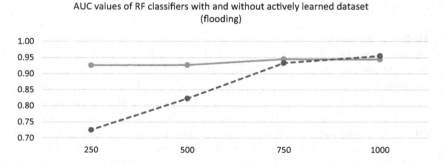

Figure 10.13 Comparison of AUC values of RF classifiers on the flooding dataset with (solid) or without (dashed) active learning

In the second phase the dataset labelled with IBk and active learning is compared to a dataset that was labelled without active and incremental learning. On the flooding dataset, after each 250 labelled posts, a RF classifier was trained and evaluated with 10-fold cross validation. The AUC values are displayed in

Figure 10.13. The evaluation revealed that the RF which was supported by active learning reaches excellent AUC values earlier. Already after 250 labelled posts, the AUC is 0.2 better than the RF without active learning. This implies the usefulness of active learning for labelling posts in disaster scenarios, reducing the amount of required labelled data.

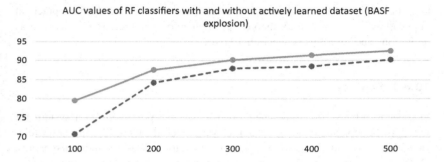

Figure 10.14 Comparison of AUC values of RF classifiers on the BASF SE dataset with (solid) or without (dashed) active learning

Within the BASF dataset we trained a RF classifier every 100 labelled posts. Even if the values are not as striking as before, in this explosion scenario, the active learning approach also seems to be useful. The AUC values can be seen in Figure 10.14. During the evaluation we noticed two particularities, which could also be the reason for the differing deviations of the two datasets. The "normal" classification already leads to good results after 200 posts, meaning that even without active learning it is fast in finding a good separation of the two classes. We also noticed that the IBk classifier performs slightly worse on this dataset, which could result in an inaccurately behaviour when choosing the active posts.

10.6 Discussion and Conclusion

Research in the field of crisis informatics revealed that social media enable emergency services to receive valuable information (e.g., eyewitness reports, pictures, or videos) contributing to situational awareness during disasters and emergencies (M. Imran et al., 2015; Reuter & Kaufhold, 2018). However, the vast amount of data generated during large-scale incidents can lead to the issue of information overload (Plotnick & Hiltz, 2016). Research indicates that supervised machine learning techniques are suitable for identifying relevant messages and filter out irrelevant messages, thus mitigating information overload (Habdank et al., 2017). However, in order to train an accurate classifier, clear criteria for relevance labelling and classification, a considerable amount of labelled data and a usable interface to facilitate the labelling process are necessary. In the following, we outline our results with regard to our research questions (section 10.6.1), practical implications (section 10.6.2), theoretical implications (section 10.6.3) as well as limitations and future work (section 10.6.4).

10.6.1 Results

In order to address the above-mentioned issues, we presented (1) a system for social media monitoring, analysis and relevance classification, (2) criteria for relevance labelling and prediction in social media during disasters and emergencies, (3) the evaluation of a well-performing algorithm for relevance classification using batch learning as well as (4) an approach and preliminary evaluation for relevance classification including active and online learning to (4.1) reduce the amount of required labelled data by real-time evaluation and (4.2) correct misclassifications of the algorithm by feedback classification. Our results contribute to answering the following research questions.

Which are suitable criteria for relevance classification and labelling in disasters and emergencies (RQ1)? Our literature review revealed that there is already a multitude of approaches dealing with relevance classification problems, which are summarized in Table 10.3. However, we found out that they often lack a clear definition of relevance as well as a set of criteria to determine relevance. Thus, we identified on the one hand abstract and interpretative (Table 10.1) and on the other hand clear and precise (Table 10.2) relevance criteria. While both types of relevance criteria are useful for labelling only the latter ones are suitable for relevance classification using supervised machine learning. The determined criteria also served as an input for the design and evaluation of a batch learning RF classifier (section 10.4).

How can existing supervised machine learning techniques for relevance classification be improved for use in real disaster and emergency environments (RQ2)? In this paper, we evaluated supervised machine learning classifiers which do not only take textual features but also relevance criteria into account. In our evaluation based on a flooding dataset, the combination of textual content (words), geographical distance (author and tweet), temporal distance and length reached the best classification quality (Table 10.10). With almost 92% accuracy, 94% precision and 85% recall the Random Forest classifier reached an excellent quality after the parameter optimization phase. Also, the classifier based on the BASF incident dataset research better results than in Habdank et al. (2017) with almost 90% accuracy, 90% precision and 90% recall. Thus, the classifiers reached a better accuracy than those in related work because the insights of these studies were incorporated into this one. For instance, Imran et al. (2013a) and Spielhofer et al. (2016) used Naïve Bayes classifiers, which yielded worse results for the present classification problem. Although Habdank et al. (2017) also used Random Forests, they are only based on textual features. Then again, Spielhofer et al. (2016) and Habdank et al. (2017) reached better recall values. However, the recall value could be improved by the further adaptation of threshold values during the parameter optimization phase.

Furthermore, many machine learning studies only focus on the evaluation itself, thus neglecting the importance of the efficiency (training time) of a classifier. However, in a productive system, the balance between effort and quality is important. Accordingly, the evaluation of the classifier also considered time expose for training (section 10.4.2.2). The previously outlined RF classifier required acceptable 18 minutes of time consumption for pre-processing, learning and 10-fold cross validation. However, if a faster classification is required, the feature subset selection phase outlined in the case of a flooding dataset, that a classifier using 148 features (instead of 10153) achieved 91.3% accuracy, 98.3% precision and 80.4% recall with a time expose of only 3 minutes, also reducing the memory consumption of the classifier.

How can the amount of labelled data required for relevance classification be reduced by active incremental learning and transparent visualization of the classifier's quality (RQ3)? We proposed methods to enhance research regarding classifying relevant messages in social media. At its core, we looked at the dilemma that, on the one hand, classifiers require a lot of training data for the present classification problem to achieve a considerable quality but, on the other hand, at the beginning of a disaster, often no labelled data is available and data has to be labelled in a limited amount of time. To address this issue, multiple researchers describe and evaluate approaches based on *domain adaptation* that use datasets from past events to reduce the amount of required labelled data from the current event (M. Imran et al., 2017; M. Imran, Mitra, & Srivastava, 2016; H. Li et al., 2017, 2015). In this work, however, we discuss approaches for improving the labelling process without incorporating data from past events.

To reduce the amount of required labelled data we implemented an active learning unit. This required rapidly trained classifiers that adapt with each new post during the labelling process. Thus, we realized it in an online learning environment which was realized via incremental classifiers (section 10.5.1). However, without further support, the labelling process is not transparent, i.e., the labeller is not able to assess the quality of the classifier during the labelling process and does not know when to finish the labelling process. Thus, we implemented an approach for predicting the classifier's quality in terms of accuracy, precision and recall, on frontend (section 10.3.4.2) and backend (section 10.5.1.1) level. Based on the merits of incremental learning, we proposed an approach for real-time evaluation.

The evaluation revealed that not all incremental learning algorithms are suitable for the present classification problem (section 10.5.1.2). However, IBk classifiers reached an acceptable quality for real-time predictions and constitute a successful learning unit for active learning. It was furthermore shown that the active learning unit was able to substantially reduce the number of labelled posts required for a well-performing classifier.

How can the dynamic retraining of relevance classifiers be supported by user feedback performance-wise using batch learning with feature subset selection (RQ4)? Even if the classifier achieves an excellent quality, misclassifications still occur. End-users, such as emergency services, might be interested in correcting these misclassifications while performing their analytical tasks to improve the quality of the classifier based on their feedback (Markham & Muddiman, 2016). Thus, we implemented a feedback classification mechanism on frontend (section 10.3.4.2) and backend (section 10.5.1.2) level. However, this kind of a posteriori classification is subject to the slow classification time of offline or batch learning algorithms, which are required to maximize the classifier's quality. In order to allow a rapid adaptation, we implemented a feature subset selection approach.

10.6.2 Practical Implications

As a practical contribution, we created a ready-to-use system for potential end-users, such as emergency mangers, and as a foundation for conducting further machine learning evaluations. Based on this, we derived the two following major practical implications.

Incorporate metadata of social media to improve the quality of relevance classification (P1). From a practical point of view, our results indicate that the consideration of metadata, as identified and summarized as precise relevance criteria (see RQ1), yield in better relevance classification results in comparison to classifiers that only consider textual features (Habdank et al., 2017; Markham & Muddiman, 2016) (see RQ2). However, the best classification results were achieved with a different set of metadata between our two datasets. Thus, characteristics of datasets, such as the number of specified locations, should be considered for determining the best classifier that still performs well also with regard to training time. To assist this process, future research could determine social media guidelines (Kaufhold, Gizikis, et al., 2019) for the interpretation of dataset metadata and its implications for the selection of classifier features. Furthermore, since professional roles, such as incident or public relation managers, may have different conceptualizations of relevant information, different classifiers using distinct metadata may be deployed to improve the actionability of information (Zade et al., 2018).

Use active and online learning to reduce classifier training time during disasters and emergencies (P2). The promising results of our active learning unit highlight that active learning can significantly reduce the training time of a well-performing classifier (see RQ3). Its combination with an approximate real-time evaluation of the classifier's quality during the labelling process assists the user in determining the stage where the classifier's quality is good enough for a first deployment during an emergency. After initial deployment, the labelling of data could be continued to deploy improved revisions of the classifier over time, also considering changes in language during developing events, such as emerging hashtags or words during collective sense-making processes (Stieglitz, Mirbabaie, & Milde, 2018) that correlate with relevant or irrelevant content. However, the overall process must be designed time as efficient as possible due to scarcity of time of emergency managers (Reuter, Ludwig, Kaufhold, et al., 2016), which suggests combining domain adaptation approaches to combine labelled data from past similar events with new data of the current event (M. Imran et al., 2018).

Further practical implications, especially with regard to the deployment of the architecture in real-world environments, have to be derived by the evaluation of the system also including the dynamic retraining of relevance classifiers (RQ4), with expert users from emergency services, which is discussed in section 10.6.4.

10.6.3 Theoretical Contributions

Our paper contributes to the area of textual content analysis, applied to the domains of emergency management and social media, and which comprises the areas of relevance classification, information quality assessment, sentiment analysis, clustering and summarization, humanitarian classification, topic modelling and named entity recognition, amongst others (Alam et al., 2019; Gründer-Fahrer et al., 2018; M. Imran et al., 2018; Kaufhold, Rupp, et al., 2020). More specifically, in the area of textual relevance classification, we made two contributions.

Identification of abstract and precise relevance criteria for labelling and classifier training (T1). In order to improve the foundations of relevance labelling and classification in future research, we identified and compiled a variety of abstract and factual criteria from existing research publications. While abstract and interpretative (summarized in Table 10.1) criteria help to develop labelling guidelines for datasets, which might vary according to the needs and structures of different personnel, roles or organizations (Hughes et al., 2014; Reuter, Ludwig, Kaufhold, et al., 2016), our results show that classifiers considering factual and precise relevance criteria (summarized in Table 10.2) are able to outperform those who focus on textual features only but not considering metadata (Habdank et al., 2017).

Novel concept for rapid relevance classification of social media posts in disasters and emergencies (T2). On the other hand, as a means of reducing information overload for emergency managers especially in large-scale emergencies, we proposed a novel concept for rapid relevance classification of social media posts, as summarized in Figure 10.11. Firstly, it comprises data labelling component with an approximate real-time evaluation of the expected classifier quality via online learning to support the user in quality assessment already in the labelling process. While research starts recognizing the value of white-box approaches to increase the users' trust and understandability of (transparent) algorithmic decisions (Kaufhold, Rupp, et al., 2020), this component contributes by making the approximate quality of the classifier transparent already during the labelling phase.

Secondly, it incorporates active learning to reduce the amount of labelled data required to achieve good quality classifiers in time-constraint settings, such as emergencies. Due to the promising results, active learning should be considered in analytical frameworks and the labelling and evaluation of more datasets is required to identify criteria determining the performance of active learning in the domains of emergency management and social media. Furthermore, a combination of active learning and domain adaptation seems worth researching too allow the use of already labelled data from similar past events and actively learn from new event data (M. Imran et al., 2018).

Finally, it allows the retrospective retraining of the classifier based misclassified instances. Despite this functionality allows the user to further tune the performance of the classifier, its effect is subject to evaluation in future. Again, we envision that a white-box approach, helping the user to understand why a specific message was classified as relevant or irrelevant, would help to identify patterns to increase the users' competence in training classifiers appropriate and according to their needs, which might result in gathering more actionable information from social media (Zade et al., 2018).

While this concept is a first theoretical contribution, further research from human-computer interaction, machine learning and social media analytics has the potential to distil this concept into a human-centred and white-box analytical framework for rapid relevance labelling and classification.

10.6.4 Limitations and Future work

This paper is subject to limitations and future research potentials, alongside those outlined in the two chapters before. Firstly, our evaluations are based on two dataset of the 2013 European floods, which was the biggest disaster event in Germany in terms of social media use (Reuter, Ludwig, Kaufhold, et al., 2015), and the 2016 BASF SE incident. Future work should examine the approach with further datasets and in different scenarios, such as small-scale emergencies (e.g., crime, house fires or multiple collisions), major events (e.g., violent demonstrations) or large-scale disasters (e.g., hurricanes, terrorist attacks or wildfires). Furthermore, the datasets only contained posts from Twitter. However, our architecture allows the creation and analysis of cross-source datasets, which could be examined in the future. Our approach, combining active and online learning, has further research potentials on both algorithmic and user interface level.

With regard to the algorithmic perspective, further offline and online machine learning algorithms as well as their impact on the classifier's quality and the active unit, considering the time effort of labelling, should be examined. For instance, Wang et al. (2009) propose approaches for incremental Random Forest algorithms. Furthermore, it was observed that *Stochastic Gradient Descent* classifiers yield similarly good results as do IBk classifiers. It is also plausible that HT and iNB classifiers achieved bad classification results due to an inability to deal with the large number of features. As our evaluation revealed, well-performing classifiers are still possible by only incorporating metadata-based features, omitting the content or words of a post. A feature vector based on metadata of posts could improve the success probability of both incremental approaches. Furthermore, it is possible to use active learning approaches beyond uncertainty sampling, as indicated by the work of Párraga Niebla et al. (2011). The works from Caragea et al.

(2016) and Nguyen et al. (2016) also show that *Convolutional Neuronal Networks* (CNN) are able to achieve excellent classification results. Currently, the SMO is limited to the processing of the posts' content and metadata but CNNs are also able to assist in image processing to assess the damage caused by a disaster, as demonstrated by Nguyen et al. (2017).

From a user interface perspective, a system such as the Social Media Observatory (SMO) allows to evaluate our algorithmic approach empirically from the perspective of human-computer interaction (HCI) (Wobbrock & Kientz, 2016). The deployment of SMO for emergency services, such as fire services and police, in experimental or real-world settings would allow for the evaluation of several aspects: a) the hedonic and pragmatic quality criteria of the overall interface (Hassenzahl et al., 2003), b) the quality of the labelling interface (Figure 10.6), also with regard to the displayed relevance criteria for labelling assistance, as well as c) the perceived usefulness of both the proactive real-time evaluation of the classifier's quality during labelling and reactive feedback classification to correct misclassifications (Figure 10.8). Although the SMO is able to show data in both static list and visualization views separately, we envision an customizable, interactive, integrated and real-time visual analytics dashboard (Keim et al., 2008; Onorati et al., 2018) to enhance emergency services' situational awareness. Here, the application of relevance classification, including the proposed feedback classification approach, could help to mitigate information overload. Based on this, its integration with other means of reducing information overload, such as alert generation (Adam et al., 2012; Avvenuti et al., 2014; Cameron et al., 2012; Párraga Niebla et al., 2011; Purohit et al., 2018; Reuter, Amelunxen, et al., 2016), event summarization (M.-T. Nguyen et al., 2015; Rudra et al., 2015; Rudra, Goyal, et al., 2018) and precise search keyword selection (Abel, Hauff, & Stronkman, 2012; Johansson et al., 2012; Vieweg, 2012b), should be examined.

At its current state, the overall architecture is optimized for a single system since the focus of this study was on algorithmic feasibility. Despite it works performant on our single-node server, in terms of scalability, it would be possible in future work to implement multiple processing instances for parallelization and database "sharding", i.e., using a distributed database. On the other hand, data access is limited by social media tokens, e.g. Twitter allows 450 queries per 15 minutes on their search API, whereof each query contains a maximum of 100 messages (up to 45.000 messages per 15 minutes effectively) (Reuter & Scholl, 2014). Using multiple tokens for one application must be carefully checked against different social media terms of services.

Design and Evaluation of a Mobile Crisis App

11

Emergencies threaten human lives and overall societal continuity, whether or not the crises and disasters are induced by nature, such as earthquakes, floods and hurricanes, or by human beings, such as accidents, terror attacks and uprisings. In such situations, not only do citizens demand information about the damage and safe behaviour, but emergency services also require high quality information to improve situational awareness. For this purpose, there are currently two kinds of apps available: General-purpose apps, such as Facebook Safety Check or Twitter Alerts, already integrate safety features. Specific crisis apps, such as Katwarn in Germany or FEMA in the US, provide information on how to behave before, during and after emergencies, and capabilities for reporting incidents or receiving disaster warnings. In this paper, we analyse authorities' and citizens' information demands and features of crisis apps. Moreover, we present the concept, implementation and evaluation of a crisis app for incident reporting and bidirectional communication between authorities and citizens. Using the app, citizens may (1) report incidents by providing a category, description, location and multimedia files and (2) receive broadcasts and responses from authorities. Finally, we outline features, requirements and contextual factors for incident reporting and bidirectional communication via mobile app.

This chapter has been published as the conference article "112.social: Design and Evaluation of a Mobile Crisis App for Bidirectional Communication between Emergency Services and Citizens" at the European Conference on Information Systems (ECIS) by Marc-André Kaufhold, Nicola Rupp, Christian Reuter, Christoph Amelunxen, and Massimo Cristaldi (Kaufhold et al., 2018).

M.-A. Kaufhold, *Information Refinement Technologies for Crisis Informatics*, https://doi.org/10.1007/978-3-658-33341-6_11

273

11.1   Introduction

Emergencies, crises and disasters threaten human lives, interfere with societal continuity and induce monetary damages all over the world. Some disaster databases and studies indicate that the frequency and intensity of natural disasters, such as the Tōhoku earthquake and tsunami in 2011, the European floods in 2013, or the hurricanes Harvey and Irma in 2017, have increased over the last decades (Eshghi & Larson, 2008; Munich Re, 2017, p. 53). Moreover, the number of man-made disasters and casualties by terrorism is increasing worldwide (Coleman, 2006; Giuliani, 2016), such as the November 2015 Paris attacks, the 2016 Munich shopping mall shooting or the 2017 London Bridge attack. However, also small occasions, such as car accidents or fires, are emergencies that have to be considered in this context.

Besides social media, which meanwhile play an important role in informing the population (Reuter & Spielhofer, 2017) and acquiring situational awareness (Reuter, Ludwig, Kaufhold, et al., 2016), mobile crisis apps can support the information needs of both emergency services and citizens. While emergency services are interested in situational updates, multimedia files and public mood (Reuter, Ludwig, Kaufhold, et al., 2016), citizens demand instructing and orientation information (Coombs, 2009; Nilges et al., 2009). With *crisis apps,* we refer to mobile apps providing specific functionality needed during crises, emergencies or disasters, such as Katwarn (Meissen et al., 2014) or FEMA (Bachmann et al., 2015). These provide information on how to behave before, during and after emergencies, as well as capabilities for reporting incidents or receiving disaster warnings (Reuter, Kaufhold, Leopold, et al., 2017b). However, although some crisis apps support the reporting of incidents (Groneberg et al., 2017), none allows the establishment of bidirectional communication threads, enabling a dynamic and timely request and exchange of multimedia files and situation updates between authorities and citizens across different phases of the emergency management cycle. Thus, we strive for three contributions: First is the development of a novel mobile crisis app for bidirectional communication among authorities and citizens (C1). Furthermore, we seek insights into two research questions: What are features and requirements for the successful reporting of incidents using a mobile app concept (C2)? What are contextual factors for the successful establishment of bidirectional communication between authorities and citizens (C3)?

The paper is structured as follows: Firstly, we present related work on authorities' and citizens' information demands during emergencies, crises and disasters as well as existing crisis apps supporting crisis response (section 11.2). Secondly, based on the design science research paradigm, we outline the requirements analysis, development, features and implementation of the mobile app 112.social (section 11.3). Thirdly, we present the evaluation of 112.social during multiple practices and field trials (section 11.4). Finally, the paper concludes with a summary, discussion and outlook (section 11.5).

11.2   Related Work and Comparison of Crisis Apps

About 2.32 billion people were using smartphones worldwide in 2017, a number which is estimated to increase to 2.87 billion by 2020 (Statista, 2017c). Citizens use smartphone apps to read and share information in different social media, such as social networks (Facebook), microblogging services (Twitter), multimedia sharing platforms (YouTube) or instant messengers (WhatsApp). These social media are not only used in everyday life, but also to stay informed during emergencies, crises or disasters (Eismann et al., 2016). By using different kinds of social media during crises, people often publish information of some value to the emergency services, such as eyewitness reports in real-time (Reuter, Hughes, et al., 2018). For the conceptual framing of this paper, we refer to the crisis communication matrix (Reuter & Kaufhold, 2018) which distinguishes authorities and citizens, both as sender and receiver, to derive four communication flows: crisis communication (A2C), self-help communities (C2C), interorganizational crisis management (A2A), and integration of citizen-generated content (C2A). Since this paper examines bidirectional communication between authorities and citizens using crisis apps, we focus on the communication flows of C2A and A2C.

11.2.1 Authorities' Demand for Citizen-Generated Content (C2A)

The multidisciplinary research field of crisis informatics has revealed interesting and important real-world uses for information and communication technology (ICT) during crises (Hagar, 2007; Palen & Anderson, 2016). Fischer, Posegga and Fischbach (2016) indicate that the communication between authorities, such as emergency services, and citizens faces technological, organisational and social barriers across the mitigation, preparedness, response, and recovery phases of a crisis. In the beginning of and during an emergency, it is vital for authorities to get as much emergency-relevant information as possible to obtain and maintain a situational overview, support decision making and carry out effective crisis communication (Coombs, 2014; Vieweg et al., 2010). Besides "getting the right information to the right person at the right time" (Hagar, 2010), emergency services have to deal with information production by diverse actors and agencies using informal and formal channels, conflicting information and the credibility of different sources.

The rise of social media and distribution of smartphones empowered the role of citizens as active participants before, during and after emergencies, which is also of use for emergency services (Reuter, Hughes, et al., 2018). The potential of benefitting from citizen-generated content lies within illustrating problematic situations through eyewitness reports, photos or videos taken with mobile phones (Alam et al., 2018; Olteanu et al., 2015). A survey with 761 emergency services workers from 32 European countries revealed situational updates (73%), photos (67%), public mood (62%), videos (59%) and specific information (56%) to be important types of information derived from social media (Reuter, Ludwig, Kaufhold, et al., 2016). To make use of such content, the field of social media analytics aims to combine, extend, and adapt methods for the analysis of social media data across the steps of discovery, collection, preparation, and analysis (Stieglitz, Mirbabaie, Ross, et al., 2018). Accordingly, several contributions aim at extracting situational awareness from citizen-generated content, highlighting the importance of geographic coordinates and timely information (M. Imran et al., 2015; Moi et al., 2015).

11.2.2 Citizens' Demand for Crisis Communication (A2C)

The citizens' needs for information differ in the phases of the emergency management cycle (Fischer et al., 2016). In the preparation phase, information for sensitisation and crisis preparation are necessary. For the implementation of preventive measures, a sense of danger is required and warnings have to be delivered urgently (Geenen, 2009; Volgger et al., 2006). This comprises information about existing and potential hazards, their probabilities, possible consequences, as well as *instructing information*, i.e. plans and instructions with best practices for emergencies (Coombs, 2009). Predictable crises have to be communicated on every available channel as early as possible, making sure that as many people as possible get the information (Volgger et al., 2006). Although social media and crisis apps offer rich opportunities (Reuter, Kaufhold, Leopold, et al., 2017b), warnings of disasters are mainly distributed through mass media (TV, radio), sirens or multichannel warning systems, where, for example, SMS, email, and RSS feeds can be combined (Klafft, 2013).

In the response phase, consistent and transparent information supply is necessary, i.e. *orientation information* for affected people to assess the situation as best as possible is required, such as weather warnings (Nilges et al., 2009) or other people's safety (Wade, 2012). Information disseminated to the public should contain basic information, e.g. recommendations for action, site-specific information, e.g. the expected duration of a local power breakdown, and configuration-specific information, e.g. for individuals with special needs (Reuter, 2014a). In the course of communication between citizens, information is also distributed, discussed, and interpreted, showing that the contact with our fellows significantly co-determines our behaviour during a crisis situation (Kaufhold & Reuter, 2016). This process of *sensemaking* can be understood as a "steady process of gaining knowledge through the transformation and integration of new information into cognitive schemata" (Mirbabaie & Zapatka, 2017), which is supported by different roles, such as information starters, amplifiers and transmitters.

11.2.3 Crisis Apps and Contextual Factors for Bidirectional Communication

Mobile crisis apps are not uncommon nowadays: Reuter & Ludwig (2013) compared 25 apps which support *different functionalities* such as the interactive display of crises on maps, sharing of information, collection of eyewitness reports, or live broadcasts by authorities or infrastructure providers. Some apps are specialized for *different types of disasters* such as earthquakes, epidemics, floods, storms and wildfires, or they provide instructing information on how to act during emergencies. Karl, Rother and Nestler (2015) compared characteristics and purposes of crisis apps, including alerts for situation awareness, sending alarms and asking for help, behavioural instructions and support, and the use by volunteers and trained first aiders. Another study compared warning apps in terms of location-based warnings, warning maps, general disaster information, information sharing, and disaster reporting (Reuter, Kaufhold, Leopold, et al., 2017b). Furthermore, Kotthaus, Ludwig and Pipek (2016) compared user comments from app stores on Katwarn and NINA, concluding that warning messages "lack in quality and timing", "malfunctions lead to high amount of user complaints" and "both apps [do not] aim at addressing users [individually]".

Finally, a research report from the German project SMARTER analysed 59 international crisis apps from 14 countries, categorising their functions into information (i.e. push notifications, maps, news, organisational information), communication (i.e. social media integration, direct 112 emergency calls, contact directory, "I'm safe" notification), preparation (i.e. emergency planning, behavioural tips, descriptions of dangers, trainings) and other (i.e. language change, app rating, feedback) (Groneberg et al., 2017). These include mobile apps that support either bidirectional communication or the reporting of incidents: The MoRep app allows relief forces to report incidents to the headquarters by providing a title, description, photo and geocoordinates of the incident (Reuter, Ludwig, & Pipek, 2016). In addition to these metadata, the Ushahidi app allows the selection of a category, date and time, as well as a news URL (Ludwig et al., 2014). Similarly, the FEMA app allows citizens to report disaster photos, enriched with geocoordinates, to authorities (Bachmann et al., 2015). While MoRep focuses inter-organisational reporting, all three apps do not allow the establishment of bidirectional communication threads. Contrarily, the Hands2Help mobile app supports the bidirectional communication between authorities and citizens; however, its focus does not lie on reporting of incidents and the dynamic exchange of situation-related information, but on coordination of demands from authorities and offers by citizen volunteers (Sackmann et al., 2014).

Groneberg et al. (2017) conclude that no documented experiences or scientific surveys about the actual use of smartphones or crisis apps are currently available, and that existing scientific publications focus on the development of smartphone apps on a conceptual level or the integration of social media into crisis and disaster response (Al-Akkad & Raffelsberger, 2014). Furthermore, literature indicates a variety of contextual factors that are worth examining for the successful establishment of bidirectional communication via crisis apps. Based on two quantitative studies, a snowball-based survey in Europe and representative study in Germany, Reuter et al. (2017b) found a low interest in installing a crisis app (16%), whereby 11% use weather apps and 6% warning apps, and suggest further research on the promotion and motivation of using crisis apps. According to an extended representative study in Germany on citizens' perception about social media and crisis apps (Reuter, Kaufhold, Spielhofer, et al., 2017), 57% expect to receive emergency warnings, 51% to get advice on how to stay safe, and 42% to share information with emergency services via crisis apps in the future. However, research indicates that the adoption of new technology by emergency services faces barriers such as lack of sufficient staff, skills, guidance and policy documents and, in terms of integrating citizen-generated content, the issues of trustworthiness and information overload (Hughes & Palen, 2014; Plotnick & Hiltz, 2016).

11.2.4 Research Gap

The presented studies indicate a growing body of mobile apps designed for crises (Groneberg et al., 2017; Reuter & Ludwig, 2013). Although crisis apps support multiple scenarios and provide a vast amount of useful functionality, only a small portion allows the reporting of incidents and none supports a bidirectional communication between authorities and citizens facilitating the dynamic and timely request and exchange of situation-related information for mutual situation awareness (Hagar, 2010; Karl et al., 2015). Although not explicitly focused on the technology of crisis apps, previous research indicates a variety of requirements (i.e. reporting of situational updates, geolocation, multimedia files and text for emergency services, and the dissemination of instructing and orientation information for citizens), but also contextual factors (i.e. the promotion of crisis apps, citizens' use motivation, authorities organisational barriers and quality issues of citizen-generated content) that have to be considered in the design of a respective crisis app. Hence, the 112.social concept intends to assist in the structured reporting of and bidirectional communication during incidents (C1). Furthermore, there is a lack of scientific studies on requirements elicitation, documentation and evaluation of the actual use of crisis apps (Groneberg et al., 2017). While there is a large-scale study on the distribution and intended use cases of crisis apps within the population (Reuter, Kaufhold, Leopold, et al., 2017b) which backs up the importance of researching the impact and potentials of crisis apps, our study aims to contribute insights from the evaluation of 112.social. Thus, we investigate features and requirements for the successful reporting of incidents using a mobile app concept (C2) as well as contextual factors for the successful establishment of bidirectional communication between authorities and citizens (C3).

11.3 Development and Architecture of the Emergency Mobile App

This chapter outlines the overall design approach and the process of requirements engineering, subsequently presenting the development of the mobile app 112.social. The system is connected to an Emergency Service Interface (ESI) and a Processing and Analysis Subsystem (PAS), which are not within the scope of this paper but are mentioned to clarify existing interfaces.

11.3.1   Overall Methodology

One goal of the project was to show the positive impact of gathering, qualifying, mining and routing citizen-generated content on the management of emergencies (EmerGent, 2017), which is realised through requirements analysis, as well as development and evaluation of artefacts for emergency services and citizens. Thus, design science research (DSR) plays a significant role, which is a problem-solving paradigm that "seeks to create innovations that define the ideas, practices, technical capabilities, and products through which the analysis, design, implementation, management and use of information systems can be effectively and efficiently accomplished" (Hevner et al., 2004). According to Hevner (2007), the central process is the building and evaluation of design artefacts and processes to improve an application domain, which comprises people as well as organisational and technical systems (*design cycle*). This design process should be grounded in and contribute to the knowledge base (*rigor cycle*). Since insisting "that all design research must be grounded on descriptive theories is unrealistic and even harmful to the field", as Hevner (2007) suggests, we integrated "several different sources of ideas for the grounding of design science research", such as requirements based on literature findings, existing artefacts and the inquiry of domain experts. Our intended contributions to the knowledge base are reflected by C1-C3 (cf. section 11.2.4). Furthermore, field testing is needed to raise requirements for the development or refinement of technology that supports emergency services in their emergency management practice (*relevance cycle*), and to decide whether additional iterations of this cycle are needed.

Table 11.1   Empirical pre-studies and workshops

Title and Focus	Year	Participants
Interviews I: Social media in emergencies (Reuter et al., 2015)	2014	11
Workshops I-III: End-user advisory board (ES) (Gizikis et al., 2017; O'Brien et al., 2016)	2014/15/17	16/18/15
Surveys I-III: Perception of emergency services and citizens (Reuter et al., 2016; Reuter et al., 2017b; Reuter and Spielhofer, 2017)	2015/16/17	761/1,034/473
Evaluation I: First round of system evaluation (Reuter, Amelunxen, et al., 2016)	2016	12

To identify requirements, we employed a requirements analysis process: (1) Scenarios and use cases from real-life operations were chosen to be illustrated and analysed; (2) these were presented to end users, development teams and experts (workshops) to understand different approaches, establish a common understanding and allow interaction with each other; (3) and to involve a broader community, we conducted online survey to collect data. All interventions (Table 11.1) were supported by prior literature reviews, including a review of mobile app concepts (Ludwig et al., 2014), to consider the state of the art and to inform the application of appropriate methodologies. The complete requirements analysis process and its results are documented in a project deliverable (Akerkar et al., 2016). Since a fine-grained specification of requirements would exceed the scope of this paper, all elicited requirements were aggregated to high-level abstractions in Table 11.2.

Table 11.2   Abstraction of requirements comprising 112.social, ESI and underlying architecture

Category	Description
Architecture	Easy-to-use, high available, maintainable, privacy-respecting, secure, stable and scalable standalone solution.
Communication	Reception, publication, response and broadcast of messages with multimedia (audio, photo, video) between authorities and citizens.
Processing	Cross-platform gathering, enrichment, relevance and quality assessment of citizen activities and alert generation.
Tailorability	Filtering of results in terms of geolocation, keywords, relevancy (mining) and information quality.
Visualization	Display of generated alerts on a list and map view, and classification of alerts according to the interoperable Common Alerting Protocol (OASIS, 2010).

11.3.2   Functionalities of the Emergency App

The architecture supports multiple information flows between authorities (A) and citizens (C). Firstly, citizens may use 112.social to forward so-called "app alerts" to the ESI (C2A). Secondly, authorities may use the ESI to disseminate messages to 112.social users via broadcast or direct reply (A2C). In this way, chat-based communication threads are established across all phases of the emergency management cycle: For instance, if a citizen reports an early warning for an emergency or a witnessed incident (preparation), authorities may disseminate behavioural tips, such as instructing or orientation information, or request additional information or updates on demand to improve their situational awareness (response), allowing citizens to reply accordingly. Finally, citizens may report damages after the incident (recovery). Figure 11.1 shows the first (I-III) and second version (IV-VI) of 112.social, which was developed after the first trial at the OSCE (section 11.4.1.1).

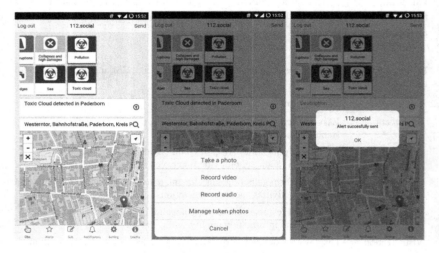

Figure 11.1   112.social-1: (I) start screen, (II) multimedia dialogue, (III) alert sent confirmation

In the upper area of the start screen (I, IV), the user first chooses the main category of the incident, such as ambulance service, police, fire, accident or severe weather, and a subcategory (i.e. nature, vehicle or building for the fire category). In this way, for instance, the public-safety answering point (PSAP) would know right away what kind of help is needed. After category selection, the user may use a text field to provide additional information. On the right side of the text field, a blue-arrow icon opens the dialogue for adding multimedia files, such as photo, video or audio files (II). It is important to stress that any kind of multimedia file have to be taken in the moment the user wants to send an app alert, which is why multimedia files cannot be loaded from the phone's memory.

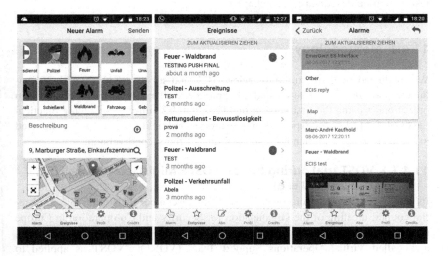

Figure 11.2   112.social-2: (IV) start screen, (V) communication threads and (VI) details

At the bottom, the user may use the map and search box to determine the location of the incident (I, IV). By default, the user's GPS location is attached, and the user cannot select locations that are more than one kilometre away from their own GPS location. These restrictions were implemented to reduce the potential of abuse and allow the precise reporting of incidents in terms of location and time. After everything is set and a multimedia file is added, the so-called "app alert" is sent to the authorities (III). Since the focus was to enable bidirectional communication, a new communication thread is opened with the dissemination of an app alert (C2A) or, alternatively, if the user receives a message from authorities (A2C). A specific tab displays communication threads (V), where unread messages are indicated, and allows the user to view specific threads (VI). In this way, after the app alert has reached the ESI, the operator can choose to send a direct reply or broadcast to every app user in a certain area around the incident scene, for instance, to ask for more or specific details. Upon a reply, the app user receives a push message on his phone and can answer it by using the app again (VI).

11.4   Evaluation of the Emergency Mobile App

The evaluation of the app followed an iterative approach, whose methodology and results will be described in this chapter.

11.4.1 Methodology

We conducted semi-structured interviews (E2) and surveys (E1, E3) with 35 evaluation participants. While E1 had an exploratory character to get feedback from practitioners using the first version of 112.social, E2 aimed at achieving practical feedback from domain experts during exercises and field trials. E3 was conducted to evaluate the design, handling and stability of the second version.

11.4.1.1   VOST trial during the 23rd OSCE Ministerial Council (E1, Dec 2016, N = 6)

In Hamburg, Germany, on the 8th and 9th of December 2016, the 23rd ministerial council meeting of the Organisation for Security and Co-operation in Europe (OSCE) took place. The participants were members of VOST112 (Table 11.3), a so-called Virtual Operations Support Team (VOST): "VOST as applied to emergency management and disaster recovery is an effort to make use of new communication technologies and social media tools, so that a team of trusted agents can lend support via the internet to those on-site who may otherwise be overwhelmed by the volume of data generated during a disaster" (VOSG, 2017).

Table 11.3 OSCE survey participants by occupation, experience (multiple selection) and device

Category	Data
Occupation	Fire Brigade (2), Paramedic (1), student (1), none (2)
Experience	Voluntary Fire Brigade (5), Accident Ambulance (3), Emergency Service (1), none (1)
Devices	Samsung Galaxy S7 (2), Apple iPhone 6 S (1), Moto G1 (1) OnePlus One (1), Samsung Galaxy S5 (1)

During the council, 13 members of VOST112 used 112.social to send pictures and information to the ESI, which was watched by a silver level member of the fire department (FD) Hamburg. Additionally, the members used a dedicated channel in the Telegram messenger to report bugs and errors. After the two days of use, a survey was conducted with six VOST112 members. Besides personal details, organisation, experience and device, the survey asked whether the participant would (1) use or even (2) recommend 112.social to other citizens, (3) how they would classify the importance of A2C and C2A communication, sending multimedia content and simultaneous sharing of information on own social media channels, and (4) if the participant would like to provide additional feedback.

11.4.1.2 Emergency Service Demonstrations and Field Trials (E2, Jan—Jun 2017, N = 21)

Firstly, during a live system demonstration, the participants interacted with the system, while in a paper-based demonstration, the interviewer introduced prepared screenshots of the system and explained its functionality. One survey was conducted in 2017 by the Scientific and Research Centre for Fire Protection— National Research Institute (CNBOP) in Poland, as well as further two interviews with members of volunteer fire departments (FD) in Germany. Secondly, the integrated system was tested at a convention in Salzburg, Austria. During the live simulation, a video of an incident (a fire) was shown, and the audience of citizens was asked to participate by using 112.social. They used the video for taking pictures of the incident scene, sending them along other valuable information to a simulated command and control (C&C) room which, among others, was manned with an incident commander (gold level firefighter, FD Dortmund) and a social media manager (bronze level firefighter, FD Ljubljana) who used the ESI to get incident information from 112.social. Finally, for longer-lasting testing periods in real-world scenarios, three field trials were conducted. The field trial of FD Dortmund was in March/April 2017, and the one of FD Hamburg took place in April/May 2017, both lasting four weeks. Another one was conducted in July 2017 with FD Hamburg during the G20 event. During these trials, the system was used by the departments of public relations and strategic planning as well as by the head of dispatchers.

Overall, we conducted 21 semi-structured interviews with emergency services (Table 11.4). In terms of personal details, the interview guideline asked for the type of organisation, main role, command level, working years, age, gender, and country. Additionally, organisational details such as the role of social media and organisational barriers were asked. The core guidelines consisted of seven questions on the (1) first impression, (2) importance of functionality for their job, (3) evaluation of "app alerts", (4) "social media alerts" and (5) "information quality", as well as the (6) most useful functionality or (7) desired functionality in the future. Interviews were audio-recorded and transcribed for further analysis. In

Table 11.4   Demonstration, field trials and workshop participants (I1-I22)

Category	Data
Roles	crew (10), head (1), incident commander (8), other (6), press (5), PSAP operator (1), PSAP supervisor (5), section leader (3)
Level	gold (3), silver (9), bronze (8), none (1)
Age	20–29 (2), 30–39 (10), 40–49 (6), 50–59 (3)
Gender	male (18), female (3)
Country	Germany (13), Poland (7), Slovenia (1)
Evaluation type	Field trial (10), live and paper-based demonstration (9), workshop exercise (2)

our subsequent analysis, we employed open coding (Strauss & Corbin, 1998), i.e., gathering data into approximate categories to reflect the issues raised by respondents based on repeated readings of the data and its organisation into similar statements.

11.4.1.3 Citizen Treasure Hunt for Functionality and Usability Evaluation (E3, Jul 2017, N = 8)

During the field trials, the app users were mostly firefighters. However, the basic idea of 112.social was to have citizens as users. Therefore, a mixed group of students and researchers from different fields tried the app during a treasure hunt on the campus of the University of Paderborn. None of these participants had a background in the field of security, qualifying them as target users, and only two of them had used the app before. The others did get a very short introduction to the general idea and approach of the app, as it could be composed in a description in the app store. As the primary focus was the functionality and usability for untrained citizens, most of them used the same device with Android (Table 11.5).

Table 11.5   Treasure hunt survey participants with occupation and device information

Category	Data
Occupation	Research assistant (4), mechanical engineer (1), student (2), none mentioned (1)
Devices	Moto G1 (2), Moto G4 (4), OnePlus 3 T (1), Samsung Galaxy S8 (1)

The treasure hunt consisted of eight "treasures" that contained answers to previously asked questions. These treasures were located on different spots of the campus and, after finding one, the participants had to report them via 112.social using every possible media file (audio, pictures and videos) at least once (C2A communication). One researcher posed as the PSAP using the ESI. After reporting a found treasure, the PSAP sent requests for further information and gave new instructions for finding the next treasure (A2C communication). In this way, the interplay of C2A and A2C communication were tested. After the treasure hunt, the citizens were asked to answer a survey. Besides personal details, organisation, experience and device, the survey asked about the (1) first impression, (2) design, (3) handling, (4) quality of features' implementation, (5) problems during the treasure hunt, (6) importance of the functionalities and (7) communication channels, whether the participant would (8) use or even (9) recommend the app to friends and, finally, (10) provide additional comments, suggestions and wishes concerning the app.

11.4.2 Results

In the following chapter, we present the results of the individual evaluation events which will then be discussed in an integrated manner in the discussion and conclusion chapter.

11.4.2.1 VOST Trial during the 23rd OSCE Ministerial Council

Issues regarding the requirements of a GPS signal and internet connection. The use of 112.social during the 23rd OSCE ministerial council was the first field trial of the app within the scope of the project. In the first version, it was necessary to sign in anew for every post (e.g. via Facebook, Google+ or Twitter), which was criticised as time consuming by the users and, consequently, fixed for the second version. One of the main errors related to the fact that the app only works with GPS, since the idea is to send the exact location of the user. This was an issue if the users were inside a building weakening the GPS signal. If the signal was not received properly, the app sometimes crashed, which made a restart with a repeated sign-in necessary. Another main issue was the fact that the app cannot work without an internet connection, thus not allowing to create an app alert and look for an internet connection afterwards. The users made clear that this is necessary for a future version as it is still common to lose the internet connection every now and then. The users would also have liked to be able to use saved pictures from their phones gallery to create an app alert. One user stated that getting the phone's camera ready was fast but getting the app ready took more time and thus, relevant information could get lost. However, the use of old pictures was prohibited by design to keep the risk of false information to a minimum and ensure the currency and authenticity of the information. In summary, all participants said that they would use the app as a citizen and recommend it to other citizens.

Table 11.6 Refined categorisation of emergencies as implemented in version two of 112.social

Rescue	Police	Fire	Crash	Weather
Unconsciousness	Crash	Forest	Car / Bike	Blocked road
Not breathing	Riot	Car	Ship / Boat	Flooded cellar
Polytrauma	Demonstration	Building	Train	Fallen tree
Severe pain	Burglary		Plane	
Heart attack / Stoke	Violence			
Birth	Shooting			
Other	Other			

Features and refinement of the categorisation of emergencies. Essentially, the users liked the basic idea of having a map, a field for typing in information, and the possibility to send multimedia files, such as audio, pictures and videos. Moreover, they were satisfied with the basic design and usability, but disliked the predefined categories. Due to the developers' location in Italy, categories like "earthquake" and "eruptions" were prevalent. In close relation with the end-user advisory board, we developed new categories which were more fitting to the targeted areas for the field trials in Germany. Our end-users decided that eight categories with subcategories were necessary, whereof the first five are depicted above (Table 11.6). According to our end-users, these five categories and subcategories were the most common. Three more seldom categories are: explosion, collapse and CBRN (hazardous material: chemical, biological, radiological and nuclear). These were perceived as important as well, but no subcategories were created. Since it was important to use icons for an intuitive understanding of categories, we worked in close cooperation with our end-users and came up with one icon for each category and subcategory (Figure 11.2., IV).

11.4.2.2 Fire Departments' Live Demonstration and Field Trial Evaluation

Importance and usefulness of app functionality. Five participants valued the C2A flow (Table 11.7), representing "app alerts", due to photos and better situational assessment (I6, I7, I20, I21): "It will help to determine an exact place of an event. Sometimes it is difficult to get to a place, even if the location has been given. Citizens using the app could help with that by giving some clues. Moreover, the photos from the scene would help a lot to estimate the current situation" (I3). Three participants valued the C2A flow on a more general level: "It's an additional way of contacting emergency services and in situations when lives are endangered all ways are welcome and increase the possibility of helping a victim" (I5). It has the potential of an information advantage: "Information" before the control room receives them or before the personnel is on the ground. Information can be received which otherwise would have to be manually searched for" (I14). Thus, in everyday life, "the information acquisition is most useful" (I17). On the other hand, two participants highlighted the relevance of A2C communication (e.g., sending a message or broadcast from ESI to 112.social): "Being in the 'hot zone' I am receiving a proper message directly on my smartphone, at least I will consider that it is serious, and I will follow the instructions" (I1). Thus, "it may be helpful to tell people what to do in emergency" and "they may feel comforted as they know that they are not alone" (I6).

Table 11.7 Indicated importance of information flows and benefit of functionality

Importance	max (4)	high (3)	low (2)	min (1)	Ø
C2A	4 (19%)	11 (52%)	4 (19%)	2 (10%)	2.81
A2C	8 (38%)	6 (29%)	4 (19%)	3 (14%)	2.90
Benefit	**huge (3)**	**moderate (2)**	**small (1)**	**none (0)**	Ø
App alerts	11 (52%)	6 (29%)	3 (14%)	1 (5%)	2.29

In an overall feedback session after the workshop, it was stated that by using the system, it was possible to raise the alert level quicker than usual. A second dispatch was possible only one and a half minutes after the first dispatch, seven minutes before the firefighters would have arrived at the scene. Especially the information coming in via 112.social was perceived as important, since it was considered to be coming from a safe source comparable to a 112 call. The social media manager also used the ESI to get in touch with 112.social users, asking for more details or giving advice how to behave best.

Relevance of multimedia files. In total, 11 participants indicated huge, six moderate, three a small, and just one no benefits at all regarding the "app alerts" functionality. The good access to smartphones (I1) allows such a reporting app to be a fundamental alternative to 112 calls (I8, I19) and thus to reduce the number of 112 calls (I9). It was perceived to be useful (I4, I5, I18, I20) since it encourages users to upload multimedia files such as photos (I3, I11-13), allows a quick classification of information due to the CAP categories (I8), a direct contact to citizens (I3) and quick response if direct exchange is required (I6, I11): "Direct contact between a citizen and authorities may be crucial in estimating the type and scope of an emergency, e.g., fire. Especially if a person sends a photo. Having several photos may help with deciding how many people and which equipment to send" (I3). According to I13, the deployment of drones would require ~60 minutes, emphasizing the importance of a quick citizens' response.

Limitations of individual processing and quantification of app alerts. However, despite the positive attitude, many reservations about the app were discussed. First, given a broad distribution of the app, the processing of individual app alerts binds personal resources (I2, I16). Thus, it would not be possible to process them individually and always react directly (I9), but multiple alerts at a certain location could be a useful indicator of an exceptional event (I15). However, quantification is not trivial (I16): "On Saturdays in the city, you can expect more than one message. In case of a big fire, you always have several calls. Of course, this is different in rural or uninhabited places [since the] distribution of the app is different" (I15). Considering that apps already exist, one user questioned the distribution and motivation of use: "And there are a lot of apps. How could we make people install and use this particular one?" (I4).

Credibility check mechanisms and selective user groups. Although doubts about the users' credibility and quality of information were expressed (I4, I15, I17), a high quality of information could be received from credible or trusted users (I15, I16). Thus, several options were discussed: To check the identity, for instance, via registration process (I15), to sanction or block users on misuse (I16, I19, I21) or to distribute the app to dedicated user groups only (I12, I17), e.g., to qualified personnel such as authorities (I11), VOST (I8), or volunteer fire brigades (I15, I18): "Expert groups, trusted people, THW relatives, potential [112.social] app users, etc.—would be a high-quality group of users" (I10).

Improving the communication between emergency services and citizens. In the interplay of ESI and 112.social, participants emphasised the need to confirm the reception of messages, e.g., if the 112.social user sends an app alert to an ESI operator (I1) or if an ESI operator sends a broadcast or direct message to 112.social users (I9). Moreover, using ESI, each incoming app alert was displayed as an individual entity, regardless whether it was a completely new alert or a reply to an already ongoing communication thread: "Provide a history of the operators own messages, and show the different communication threads with individual users" (I9). It furthermore should be easily visible whether someone "already replied to the message" (I9). Also, one participant emphasised that the system should support more categories since "the reporting of injuries and deaths should have top priority" (I9) and suggests implementing this information into the CAP protocol and highlighting it appropriately in the interface.

11.4.2.3 Citizen Treasure Hunt for Functionality and Usability Evaluation

Issues in terms of design, handling and usability. More than 90% of the users rated the first impression as "neutral" and, concerning app design, the answers ranged from bad to very good, with most participants classifying it as "good" (Table 11.8). Addressing the overall handling of the app, more than 60% voted "bad". The analysis of detailed feedback provided indications for the negative feedback on the overall handling: Firstly, the users criticised that the button for adding multimedia files was too small, which was especially a problem when the app was used for the first time. Secondly, they desired the possibility to send text only, since it is currently required to add a multimedia file. Considering a case where emergency services request specific information that cannot be documented by multimedia files, this design decision would impair the citizens' response time. Furthermore, the push notifications were partly seen as confusing and the users found it difficult to respond to those notifications. Finally, a common problem was that the keyword overlapped with the text field, so that some users were not able to see what they were typing at all.

Technical problems in terms of connection, functionality, performance and stability. Unfortunately, the connectivity on the campus was erratic. This made the app crash or slowed down the sharing of multimedia files a lot. The users were missing some resilience of the app, being able to pick up the point in the process where they had lost the connection. Instead of this they had to sign in again after every loss of connection. Another problem was the GPS since the signal was not available inside a building and also sometimes lost even outside. The response time between 5 and 20 seconds after clicking the button for adding a multimedia file was unsatisfying for the users. In summary, as depicted in Table 11.9, the citizens stated that the general approach of the app was between "important" and "very important". Concerning the kind of multimedia files, the opinions differed. Especially for the audio files, citizens could not see a real use case. Most value was ascribed to the attachment of pictures and GPS tracking.

Table 11.8 Indicated quality of functionalities

Evaluation of functionality	Very good	Good	Bad	Very bad	I don't know
Receiving messages from the headquarter	5 (62.5%)	1 (12.5%)	2 (25%)	0	0
Sending messages to the headquarter	3 (37.5%)	2 (25%)	2 (25%)	0	1 (12.5%)
Allocation of categories and subcategories	2 (25%)	5 (62.5%)	0	1 (12.5%)	0
Adding a description	2 (25%)	3 (37.5%)	3 (37.5%)	0	0
Recording and sending multimedia files	1 (12.5%)	4 (50%)	3 (37.5%)	0	0
GPS tracking by the app	1 (12.5%)	1 (12.5%)	2 (37.5%)	0	4 (50%)

Table 11.9 Indicated importance of functionalities

Importance of functionality	Very high	High	Low	Very low	I don't know
GPS tracking through the app	6 (75%)	2 (25%)	0	0	0
Attaching pictures	5 (62.5%)	2 (25%)	1 (12.5%)	0	0
Adding a description	3 (37.5%)	5 (62.5%)	0	0	0
Attaching videos	3 (37.5%)	3 (37.5%)	1 (12.5%)	1 (12.5%)	0
Classification of the event into categories	2 (25%)	5 (62.5%)	1 (12.5%)	0	0
Attaching audio files	2 (25%)	0	4 (50%)	2 (25%)	0
Setting keywords and areas	1 (12.5%)	2 (25%)	1 (12.5%)	0	4

11.5 Discussion and Conclusion

Summary. Given the widespread use of smartphones in western societies (Reuter, Kaufhold, Leopold, et al., 2017b), mobile applications provide novel opportunities for bidirectional communication between authorities and citizens. In this paper, we reviewed related work on authorities' and citizens' information demands during emergencies and performed an analysis and comparison of existing mobile crisis apps (section 11.2). Based on a requirements analysis, we presented the development of the mobile app 112.social, which intends to support the bidirectional communication between authorities and citizens during emergencies (section 11.3). Finally, we evaluated 112.social during a council in 2016 (E1), several field trials, demonstrations and a workshop with emergency managers and citizens in 2017 (E2), and a technical evaluation of design, handling and stability in 2017 (E3), using semi-structured interviews and surveys (section 11.4).

Development of a novel mobile crisis app for bidirectional communication among authorities and citizens (C1). Based on the conceptual framing of the crisis communication matrix (Reuter & Kaufhold, 2018), our first contribution represents the artefact 112.social for facilitating the bidirectional communication between authorities and citizens. It enables citizens to send original app alerts (i.e. to report an incident or send situational updates), which are defined by a category, subcategory, description, geolocation, multimedia files and, indirectly, time of an incident (C2A). Furthermore, authorities can broadcast messages to a larger audience of citizens or reply to citizens' app alerts (A2C), establishing chat-based communication threads across different phases of the emergency management cycle. Although existing literature documents a large variety of different functionality that is not implemented within this concept (Groneberg et al., 2017; Reuter & Ludwig, 2013), to our best knowledge, none of the existing app concepts realised or was evaluated in terms of a bidirectional communication feature between authorities and citizens alongside a reporting mechanism.

What are features and requirements for the successful reporting of incidents using a mobile app concept (C2)? The presented app was developed to research the feasibility of establishing a bidirectional communication between authorities and citizens using app alerts. Our evaluation participants identified a variety of beneficial use cases. From the authorities' perspective, these app alerts were designed to capture metadata relevant for enhancing situational awareness (Table 11.10, left) and assist decision-making (Vieweg et al., 2010). The proposed *categorisation*, based on the CAP specification, was perceived as a useful feature for quick information assessment, although additional information was requested, for instance, to indicate deaths or injuries. Thus, further discussion of

Table 11.10   Summary of features, requirements (C2) and contextual factors (C3)

Features and requirements (C2)	Contextual factors (C3)
• A quick categorisation of incidents improves reaction time, but the categories have to be adjusted to the needs of authorities. • Structured textual descriptions and multimedia files, such as photos, may supplement authorities' situational assessment. • Precise location information is required for situational awareness, but also tolerance in contexts where the determination of the users' geolocation is impaired. • Timely (multimedia) information is required to assess the relevance of the alert, but also tolerance for adding (older) files that were created outside the app.	• The incident's scale and time affect the interpretation of app alerts. • The connectivity on-site determines the operationality of the app. • Promotion is required to ensure the app's distribution for cases of emergencies. • The users' motivation and technological access are required to increase the potential use. • The credibility of information determines the usefulness of app alerts. • Authorities' barriers regarding law, personnel and time have to be considered.

the categorisation with emergency services seems sensible to encourage the refinement of specifications such as CAP. In line with previous literature (Reuter, Ludwig, Kaufhold, et al., 2016), participants highlighted the relevance of adding *multimedia* files such as photos and videos.

However, during the council (E1) and paperchase (E3) evaluations, feedback from the perspective of citizens regarding the usability of 112.social revealed interesting design trade-offs where too strong regulations for information quality assurance led to user resistance: Firstly, participants were sceptical about the mandatory attachment of at least one media file and desired the option to add multimedia files that were not created with the app. Hence, further studies could investigate authorities' willingness to compromise on the handling of multimedia files. Secondly, the importance of *location* information for situational awareness (M. Imran et al., 2015) was confirmed in this study. However, the requirement of operating 112.social with GPS signal and internet access, although intended to ensure the accuracy of geolocation information, led to regular crashes among the users. This suggests that the preparation of app alerts should be possible without internet connection. Furthermore, infrastructure-less technologies such as off-grid ad-hoc networks could be explored as an opportunity to move information to devices or into zones with established connectivity (Al-Akkad et al., 2014; Alvarez et al., 2016). Furthermore, general issues of the design and handling of 112.social might be addressed with a first use tutorial and the redesign of concerned interface elements.

What are contextual factors for the successful establishment of bidirectional communication between authorities and citizens (C3)? In favour of app alerts, most emergency services' participants highlighted the good smartphone access nowadays as an enabler for bidirectional communication among authorities and citizens. However, contextual factors have to be considered (Table 11.10, right). While app alerts were perceived as an alternative to 112 calls in the future, participants emphasised that the processing of individual app alerts may be too time-consuming in large-*scale* emergencies, which is in line with prior research reporting on the limited resources of emergency services in terms of personnel and time (Hiltz et al., 2014). Furthermore, studies indicate that only 16% of citizens in Europe have downloaded smartphone apps for emergencies, suggesting that there is a need to examine the *promotion* of emergency apps and the users' *motivations* for installing such apps (Reuter, Kaufhold, Leopold, et al., 2017b). Despite varying *technological access* in the population, such a concept may be developed into a worthwhile complement to existing mass media or multichannel warning systems (Klafft, 2013).

Besides findings that comply with observations from previous research, the interviews with emergency services revealed novel insights: multiple app alerts at a certain location were estimated to be a useful indicator of an exceptional event, although it has to be contextualised into different factors such as *population* density of a certain spot, the *time* of the current day or week, available *connectivity*, or specific ongoing events. Furthermore, participants emphasised that the *credibility* of information depends on the credibility of the users distributing them, although by tendency, information provided via such a communication app was perceived more credible than arbitrary social media content. Thus, the idea was preferred to implement credibility check mechanisms and to hand out the app to emergency services, volunteers, VOST or qualified citizens only for mobile reporting (Ludwig et al., 2013). For instance, fire departments could distribute 112.social in voluntary fire brigades or to trusted relatives (Kaufhold & Reuter, 2017).

Limitations and Outlook. Our results suggest the conduction of an additional design cycle (Hevner, 2007) followed by field testing to evaluate the practical relevance and value of 112.social. Although the evaluations were conducted with emergency service staff, VOSTs and a small sample of citizens, large-scale evaluations with citizens based on a more representative sample in exercises, serious games (Link et al., 2014) or real-world settings would allow a more rigorous research contributions in terms of (1) citizens' perceived usability and utility,

(2) emergency services' handling of large numbers of app alerts during large-scale emergencies and (3) the technological maturity and scalability of 112.social. Furthermore, the evaluation was mainly conducted with fire services, limiting the applicability of results to other types of organisations. After implementing the gathered user feedback, further evaluations could examine requirements and specifics of danger prediction and prevention by the police interacting with 112.social users. In future, concepts for keyword-based subscriptions or location-based broadcastings will be implemented and tested. Since structured textual information potentially improves situational awareness (Starbird & Stamberger, 2010), concepts supporting the structuring of information could be examined. Furthermore, future research could consider more enhanced concepts for integrating volunteered geographic information (VGI) into emergency response (de Albuquerque et al., 2016).

Part IV
Conclusion and Outlook

Information Refinement Technologies for Crisis Response

12

In part I of the thesis, information refinement was defined as the process of app-lied practices and used technologies to refine obtained information according to contextual, i.e. event-based, human, organizational, societal, and technological, boundary conditions in order to improve crisis management, response and, as a desired consequence, social wellbeing (section 2.4). Based on the findings of the parts II and III, this chapter proposes an information refinement framework for emergency services (section 12.1). In the following, it will discuss the event-based (section 12.2), organisational (section 12.3), societal (section 12.4), and technological (section 12.5) perspective of the established information refinement framework.

12.1 Overview of the Information Refinement Framework

The information refinement framework combines event-based, organisational, and societal boundary conditions influencing information refinement practices and technologies, which are characterized by information channels, access, content, analysis, filtering, and evaluation. The framework fulfils multiple purposes: *it should raise awareness for existing boundary conditions and technological poten-tials for information refinement in emergency services practice, it can be used to derive analytical features for research-related activities, but it should also serve as a guideline for the design and implementation for objective-oriented informa-tion refinement systems.* The framework was inspired by the goal of transforming the high volume of noisy data into a low volume of rich content that is use-ful to emergency personnel within the EmerGent project (Moi et al., 2015). The involved information processing architecture (chapter 9) comprised the steps of

© The Author(s), under exclusive license to Springer Fachmedien Wiesbaden GmbH, 299
part of Springer Nature 2021
M.-A. Kaufhold, *Information Refinement Technologies for Crisis Informatics*,
https://doi.org/10.1007/978-3-658-33341-6_12

information gathering, mining, quality, and grouping (referred to as alert generator), which gradually *refined* the available information according to emergency services' needs. However, the architecture focused the technological view, not integrating event-based, organisational, and societal boundary conditions.

Furthermore, the information refinement framework was inspired by existing frameworks for social media analytics. For instance, the analytical framework of Fan and Gordon (2014) comprises the steps of *capture, understand*, and *present* while Stieglitz et al. (2018) distinguish the *discovery, tracking, preparation*, and *analysis* of social media content. These are general-purpose frameworks adaptable to diverse domains but are not designed to capture the event-based, organisational, societal, and technological challenges of information refinement in crisis informatics. In order to increase sensitivity to the research field of crisis informatics, the information challenges of a crisis as identified by Hager (2010) were reflected for the design of the framework. Finally, the first four articles of this thesis (chapters 4–7) focused on event-based, organisational, and societal results and the second four articles (chapters 8–11) on technological results regarding information refinement. However, the content of the following chapters is not limited to the findings of this dissertation but also integrates results from the broader scope (outlined in section 3.4) of the underlying projects EmerGent, MAKI, KontiKat, and ATHENE. Furthermore, the identified boundary conditions cannot be considered exhaustive but reflect the findings and research that I traversed during my thesis.

Table 12.1 Boundary conditions influencing the information refinement process

Boundary Conditions		
Event	**Organisation**	**Society**
Dynamics	Mission	Attitudes
Location	Roles	Behaviour
Phase	Collaboration	Roles
Type	Perception	Expectations

Information Refinement					
(1) Channels	**(2) Access**	**(3) Content**	**(4) Analysis**	**(5) Filtering**	**(6) Evaluation**
Features	Tracking	User-generated content	Preprocessing	Backend	Explainability
Purpose	Storage	Metadata	Methods	Frontend	Scope
Interaction					

The information refinement framework is summarized in Table 12.1. The boundary conditions of the crisis, organisation, and society influence the process of information refinement. For the sake of overview, the illustration only shows the *categories* (e.g. for crisis: development, phase, and type) but not the *features* (e.g. for dynamics: geospatial and temporal development) and concrete *conditions* (e.g. for geospatial development: focalised or diffused). The features and conditions of each category will be described in the following chapters.

12.2   Event Perspective

The capabilities and potentials of mobile technologies depend on event-based conditions, such as the phase and type of a disaster. Chapters 4, 7, 8, and 11 contribute findings to the event perspective of information refinement. Furthermore, the identified conditions are summarized in Table 12.2.

Table 12.2   Event-based categories, features, and conditions for information refinement

Category	Features	Conditions
Dynamics	Geospatial development	Focalised, diffused
	Temporal development	Instantaneous, progressive
Location	Capabilities	Incident management, social media management
	Infrastructure	Access, connectivity, distribution
Phase	Emergency management cycle	Mitigation, preparation, response, recovery
	Emotional phases of citizens	Threat phase, heroic phase, honeymoon phase
Type	Natural hazard	Meteorological, hydrological, geophysical, climatological, biological
	Human-induced hazard	accidental, intentional

12.2.1 Characteristics and Types of Crises

For successful crisis management, emergency services must consider the development, and types of crises and their implications for the information space. Chapter 4 provided an overview of case study research in crisis informatics across the last fifteen years, summarizing that social media was used in any major past event (Reuter & Kaufhold, 2018). When characterising the types of crises, research distinguishes hazards of natural (e.g. meteorological, hydrological, geophysical, climatological, or biological) and human-induced (e.g. intentional or accidental) origin, which also result in different kinds of threats or damages (Olteanu et al., 2015). Besides hazard type (Table 12.3), Olteanu et al. (2015) outline the temporal development and geographic spread as substantial characteristics of an event. With regard to temporal development, they distinguish *instantaneous events*, such as earthquakes or shootings, that "do not allow pre-disaster mobilization of workers or pre-impact evacuation of those in danger", and *progressive events*, such as floods or hurricanes, that are "preceded by a warning period" (Olteanu et al., 2015). The geographical spread, then again, is characterizes as either *focalized*, such as a train accident, or *diffused*, such as a large earthquake.

Table 12.3 Hazard categories and sub-categories (left). Reproduced from Olteanu et al. (2015), and categorisation used in 112.social (chapter 11)

Category	Sub-category	Examples	Categorisation of 112.social
Natural	Meteorological	Tornado, hurricane	
	Hydrological	Flood, landslide	
	Geophysical	Earthquake, volcano	
	Climatological	Wildfire, heat/cold wave	
	Biological	Epidemic, infestation	
Human-Induced	Intentional	Shooting, bombing	
	Accidental	Derailment, building collapse	

The importance of characteristics and types of crises was emphasized in the design and evaluation of 112.social mobile app (chapter 11). When citizens use the app for reporting incidents, they have to select a category (e.g. crash, fire, police, rescue, or weather) and subcategory (e.g. for the category fire: building, car, or forest) and attach geographical information in order to describe the incident as precise as possible. The report was stored according to the Common Alerting Protocol (CAP) specification, which is "an open, non-proprietary digital message format for all types of alerts and notifications" (OASIS, 2010) allowing to set an incident category. Based on the category, the citizen report should then be sent to the appropriate authority, such as police or fire departments. The evaluation revealed that the categorisation of 112.social should be adapted to the region of deployment, e.g. earthquakes and volcano eruptions were relevant types of disaster in Sicily (Italy), and the participating emergency services. Furthermore, the availability of *infrastructure*, including as the access, connectivity, and distribution of devices, might vary across events or regions. During a blackout or failure of radio masts, for instance, transmission might be realised using infrastructure-less technology, such as off-grid ad hoc networks (Alvarez et al., 2016).

12.2.2 Information Use and Expectation in Different Phases of a Crisis

The emergency services' integration of citizen-generated content for crisis response and crisis communication must consider the *phases* of a crisis, emergency, or disaster. Despite the existence of variations, the emergency management cycle (EMC) mostly comprises the phases of mitigation, preparedness, response, and recovery, whereas each phase is characterized by distinct situational patterns, tasks or risks (Baird, 2010). Still, plenty of models and research projects in the domain of social media analytics omit the phase of *mitigation*, thus reducing the cycle to (1) *preparedness* (pre-crisis; before the emergency), (2) *response* (crisis; during the emergency), and (3) *recovery*, (post-crisis; after the emergency) (Kaufhold, Gizikis, et al., 2019). While the EMC focuses on professional crisis management and response, in the domain of psychology, an EMC for affected people was elaborated by McMahon (2011): in the *threat phase*, people are under uncertainty and their emotional state is vague, whereas the *heroic phase* is characterized by high activity as well as altruism, and the *honeymoon phase* reveals strong empathy and community coherence (Gründer-Fahrer et al., 2018). This has implications for the integration of citizen-generated content: Existing research outlines that citizens publish varying types of information across different phases of a crisis with different sentiment and relevance for emergency services (Alam et al., 2019; Reuter et al., 2013). The EMC also influenced the research conducted in this dissertation: In an effort to increase the quality of citizen-generated content, Chapter 8 contributes the design and evaluation of concise social media guidelines (Table 12.4) for citizens' use of social media before, during and after emergencies (Kaufhold, Gizikis, et al., 2019).

Table 12.4 Compilation of some social media guidelines, reduced version of Table 8.3

Category	Guidelines
General Aspects	• Interact with respect and courtesy (Bundesanstalt Technisches Hilfswerk, 2011). • You are responsible for your writing, think of possible consequences (Bundesanstalt Technisches Hilfswerk, 2011).
Preparation	• Know the social media accounts of your local and national ES and follow them. This will help find real-time information during an emergency (Helsloot et al., 2015). • Read what to expect from ES in social media. Are they always online? Do they reply to posts in social media (Reuter, Ludwig, Kaufhold, et al., 2016)?
Response	• Remember you can use social media for information updates, but it does not replace emergency calls (Helsloot et al., 2015). If in danger, always call 112 first. • Share only official and reliable information and avoid spreading rumours! The spreading of false information can threaten the smooth deployment of rescue teams and put you and your relatives at additional risk (Perng et al., 2012).
Recovery	• Restore missing contact and ask for welfare of family and friends (Qu et al., 2011). • Help others reconstructing/handling the event (Mirbabaie & Zapatka, 2017).

Furthermore, as outlined in chapter 7, citizens have different information demands across phases of mitigation (e.g. information on precautionary measures), preparation (e.g. early and accessible warnings, information on potential threats, consequences and required behaviour), response (e.g. targeted information on the origin, duration and consequence of the crisis, instruction on behaviour and the safety status of close ones), and recovery (e.g. establishing contact with relatives and asking for further help to enable the recovery) (Grinko et al., 2019). These demands have implications for crisis communication: In order to assist emergencies services in overcoming the complexity of crisis communication, the EmerGent project developed comprehensive guidelines (outlined in Figure 12.1) to establish a strategy and use of social media before, during, and after emergencies (Gizikis, O'Brien, et al., 2017).

Guidelines for Emergency Services

Prepare to start using social media	Before an emergency	During an emergency	After an emergency
1. Consider the legal implications 2. Consider the needs in human and financial resources 3. Prepare a social media strategy 4. Clearly communicate the social media strategy and provide staff training 5. Explore what ICT tools are available for social media monitoring and analysis 6. Use of apps for direct communication (A2C & C2A) 7. Plan the next steps to start using social media	1. Provide information about your organisation, its operations and emergency prevention and preparation 2. Raise awareness on the use of social media 3. Use of ICT tools for social media monitoring and analysis 4. Team up with other groups and organisations 5. Publish alerts for the risk of an upcoming emergency	1. Understand how social media is used by citizens during emergencies 2. Establish communication with the public 3. Request information from the public 4. Use of ICT tools for social media monitoring and analysis 5. Respond to false information & rumours 6. Collaborate with emergent group initiatives, e.g. VOST	1. Continue the communication with the citizens 2. Evaluate your social media use during the emergency **Guidelines for citizens** Additional Content: • Data Protection & Privacy Guidelines for Processing Data • Publication and dissemination of guidelines

Figure 12.1 Social media guidelines for emergency services. From Gizikis, O'Brien, et al., (2017)

12.3 Organisational Perspective

The adoption of mobile technologies and social media depends on the goals, capabilities, and willingness of emergency services. Chapter 4 and 8 contribute findings to the organizational perspective of information refinement. Furthermore, the identified conditions are summarized in Table 12.5.

12.3.1 Goals, Objective, Strategies and Tactics of Emergency Services

Regardless whether emergency services have a concrete mission or vision statement, they need to fulfil certain *goals*, which can be translated into (multiple) measurable *objectives*. To achieve these objectives, emergencies must employ *strategies* that include conducive capabilities and methods. Then, strategies must define *tactics* which includes tasks that are performed by persons using specific artefacts to implement the strategy. Using a crisis interaction matrix (Reuter et al.,

Table 12.5 Organisational categories, features, and conditions for information refinement

Category	Features	Conditions
Mission	Goals and objectives	Crisis communication, integration of citizen-generated content
	Strategies and tactics	Information dissemination, data monitoring and analysis, conversations and coordinated action
Roles	Incident response	Incident management, public relations
	Virtual response	Virtual operations support teams
Collaboration	Interorganizational	Authorities, media, NGOs
	Public	Self-help communities, virtual and technical communities
Perception	Law compliance	Data protection, data privacy
	Organizational culture	Management support, operational support
	Facilitating resources	Expertise, personnel, technology, training

2012), chapter 4 outlined three use cases for emergency services' use of mobile technologies and social media, namely *crisis communication* and the *integration of citizen-generated content* (Reuter & Kaufhold, 2018). While crisis communication intends to inform and warn the population about emergencies, which usually takes place using modular warning systems such as MoWaS in Germany (Klafft, 2013), the integration of citizen-generated content strives for enhancing situational awareness (Reuter, Hughes, et al., 2018). For this, supportive technology, as it will be discussed in section 12.5, can assist in performing tasks such as damage and situational assessments, event detection or prediction, fake news or hate speech detection, sentiment analysis, and topic discovery, amongst others (Alam et al., 2019).

As these use cases can be considered as objectives, they require appropriate strategies for implementation, such as information dissemination, data monitoring and analysis, or conversations and coordinated action (Wukich, 2015). Here, chapter 8 presented the design and implementation of *guidelines for social media use in emergencies* (Kaufhold, Gizikis, et al., 2019). The design included guidelines for both emergency services and citizens on how to use social media before, during and after emergencies. While the guidelines for emergency services were intended to establish the implementation of social media strategies, the citizens guidelines were designed as a concise artefact to facilitate the spread of high-quality information in social media. The *citizen guidelines* were evaluated with a representative sample of German citizens, whereof 72% rated them as helpful consistently across the phases before, during and after emergencies. The more individuals were using smartphones or social media, the more they tended to agree with the proposed social media guidelines, but demographic variables did not have a strong influence on the responses. Thus, social media guidelines might be a measure to increase citizens media literature and reduce a chaotic and unstructured use of social media in emergencies, facilitating data analysis by emergency services. However, as the *social media guidelines for emergency services* were finalized at the end of the EmerGent project, an evaluation was not conducted anymore.

12.3.2 Persons, Roles and Cooperation With External Stakeholders

In order to implement strategies, personnel must fulfil tasks using specific artefacts. These persons and tasks are often bound to specific roles, such as the incident manager or the public relations officer. While a member of the incident management team might be primarily interested in information on situational awareness, the public relations officer needs to inform or warn the public about emergencies on a more general level. Chapter 4 of this dissertation developed a role typology matrix which differentiates the affiliation (authority or citizen/public) and realm (real or virtual) of emergency-involved actors. On the authority level, it differentiates the (on-site) *incident management team* but also the implementation of *virtual operations support teams* (VOST), which often constitute of trusted volunteers monitoring the virtual realm, especially social media activities, to assist formal emergency management (Reuter & Kaufhold, 2018). Furthermore, the role typology matrix includes the public as emergent groups, including affected citizens and on-site volunteers, as well as *virtual and technical communities* (V&TC) which operate in the virtual realm to assist crisis response by contributing with expertise in geographic information systems, database management, social media, or online campaigns (van Gorp, 2014).

The need for *collaboration* with external stakeholders is also emphasized in the crisis interaction matrix by *inter-organizational crisis management*, for instance for mutual awareness among fire and police departments, and the potential collaboration with public *self-help communities*, which act both in the real and virtual realm (Reuter et al., 2013) and contribute to emergency response by diverse activities, such as the organization of material donations, search for missing people or stacking sandbag for dikes, amongst others. The collaboration with external stakeholders, such as citizen volunteers, media organisations, or NGOs, was not the focus of this dissertation but their capabilities can assist emergency services in the process of refining information.

12.3.3 Perception and Challenges of Technology Adoption

The adoption of technology, such as mobile apps and social media, relies on the organizational *perception* and expected organizational, social, and technological challenges. On the one hand, chapter 4 indicates that emergency services generally have a positive attitude towards social media and already use it to share information with the public on how to avoid (preparation) and how to behave during (response) accidents or emergencies, but less to receive messages or even establish bidirectional communication with citizens (Reuter & Kaufhold, 2018). However, although emergency services acknowledged the usefulness of social media to establish situational awareness, particularly in terms of situation updates, photos, videos and information on public mood, only a small portion actually used it for that purpose (Reuter, Ludwig, Kaufhold, et al., 2016). Barriers for the adoption of novel ICT include a lack of expertise, personnel and training, organizational culture or scepticism at management and operational levels, as well as perceived issues of data protection, credibility and issues regarding law compliance (Plotnick & Hiltz, 2016). Despite these organizational and social barriers, however, an increasing number emergency services reported the integration of social media analysis and communication into their organization between 2014 and 2017 (Reuter, Kaufhold, et al., 2020). Furthermore, often reported technological barriers comprise the issues of information overload and quality (Plotnick & Hiltz, 2018), which are tackled by the ESI (chapter 9), SMO (chapter 10) and 112.social (chapter 11) artefacts, which will be further discussed in section 12.5 (Kaufhold, Rupp, et al., 2020).

12.4   Societal Perspective

As the use mobile technologies and social media by emergency services has an impact on society, citizens use, behaviour and roles in ICT as well as their perceptions and expectations towards emergency services must be considered. The chapters 4, 5, 6 and 7 comprise findings to the societal perspective of information refinement. Furthermore, the identified conditions are summarized in Table 12.6.

Table 12.6 Societal categories, features, and conditions for information refinement

Category	Features	Conditions
Attitudes	Benefits	Up-to-date information, reliability of crisis apps, increased perceived safety
	Barriers	Data privacy, difficulty of use, lack of threat awareness, rumours, unreliability of social media
	Risk culture	Authority-centred, individual-centred, fatalistic
Behaviour	Use of channels	TV, personal conversations, radio, social media, crisis apps, phone call, text messages, magazines, websites, local information
	Operations	Identification, amplification, routing, verification, structuring, synthesising information
	Challenges	Chaotic, fake, inconsistent, redundant, unorganised information dissemination
Roles	Emergent groups	Affected citizens, helper, unbound helpers
	Digital volunteers	Moderator, reporter, retweeter, repeater, reader
Expectations	Dissemination	Instructions, warnings and safety tips, duration and reason of the incident, information source
	Monitoring	Continuous or limited availability
	Response	Fast, delayed or no response

12.4.1 Citizens' Behaviour and Roles in the Digital Realm

As the information space of mobile and social media is largely consisting of citizen-generated content, understanding their behaviour and roles is pivotal for information refinement. Using the crisis communication matrix, chapter 4 reports findings on the behaviour and roles of *self-help communities* (Reuter & Kaufhold, 2018). Here, mobile technologies and social media allow for a deeper integration of the real and virtual world for crisis response: while *emergent groups* usually act in the form of neighbourly help and work on-site, digital volunteers originate from the Internet and work mainly online (Reuter et al., 2013). However, social media facilitate emergent groups to reach out for unbound helpers from unaffected regions (Detjen et al., 2016) and so-called moderators were reported to coordinate activities in both the real and virtual realm, for instance by identifying demands on-site and reporting them to social media help communities (Kaufhold & Reuter, 2016). Further identified roles of the virtual realm include the digital volunteer, helper, reporter, retweeter, repeater, and reader, each with distinct behaviour (Reuter et al., 2013; Starbird & Palen, 2011). The role typology matrix of chapter 4 contributes to research by contextualizing existing social media roles into a new perspective comprising both authorities and citizens as well as the real and virtual realm. Besides individual processing, Starbird (2013) analysed information processing through the lens of collective intelligence using distributed cognition as a methodological framework. The analysis of Twitter data revealed the citizen activities of identifying and amplifying actionable information, routing information, verifying information, structuring information, and synthesizing information into resources. In this way, individual actors or communities fulfilling different roles shape the information space in crisis and V&TC's are capable of supporting emergency services with expertise in data processing and technologies (van Gorp, 2014).

Despite the good intents of digital volunteerism, the actions of amateur or semi-professional actors pose *challenges* for emergency services and the information refinement process. As information related to crises is likely to be posted across a multitude of different information channels, the emergence of inconsistent or redundant information is likely (Reuter, Ludwig, Kaufhold, et al., 2015), also leading to rumours, uncertainty and mistakes due to chaotic and disorganized work of volunteers (Valecha et al., 2013). In order to mitigate the issue of chaotic social media use proactively, chapter 8 presented the development of social media guidelines (SMG) for citizens, comprising tips on how to best use social media before, during and after emergencies (Kaufhold, Gizikis, et al., 2019). Furthermore, the technical artefacts ESI (chapter 9), SMO (chapter 10) and 112.social (chapter 11) contribute to the issues of information quantity and quality.

12.4.2 Attitudes and Reported Use of Mobile Technologies and Social Media in Emergencies

As already indicated in chapter 4, there is a wealth of case studies documenting the use of social media in larger emergencies (Reuter & Kaufhold, 2018). While these offer a rich qualitative understanding of each case, there were less quantitative surveys using representative samples, which is why three chapters are contributing to that gap. Chapter 5 outlines that most of the German population still use TV, personal conversations with friends, family and neighbours, and radio as the main information source in crises with each over 70% usage, while social media (55%) and crisis apps (25%) had significantly lower proportions (Kaufhold, Grinko, et al., 2019). While social media was perceived more useful than crisis apps, this changed in a following study in favour of crisis apps (Kaufhold, Haunschild, et al., 2020). This effect could be attributed to the rising awareness and distribution of crisis apps in Germany, such as Katwarn or NINA, and their perception as an official authority channel. Furthermore, chapter 6 outlines that 45% of German citizens have already used social media during an emergency to share or look for information (Reuter, Kaufhold, Spielhofer, et al., 2017). In case of sharing information, most of them published information on weather (63%), road/traffic conditions (59%) or feelings/emotions (46%). Social media was *valued* as being faster than other more formal channels as well as for spreading and obtaining up-to-date information. However, participants also reflected *barriers* of social media use, including the fear for false rumours (73%), unreliable information (65%), data privacy issues (62%) and the issue that social media might not work in emergencies, for instance during a telecommunication breakdown.

Furthermore, 16% indicated that they already have downloaded a crisis apps, especially Katwarn (6%) or NINA (6%). In future, they would like to use crisis apps primarily to receive warnings (57%) and safety tips (51%). A slightly lower percentage of citizens was motivated to share information with emergency services (42%), which would allow for the establishment of bidirectional communication. The research on crisis apps is complemented by findings of chapter 8, which reported an increase of crisis apps use in Germany from 16% to 25% (Grinko et al., 2019). Over two thirds of participants favoured installing only one crisis apps, valuing a warning configurator, recommendations for actions in crises and support for personal preparation as most import features. The availability of up-to-date information, reliability and increased perceived safety were reported *enablers*, while lack of knowledge, threat awareness and advantage over other channels perceived *barriers* of crisis apps use. In a follow-up study, it became apparent that crisis apps approval increased and crisis functionality, such as telephone emergency calls (with geolocation) and disaster warnings were still the most important features, participants also desired police-related such as fraud offences, search for missing people, or cybercrime (Kaufhold, Haunschild, et al., 2020). While already the search for missing people indicates a willingness to be not only a passive information consumer, 76% also indicated that they would submit image or video data to emergency services in case of a life-threatening event, which would allow for the establishment of bidirectional communication such as proposed by 112.social (chapter 11). Furthermore, German citizens saw crisis apps rather as a complement than an replacement for existing channels (Kaufhold, Haunschild, et al., 2020), which supports the cross-media approach realized in ESI (chapter 9).

12.4.3 Expectations Towards Authorities, Emergency Services, and Other Organisations

The conducted empirical studies revealed citizens expectations towards the use of media by authorities, emergency services or organisations. In chapter 5, in case of an infrastructure breakdown, at least 70% of participants desired instructions and information (including its source) on the duration and reason of the incident by infrastructure operators (Kaufhold, Grinko, et al., 2019). Furthermore, chapter 6 outlined that in 2017, 67% of German citizens expected emergency services to monitor social media and 47% also expected them to answer on their posts within one hour, although 43% acknowledged that emergency services are too busy. In a following survey in 2019, citizens expectations remained high, although a slightly higher percentage identified emergency services as too busy monitoring social media (Haunschild et al., 2020). The prevalence of high expectations might be attributed to the authority-centred *risk culture* in Germany, which emphasizes the obligation of authorities to solve crisis situations (Reuter, Kaufhold, et al., 2019). This is in contrast to individual-centred risk cultures, such as the Netherlands, where citizens feel themselves more responsible for overcoming crises, resulting in less expectations towards emergency services. This requirement, however, led to the development of ESI (chapter 9), which combines analysis and communication capabilities for emergency services. Furthermore, expectation management could be considered as a practice of information refinement by clearly communicating dissemination, monitoring, and response practices: For instance, a case study with the fire department of Frankfurt revealed that they communicate the purpose of a channel directly (e.g. Facebook, Instagram, and YouTube for publicity and recruitment, Twitter for operational communication, and the website for safety information) and emphasise that their social media accounts are not monitored 24/7 (Kaufhold & Reuter, 2017).

12.5   Technological Perspective

The requirements elicited in the theoretical and empirical chapters but also in the larger scope of the EmerGent, KontiKat, and MAKI projects led to the design and evaluation of the three artefacts ESI (chapter 9), SMO (chapter 10) and 112.social (chapter 11). These considered both event, organizational and societal boundary conditions and implemented different aspects of information refinement, including the channels, access, content, analysis, filtering, and evaluation of information. Furthermore, the identified conditions are summarized in Table 12.7.

12.5.1 Channels: What Are relevant Channels to Fulfil Objectives?

Before *interacting* with information, emergency services need to choose which channels are relevant to their objectives, or *intent*, and develop supportive *strategies*. Wukich (2015) differentiates the strategies and tactics of information dissemination (one-way from agency to public), data monitoring & analysis (one-way from public to agency) and conversations & coordinated action (two-way or multiway) for social media use in emergencies. For example, the fire department of Frankfurt implements the *information dissemination* strategy by limiting all *operational communication* to Twitter in a unidirectional manner, while Facebook, Instagram and YouTube are used to maintain *public image* and *recruit new members*, which also involves *conversations* with citizens (Kaufhold & Reuter, 2017). The *data monitoring and analysis strategy* was realized using the free tool TweetDeck, a dashboard application for the management of Twitter accounts, due to its real-time capability in terms of monitoring keywords and hashtags. However, the optimal social media tool should "provide a good overall usability, capabilities to pre-structure actions and content, the flexibility to adapt it to the current emergency, support for all relevant social media, and possibilities to capture the mood of citizens" (Kaufhold & Reuter, 2017, p. 607). Thus, the integration of multiple social media channels allows the analysis of different use patterns observed in social media: In the 2013 European floods, due to the availability of close or semi-open spaces (such as groups or pages), Facebook was used for a more private community *exchange* and the topic centred around actions and volunteering, emotional support and thanks, and donations in kind (Gründer-Fahrer et al., 2018; Kaufhold & Reuter, 2016). In contrast, Twitter was rather used for the *reporting* of alarm or water levels, the current situation, situational overviews incl. actions, and status updates.

Social media, however, are *general-purpose services* for everyday use but not tailored to the requirements of crisis managements (Tan et al., 2017). Thus, the next steps of the information refinement framework highlight approaches to overcome the sometimes chaotic and unstructured exchange of social big data (Kaufhold, Gizikis, et al., 2019). A different approach constitutes the development or use of *crisis-specific apps and services*, such as Ushahidi for crowdsourcing (Okolloh, 2009) or available crisis apps (Groneberg et al., 2017). While crisis apps such as Katwarn and NINA are well established in Germany, their functionality focuses on crisis communication, i.e. warning the population, but not on improving emergency services situational awareness (Reuter, Kaufhold, Leopold, et al., 2017a). Thus, chapter 11 presented the design and evaluation of 112.social

Table 12.7 Technological steps, categories, features, and conditions for information refinement

Category	Features	Conditions
Channels		
Features	Exchange	Actions and volunteering, emotional support and thanks, donations in kind
	Reporting	Alarm or water levels, current situation, situational overview incl. actions, status updates
Purpose	General services	Facebook, Instagram, Twitter, YouTube
	Crisis-related services	Katwarn, NINA, Ushahidi
Interaction	Intent	Operational communication, public image, employee recruitment
	Strategy	Information dissemination, data monitoring and analysis, conversations and coordinated action
Access		
Tracking	Approach	Keyword-, actor-, URL-related
	Method	Application programming interfaces, website scraping
Storage	Database	Document, graph, key-value, relational database
	Format	Structured and unstructured data
Content		
User-generated content	Production	Generative, synthetic, derivative, innovative
	Dimension	Objectivity, type, source, credibility, time, location
Metadata	Message-related attributes	Date, time, title, comments, replies, number of views, likes, dislikes, retweets, shares

(continued)

Table 12.7 (continued)

Category	Features	Conditions
	Person-related attributes	Age, birthday, gender, location, education, number of uploads/watches/total posts, name
Analysis		
Preprocessing	Cleaning	Duplicate or stop word removal
	Annotation	Part-of-speech tagging
	Normalisation	Stemming, lemmatisation
Methods	Aggregation	Clustering, event detection, topic modelling
	Classification	Credibility, damage, humanitarian, relevance, sentiment
	Extraction	Data enrichment, named entity recognition
	Relationships	Social network analysis
Filtering		
Backend	Classification	Credibility, relevance
	Chunking or message grouping	Distance metrics, information summarisation
Frontend	Attribute filtering	Keywords, language, locations, platform
	Interactive filtering	Charts, classifications
Evaluation		
Explainability	Content	Inconsistency, redundancy
	Transparency	Indicators, white boxing
Scope	Goals and objectives	Configurability, integration
	Interpretation	Accuracy, access, use, worldview

(Figure 12.2), which is a novel crisis app for bidirectional communication between emergency services and citizens (Kaufhold et al., 2018). In contrast to existing crisis apps, it enables citizens to report emergencies in a structured manner by defining a category and subcategory, adding a description, multimedia files (audio, photo, or video) and GPS location. Based on a reported incident, a chat-based approach allows emergency services to request additional information from citizens on-site. Furthermore, besides targeted communication, a broadcast feature for warning app users was implemented.

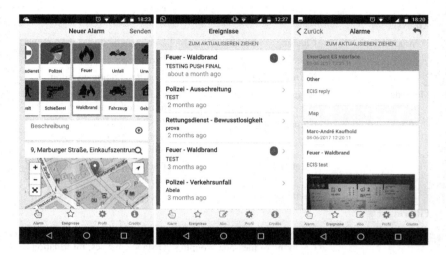

Figure 12.2 The 112.social mobile crisis app for reporting (left), overview (centre) and bidirectional communication (right) of incidents

12.5.2 Access: Which Data Is Available and How It Is Accessible?

After selecting relevant channels for analysis and communication, emergency services must establish access to different social media. In social media, information access can be distinguished between normal end-user and developer usage. For end-users, social media provide different degrees of public and private spaces. For instance, while most communication on Twitter is public, which is even accessible without user credentials, Facebook offers a variety of restricted and private spaces, such as groups, pages or the private timeline which is usually only visible to friends (Reuter & Scholl, 2014). However, the manual monitoring of diverse social media and their public or private substructures using heterogeneous interfaces might be a resource-consuming task in terms of organizational capabilities

(Reuter, Ludwig, Kaufhold, et al., 2016). Another approach lies in the development of supportive tools that provide access to relevant social media and are tailored according to the needs of emergency services (Kaufhold, Rupp, et al., 2020). In order to establish access for developers, most social media platforms provide programmatic access to their data via Application Programming Interfaces (APIs), which usually comprise search APIs allowing to query past messages and streaming APIs allowing to subscribe to real-time data feeds (M. Imran et al., 2015). These APIs typically allow the expression of constraints, such as *"(i) a time period; (ii) a geographical region for messages that have GPS coordinates (which are currently the minority); or (iii) a set of keywords that must be present in the messages [...]"* (Imran et al., 2015, p. 9).

Table 12.8 Challenges of using various social media platform APIs

Challenge	Description
Ease of data access	The varying ease of data access facilitates a bias for practitioners and researchers (M. Imran et al., 2015).
Exchange format	Platform APIs, such as Facebook Graph or Twitter Search APIs, provide data in different exchange formats using different sets of metadata (Kaufhold, Reuter, et al., 2019).
Geographical information	Only a small subset of social media messages contain geographical information, especially GPS coordinates (Schulz et al., 2013).
Policy changes	Due to regular policy changes, access to data changes over time (Kaufhold, Reuter, et al., 2019).
Query language	The query language for keywords, including Boolean operations, varies across social media (M. Imran et al., 2015).
Quota restrictions	Platform APIs usually limit the number of queries via search APIs or amount of real-time data gathered via streaming APIs (Reuter & Scholl, 2014).

In the light that emergency services might be interested in *tracking* data across different types of social media to enhance situational awareness (Reuter, Ludwig, Kaufhold, et al., 2016), different tracking *approaches*, such as *keyword-*, *actor-*, or *URL*-related tracking (Stieglitz, Mirbabaie, Ross, et al., 2018), and challenges must be considered for the development of supportive social media technologies using platform APIs, including the varying ease of data access, different exchange formats, regular policy changes, different search query languages and quota restrictions (Table 12.8). As an alternative to APIs, developers may use

HTML scraping or parsing techniques to gather social media data (Stieglitz et al., 2014); however, this approach also suffers from changes of the underlying HTML structure and is against the terms of services of most social media platforms. Furthermore, the uneven access to social media data potentially fosters research bias. In the past, a lot of research has focused on Twitter data, which is facilitated by the ease of data access in comparison to other social media, such as Facebook, which offer more private spaces that are inaccessible to public APIs (Reuter, Hughes, et al., 2018).

The analysis of tracked data often requires the storage using databases and a specific data format. Depending on the characteristics, prerequisites, and scope of planned analyses (section 12.5.4), different types of *databases*, such as document, graph, key-value, or relational databases, may perform better than other with regard to reading, writing, and deleting database entities or fetching keys (Y. Li & Manoharan, 2013). Furthermore, a data *format* is required that is capable of storing the *structured* and *unstructured* information from social media (Stieglitz, Mirbabaie, Ross, et al., 2018). In order to allow a unified analysis of heterogenous data, the ESI (chapter 9) and SMO (chapter 10) artefacts convert the heterogenous source data into the Activity Streams 2.0 specification (World Wide Web Consortium, 2016), which is used for both the exchange (JSON) and document-oriented storage (MongoDB) of data. Each *activity*, representing a post from 112.social (chapter 11) or social media, is composed of at least a unique *id*, *actor* (comprising user data and metadata) and *object* (comprising content and related metadata). If exchange format and policy changes occur, the preprocessing and conversion of data into Activity Streams 2.0 must be adapted, but interoperability remains. However, as the specification was not designed to cover all potential metadata from social media, an *enrichedData* object was added as an extension covering all available and relevant information that cannot be mapped to the base specification.

12.5.3 Content: What Are Characteristics of Emergency Information?

Once access to information is established, the structure and analysis of content come to the foreground. By analysing Twitter data, for instance, Starbird et al. (2010) identified four types of *user-generated content*, or information: *generative* information refers to original raw material, *synthetic* information integrates external information, such as other tweets, web and news sources, *derivative* information emerges as a result of informational interaction and *innovative* information is distinguished by inclusion of cross-domain expertise and interpretation. Furthermore, Imran et al. (2015) classified information according to six dimensions of factual, objective or emotional content, information provided, source, credibility, time, and location. As an addition, Olteanu et al. (2015) distinguish information as either informative, such as on-topic information contributing to situational awareness, or not informative, such as trolling or off-topic information. The identified characteristics of information are summarized in Table 12.9.

Table 12.9 Dimensions of information. Own illustration based on Imran et al. (2015), Olteanu et al. (2015) and Starbird et al. (2010)

Dimension	Manifestations
By information production	Generative, synthetic, derivative, innovative
By factual, subjective, or emotional content	Factual information, opinions, sympathy, antipathy, jokes
By informativeness	Informative, non-informative
By information provided	Caution and advice, affected people, infrastructure/utilities, needs and donations, other useful information
By information source	Eyewitnesses/bystanders, government, NGOs, news media
By credibility	Credible information, rumours
By time	Pre-phase/preparedness, impact-phase/response, post-phase/recovery
By location	Ground-zero, near-by areas

By looking at dimensions such as location and time, it becomes apparent that information in social media is not only based on the message content, but also the attached *metadata*. For instance, Moi et al. (2015) distinguish between *message-related attributes*, such as date, time, title, comments, replies, number of views, likes, dislikes, retweets, or shares, and *person-related attributes*, such as age, age range, birthday, gender, location, education, number of uploads/watches/total posts, real name, or relationship status. The information or metadata relevant to emergency services varies according to the conducted task or organisational role. While a public relations manager might be interested in credibility-related information to communicate against rumours, incident managers require informative information, including location and time metadata, to enhance situational awareness (Kaufhold, Rupp, et al., 2020). In addition, as section 12.5.4 will outline, different analysis techniques can be utilized to gather and refine required information.

While these approaches facilitate information refinement reactively, emergency services could influence content creation proactively by communication relevant information to citizens and reducing the production of misinformation. For instance "Tweak the Tweet" was proposed as a microsyntax to support citizens in the reporting of emergency-relevant information (Starbird & Stamberger, 2010). Furthermore, chapter 8 of this thesis contributed by proposing citizen guidelines for social media use in general and in emergencies (Kaufhold, Gizikis, et al., 2019). The guidelines comprise tips on the *preparation* (e.g. to identify channels that might post relevant information for affected citizens), *response* (e.g. how to disseminate high-quality content and connect with volunteer initiatives) and *recovery* (e.g. to give feedback and participate in recovery activities) phases of an emergency. However, as the design of the guidelines were primarily by workshops with fire department personnel (Gizikis, Susaeta, et al., 2017; O'Brien et al., 2016), other emergency services such as police and technical rescue might need to tailor the guidelines according to their needs and objectives before disseminating them publicly.

12.5.4 Analysis: How Can Technology Assist in the Analysis of Information?

As soon as emergency services determined the relevant content, they also need to design the way they want to analyse the data. Until now, the research field of crisis informatics produced a wealth of algorithmic approaches for the analysis of social media content (Alam et al., 2019) often using *methods* and techniques from natural language processing (NLP), machine learning (ML) or statistics. Without any claim of completeness, Table 12.10 outlines algorithms that can be roughly clustered into the objectives of *aggregation* (e.g. clustering, event detection, topic modelling), *classification* (e.g. credibility, damage, humanitarian, relevance, sentiment), *extraction* (e.g. data enrichment, named entity recognition), and *relationship* analysis (e.g. social network analysis). In most cases, some steps of data *preprocessing* are required (Stieglitz, Mirbabaie, Ross, et al., 2018), including the *cleaning* (e.g. duplicate or stop word removal), *annotation* (e.g. part-of-speech tagging), and *normalisation* (e.g. stemming, lemmatization) of data before analysis can take place. The selection of a single or combination of algorithms, then, relies on the identified objectives and the content relevant to reach these objectives. For instance, if emergency services want to reduce information overload in a large-scale emergency, but also evaluate the credibility of the remaining information, they could use a tool employing a combination of algorithms for relevance classification and credibility assessment. The results of algorithmic computation in the backend must be visualized in a useful manner, allowing to filter (section 12.5.5) and evaluate (section 12.5.6) produced or refined information.

Especially in large-scale emergencies, it is often not possible for emergency services to scan through all potentially interesting social media posts, even if search queries for monitoring are well designed, nor have they the capability to create large training data sets (for supervised ML) quickly under the time-critical constraints of crises (M. Imran et al., 2018). Chapter 10 of the thesis contributes by the design of an algorithm for rapid relevance classification which is also embedded in the visual interface of the Social Media Observatory (SMO) (Kaufhold, Bayer, et al., 2020). The novel algorithm uses Random Forest, which proved as an efficient algorithm for detecting relevant social media information in emergencies (Habdank et al., 2017), to binarily classify social media posts as relevant or irrelevant to a trained crisis scenario, such as fire- or flood-related incidents (Figure 12.3). Using the relevance classifier, users of the SMO can hide messages classified as irrelevant and this way reduce the volume of displayed information and thus mitigate information overload. Even with a well-trained model, there is a risk of false negatives (e.g. relevant information labelled as irrelevant), which requires on backend level the optimization of the algorithm's accuracy, but on interface level the use of the relevance classifier must not be mandatory.

In this way, emergency services can browse through irrelevantly labelled social media posts if personnel and time resources allow. Furthermore, the approach comprises a *feedback classification* mechanism which allows to reactively correct misclassifications of the algorithm, allowing to continuously finetune the classifier. Furthermore, while previous approaches determined relevance by the content of the message (Kaufhold, Bayer, et al., 2020), the approach of chapter 10 includes metadata such as location or time, which improved the accuracy of relevance classification. Furthermore, by the use of *active learning*, which followed the assumption of uncertainty sampling by labelling posts where the classifier has the lowest confidence (Lewis & Catlett, 2014), this approach was able to reduce the training time of Random Forest classifiers, requiring only up to a quarter of labelled data compared to a batch learning approach. However, the evaluation revealed that the optimal set of metadata and also the effectivity of active learning varied across datasets; the examination of both characteristics are future research potentials to further optimize the effectivity of models as well as their efficient creation.

Table 12.10 Exemplary application of machine learning techniques in crisis informatics. Own overview based on the referenced literature

Analysis technique	Image	Text
Clustering: Categorization of text documents into similar groups using distance and similarity metrics, often combined with approaches for information summarization (Fahad et al., 2014; Rudra et al., 2015).		x
Credibility assessment: The identification of credible information comprises classification- and multifeature-based, graph- and propagation-based, indicator-based and survey-based detection approaches (Hartwig & Reuter, 2019; Viviani & Pasi, 2017).		x
Damage assessment: Using a supervised machine learning to assess the severity of damage observed in an image, such as 'severe', 'mild', and 'none' (Alam et al., 2019).	x	
Data enrichment: Refers to the computation of additional metadata that is not delivered by any or every social media provider APIs (Moi et al., 2015).		x
Duplicate detection: Hash techniques for the identification of exact or near duplicate images with little modifications, such as cropping, resizing, padding background, changing intensity, or embedding text (Alam et al., 2018).	x	
Event detection: Detection and tracking of crisis events or subevents using techniques such as burst detection, dictionaries, or supervised classification (M. Imran et al., 2015).		x
Humanitarian classification: Semi-supervised classification of messages into humanitarian categories such as personal update, infrastructure and utilities damage, affected individual, caution and advice, donation and volunteering, missing and found people, sympathy and support, injured or dead people, and other useful information (Alam et al., 2019).		x
Named entity recognition: Natural language processing, statistical or machine learning techniques for the extraction of entities, such as locations, persons or organizations, from text (Ritter et al., 2011).		x
Relevance classification: Binary classification of relevant and irrelevant images (Alam et al., 2018) or crisis-related messages for crises often using supervised machine learning models (Habdank et al., 2017).	x	x
Sentiment classification: Natural language processing, statistical or machine learning techniques for the determination of emotional polarity, e.g., positive, neutral or negative, of a document, message or sentence (Pang & Lee, 2008).		x

(continued)

Table 12.10 (continued)

Analysis technique	Image	Text
Social network analysis: Methods and techniques designed to discover patterns of interaction between actors in social networks, including the discovery or identification of most central actors using statistical measures, hubs and authorities using link analysis, and communities using community detection techniques (Oliveira & Gama, 2012).		x
Topic modelling: Identification of topics crisis-specific topics from textual data using modelling techniques such as (unsupervised) Latent Dirichlet Allocation (Blei et al., 2003; Gründer-Fahrer et al., 2018).		x

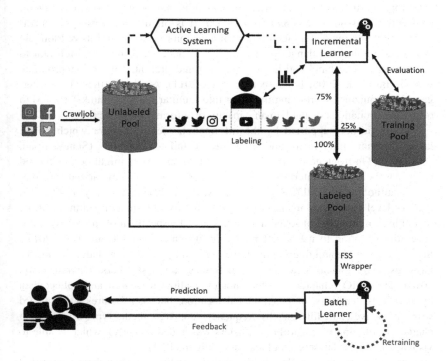

Figure 12.3 The relevance classification process of the SMO comprising active learning, real-time evaluation, and feedback classification

12.5.5 Filtering: How to Detect and Filter for the Relevant Information?

Even if the information refinement process so far improved the quality of information, e.g. by selecting the right channels, establishing access, defining the relevant content, and using techniques for analysing the data, the volume of data could still be too high to process or emergency services might be interested in different subsets of data, such as content related to a specific class, sentiment or time. The application of filtering techniques can take place both at backend or frontend level (Kaufhold, Rupp, et al., 2020). When up to hundreds of thousands of social media messages are generated in short timeframes during large-scale disasters or catastrophes, such as floods, hurricanes, or wildfires, emergency services have to deal with the issue of *information overload*. It is often defined as "[too much] information presented at a rate too fast for a person to process" (Hiltz & Plotnick, 2013) and implies the danger of getting lost in data which may be irrelevant to the current task at hand and being processed and presented in an inappropriate way (Keim et al., 2008; Landesberger et al., 2011). Already Miller (1956) suggested "organizing or grouping the input into familiar units or chunks" (p. 93) to overcome limitations of the human capacity of information processing.

This is further supported by the information seeking mantra which postulates "overview first, zoom and filter, then details-on-demand" (Shneiderman, 1996, p. 2). On frontend or *visualization* level, the overview might be established using dashboard-style interfaces with show aggregated or summarized information (Kaufhold et al., 2017; Rudra, Ganguly, et al., 2018). A survey of 477 U.S. country-level emergency managers suggests that the information summarization, 'chunking' or grouping of social media messages by specific tools positively influences the intention to use social media during emergencies (Rao et al., 2017). Secondly, zooming and filtering might be realized by interactive charts, keywords, hashtags, geographical locations or timeframes, amongst others (Onorati et al., 2018). The SMO (chapter 11) does not only allow to search and collect data by keywords, hashtags, or geographical location, but also features a dashboard with interactive chart filtering (by frequency, language, media type, or sentiment charts; the filtering by algorithmic clusters was added recently), which is able to visualise multiple datasets simultaneously (Figure 12.4).

Furthermore, details on demand implies the ability to switch from aggregated to original data, e.g., allowing to see the individual social media messages that were grouped to a single information chunk in the interface (Reuter, Amelunxen, et al., 2016). Moreover, based on supervised machine learning, configurable relevance classifiers might assist in filtering out irrelevant information and thus reduce

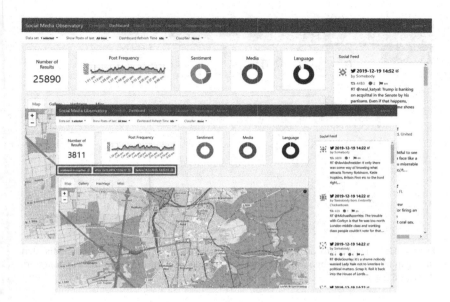

Figure 12.4 Dashboard of the SMO without (background) and with (foreground) enabled interactive chart filtering, showing only information of negative sentiment and in a specific timeframe

the volume of data by only showing relevant data (Habdank et al., 2017). Finally, for situational awareness emergency services have to select the most accurate information or those data with the highest information quality (Shankaranarayanan & Blake, 2017). The core of ESI (chapter 9) constitutes a dashboard which visualises mobile app and social media alerts (Figure 12.5). By clicking on social media alerts, which are "a set of classified messages sharing a similar context, which is defined by event type, keywords, language, location, platform, quality, relevancy, and time" (Kaufhold, Rupp, et al., 2020, p. 328), the emergency manager can see the individual social media posts of an alert, thus retrieving details on demand. The interface furthermore features configurable keyword, relevance (on; off), and quality (low; medium; high) filters. The relevance and quality filters are then based on backend implementations of information mining and information quality components. The information mining component used a Naïve Bayes model for classification social media posts as relevant or irrelevant, which was improved in the later SMO implementation (see chapters 10 and section 12.5.4).

Figure 12.5 Dashboard of the ESI featuring mobile app and grouped social media alert in a map and list view; upon the click on a social media alert, the individual messages are displayed (right)

For the information quality component, a framework for algorithmically measuring information quality in social media was developed in the EmerGent project (Moi et al., 2017). Information quality is understood as a multidimensional concept that is often defined as *fitness for use*, which includes that the right information is available in the right format and quantity at the right time for the right person (Gonzalez & Bharosa, 2009; Kandari, 2010; R. Y. Wang & Strong, 1996). The framework consists of seven information quality criteria: believability, completeness, impact, relevancy, reputation, timeliness, and understandability. These were measured by diverse indicators based on social media data (Table 12.11). As outlined in chapter 9, "the dependencies between criteria and indicators are modelled as nodes of a Directed Acyclic Graph (DAG)" and "the output of each indicator node lies within [0,1] and criterion nodes collect and aggregate the output of indicator nodes dependent on them" (Kaufhold, Rupp, et al., 2020, p. 328). Then, the information quality (IQ) score of each social media post is measured by the weighted mean of all seven criteria scores, whereby the weighting is realized using a backpropagation algorithm (Werbos, 1994).

However, emphasising the subjective component, Hilligoss and Rieh (2008, p. 1469) describe information quality as "people's subjective judgement of goodness and usefulness of information in a certain information use setting with

Table 12.11 Social media information quality framework with criteria (left) and indicators (right). Reproduced from Moi et al. (2017)

Criteria	Indicators
Believability	Existence of URLs, locality, proximity, existence of media files
Completeness	Existence of URLs, number of characters, number of hashtags, what information present, when information present, where information present
Impact	Number of comments, number of shares, involvement, number of likes, number of views
Relevancy	Existence of emergency words, relative frequency of emergency words, amount of contained crawl keywords, number of relevant entities, number of sentences with relevant entities, relative frequency of relevant entities
Reputation	Number of followers, number of statuses, verified account, trusted account
Timeliness	Closeness, post age, first occurrence of an emergency word
Understandability	Average length of words, readability, existence of media files, information noise, appropriate language

respect to their own expectations of information or regard to other information reasonable". Thus, the information quality framework also supported the manual weighting of criteria, allowing emergency services to finetune the weights according to their organisational requirements (Moi et al., 2017). The filtering capabilities of the ESI were further extended by the information gathering component, allowing to formulate and adapt simple or complex search queries using keywords and operators. Finally, by the integration of an alert generator, the filters could not only be applied to individual posts but also alerts which consisted of similar messages grouped together. In this way, multiple layers of configurable filters were implemented in the backend but also made configurable at the interface level by users, allowing emergency services tailoring them according to their current task at hand. As the ESI was developed in the international Emer-Gent consortium, the components were owned by different partners and thus were not integrated into the SMO, which however would have allowed an even more comprehensive combination of backend and frontend filtering techniques.

12.5.6   Evaluation: How Can the Results Inform My Decision and Action?

Having analysed and filtered the information, the user still needs to evaluate the presented information to, in case of emergency services, establish situational awareness or even inform decision making. While the evaluation might be dependent on discussed event-based, organisational, or societal factors, technology can assist the user's *interpretation* of data by rising awareness or supporting *explainability*. First, considering that different information channels are used for different purposes (Hughes et al., 2014), the uneven access to information might also lead to biases in practice when using (cross media) social media analytics tools. The mismatch between access and use, for example, become apparent in the interplay of Facebook and Twitter. While in 2017, 52% of the German population used Facebook and only 9% used Twitter (Reuter, Kaufhold, Spielhofer, et al., 2017), the access to Facebook data is much more restricted than Twitter. As a consequence, (cross-platform) third-party applications for social media analytics and associated research rely more on Twitter data, which however does not represent the populations' social media use (Reuter, Backfried, et al., 2018; Reuter, Hughes, et al., 2018). Furthermore, as platform APIs only return a limited amount of data restricted by sampling and quota limitations (Reuter & Scholl, 2014), besides the manual adjustment based on applied filters, the *provided worldview* of third-party tools is inherently limited. By rising awareness about such limitations, technology might shape emergency managers on how to contextualize the results presented by social media tools.

Second, while machine learning techniques are capable of filtering out irrelevant or low-quality content (Moi et al., 2015), algorithms must not only achieve a high accuracy in decision-making, but emergency managers also need to understand algorithmic learning and decision-making to establish trust in such system during critical and life-threatening situations (Kaufhold, Rupp, et al., 2020). However, current black-box implementations, such as the ones' realized in the ESI (chapter 9), restrict the explainability and transparency of algorithmic decisions, which furthermore increases the effort of configuring the system so that its behaviour becomes more useful to emergency services' needs (Burnett et al., 2017). Thus, white-box approaches should be examined in future research to increase the *explainability* and *transparency* of algorithmic decisions and thus the users' acceptance and trust of complex algorithmic systems (Delibaši et al., 2013; Romero et al., 2013). Moreover, in the evaluation of ESI, emergency managers had no influence on the configuration of algorithms. However, users were requesting the ability to configure the weighting of the information quality component according to organisational needs (Kaufhold, Rupp, et al., 2020). This *usable configurability* is required to combine the demands of both easy-to-use and integrated systems with a configurability of complex components regarding the users' and organisations' objectives.

One transparency-enhancing approach constitutes the real-time evaluation of labelled data in the SMO (chapter 10). While it did not affect the algorithmic decision, i.e. explaining why the message was classified as relevant or not, each post labelled using the relevance labelling interface is submitted to an efficient intermediate model based on incremental learning (Kaufhold, Bayer, et al., 2020). In this way, the accuracy, precision, and recall for the final model can be predicted in real-time during the labelling process, making it easier to assess when a classifier with acceptable quality is achieved (Figure 12.6).

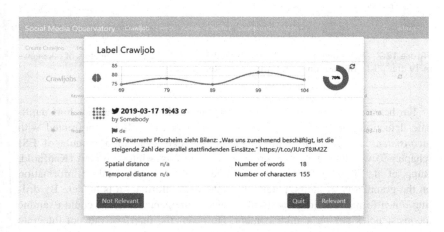

Figure 12.6 Relevance labelling of the SMO, which predicts the accuracy, precision, and recall of a model by the increasing number of labelled posts using incremental learning

Another interesting approach constitutes TrustyTweet, which is an indicator-based approach to assist users in dealing with fake news on Twitter (Hartwig & Reuter, 2019). When different fake news *indicators* are prevalent in a Twitter message, such as consecutive capitalization, excessive usage of punctuation or emotions, default account images, or the absence of an official account verifications seal, amongst others, the browser plugin attaches a warning to the affected message (Figure 12.7). By clicking at the message, a user receives information on *what* fake news indictors are prevalent and gives an explanation *why* these characteristics are considered as an indicator for fake news.

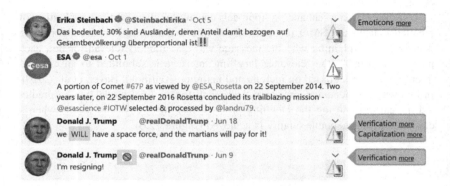

Figure 12.7   The browser plugin TrustyTweet detects and visualises indicators of fake news in Twitter posts. Illustration from Hartwig and Reuter (2019)

Furthermore, despite different steps of information refinement, such as duplicate detection and removal, emergency managers can still be confronted with *inconsistent* and *redundant* information. The alert generation module of ESI (chapter 9) was one attempt to group similar messages into one alert (Kaufhold, Rupp, et al., 2020). In this way, if emergency services just analyse information on the granularity of alerts, potentially redundant information is hidden. By drilling *down* to the individual posts of an alert, emergency services could examine inconsistencies among similar content. However, the interface did not integrate functionality designed for the treatment of inconsistent information, which could be examined in terms of explainability in a different branch of future research. Finally, the artefacts discussed in this thesis were designed for single users of an organisation, not focusing on the inter- and intra-organisational collaboration of emergency services (Reuter & Kaufhold, 2018). Thus, the integration of insights gained from mobile technologies and social media into evidence from disconnected channels or systems must be designed. On a technical level, the backend architecture of 112.social (chapter 11) and ESI (chapter 9) used the Common Alerting Protocol (CAP) to facilitate the *integration* of generated mobile app and social media alerts into command and control room systems (Kaufhold et al., 2018; Kaufhold, Rupp, et al., 2020). Still, the integration of insights must also be designed from the organisational perspective, for instance, if Virtual Operations Support Teams (VOST) assist formal crisis response by information refinement activities and systems (Fathi et al., 2020).

Conclusion and Future Work

13

Almost two decades of mobile technology and social media usage in crises led to the emergence and steadily foundation of crisis informatics, which is a research field combining computing and social science knowledge of disasters to improve ICT-supported crisis management and response (Palen & Anderson, 2016). In order to utilise these sources of social big data for the situational awareness, crisis communication, and decision making of emergency services, available information must be refined according to event-based factors, organizational requirements, societal boundary conditions, and technical feasibility. This dissertation combined the methodological framework of design case studies with principles of design science research to research technologies for information refinement focusing mobile crisis apps and social media (Hevner et al., 2004; Wulf et al., 2011). These extended design case studies comprised the four phases of theoretical reviews, empirical pre-studies, design of supportive ICT, and evaluation of ICT appropriation. This conclusion will outline main findings of the dissertation by answering the initial research questions (section 13.1). Furthermore, it will delimit and discuss empirical and theoretical contributions (section 13.2) followed by design and practical contributions (section 13.3) using the HCI research contribution typology of Wobbrock and Kientz (2016), which differentiates empirical, artefact, methodological, theoretical, dataset, survey, and opinion contributions. The conclusion finishes with the discussion of limitations and future work (section 13.4).

13.1 Main Findings

The first research question asked: ***What are citizens' and emergency services' expectations, perceptions and use patterns with regard to media, especially social***

© The Author(s), under exclusive license to Springer Fachmedien Wiesbaden GmbH, part of Springer Nature 2021
M.-A. Kaufhold, *Information Refinement Technologies for Crisis Informatics*,
https://doi.org/10.1007/978-3-658-33341-6_13

media and mobile apps in crises (RQ1)? In short, citizens in Germany *expect* emergency services and infrastructure operators to monitor and disseminate crisis-relevant information in new media, such as social media and mobile crisis apps. Although they still indicate a preference for using traditional media in crises, such as TV, radio, and personal conversation, they *value* up-to-date information, the reliability of crisis apps, and increased perceived safety using new media. However, they also articulated *barriers* of use regarding data privacy, difficulty of use, lack of threat awareness, rumours, and perceived unreliability of social media. When citizens act as digital volunteers in social media, they conduct diverse *information operations*, especially skilled virtual and technical communities (V&TCs), in an effort to assist in crisis response. However, the chaotic or voluminous use of social media can lead to inconsistent, redundant, or unorganised dissemination of information. Emergency services have a positive attitude towards social media, but still tend to use new media rather for unidirectional *crisis communication* than the integration of *citizen-generated content* for situational awareness or *bidirectional communication*. For an enhanced use of social media, they must ensure law compliance, establish a facilitating organisational culture, and have proper resources available, such as expertise and supportive software (e.g. tools for social media analytics). In order to support the incident management, an increasing number of emergency services establishes virtual operations support teams (VOST), comprising of trusted volunteers (e.g. members of volunteer fire brigades supporting fire departments), to support social media monitoring and communication. More detailed findings are summarized in the following paragraphs:

Usage, role, and perception patterns. The review of crisis informatics research (chapter 4) revealed usage, role, and perception patterns of both emergency services and citizens for social media use in emergencies (Reuter & Kaufhold, 2018). It comprises a *crisis communication matrix* highlighting both emergency services and citizens as receiver and sender of information. It furthermore elaborates a *role typology matrix* which outlines official and public roles in both the real and virtual realm. The findings suggest that emergency services utilize social media to a much higher degree for crisis communication than integrating citizen-generated information for situational awareness. In general, studies showed their positive attitude towards social media, emphasizing the potentials of situation updates, multimedia files and public mood information. However, legal (e.g. data protection, credibility, and issues regarding law compliance), organisational (e.g. lack of expertise, personnel, and training) and technological (e.g. supportive software) barriers are reported barriers for the utilization of social media. The deployment of VOST comprising of trusted volunteers was identified as measure to mitigate some legal and organisational issues. Furthermore, the review highlighted that citizens use social media to organise as self-help communities, which assist emergency services (e.g. dyke construction) or take over tasks that are not parts of their primary goals and objectives (e.g. material and monetary donations) as mostly autonomous units. In this way, social media does not only support the recruitment of unbound helpers from regions that are not affected by the crisis, so-called V&TCs use their expertise to contribute with technical solutions for community-based crisis response.

Insights on media expectations, perceptions, and use. The first representative survey (chapter 5), conducted in 2017 with 1,024 participants from Germany, found out that traditional media such as TV and radio but also personal conversations with friends, family, and neighbours were the main information source in crises for over 70%, while social media (55%) and crisis apps (25%) had notably lower proportions (Kaufhold, Grinko, et al., 2019). In case of a breakdown, over 70% expected infrastructure operators to publish instructions and information on the reason and duration of the incident and the information source. They would furthermore like to receive that information via traditional media, website, or social media (59%) but expect bidirectional communication (41%) to a lower extent. In terms of awareness, German citizens knew their local police station (70%) and administration (68%) better than ambulance stations or firehouses (both 55%). In an acute crisis, according to our qualitative results, citizens showed prosocial behaviour, prioritising communication with relatives and others and seeking information about the crisis in order to ensure firstly personal and secondly material security.

Insights on social media expectations, perceptions, and use. A further representative analysis (chapter 6), based on a survey conducted in 2016 with 1,069 participants from Germany, found out that 52% of the German population use Facebook and 9% use Twitter daily (Reuter, Kaufhold, Spielhofer, et al., 2017). Of all participants, 44% reported that they already searched or shared information over social media in emergencies and, as qualitative analysis revealed, valued social media for being faster than other channels, allowing the spreading and obtaining of up-to-date information. On the other hand, besides some general scepticism, over 60% of German citizens emphasized the barriers of false rumours, unreliable information, data privacy, and the issue that social media might not work in emergency services. Despite only 16% reported that they have already downloaded a crisis app, they indicated future use intentions: more than half of the population would like to receive warnings and safety tips, while 42% would also share information to emergency services using crisis apps.

Insights on crisis apps expectations, perceptions, and use. The findings on crisis apps are supplemented by the third representative analysis (chapter 7), which was based on a survey conducted in 2017 with 1,024 participants from Germany (Grinko et al., 2019). In contrast to 2016, the use of crisis apps increased to 25%, mostly represented by Katwarn (6%) and NINA (4%). Most participants prefer to install a single crisis app (68%) while only 21% would install multiple apps. Almost half of participants welcomes the idea of crisis apps with other everyday utilities, while 44% would like to have such app preinstalled on their phone. In terms of functionality, citizens would like to configure the reception of crisis warning messages according to their preferences (73%), receive recommendations for action (71%) and desire support for their personal crisis preparation (67%). The qualitative analysis revealed perceived advantages, such as current information, fast and efficient warnings, or enhanced personal safety, and disadvantages, such as no need, technical prerequisites, or the availability of other media.

Moreover, the second research question asked: *What are implications for the design of social media technologies and mobile crisis apps to support information refinement for crisis response? (RQ2)?* In short, the design of tools for mobile and social media analytics requires the selection of *channels*, establishment of *access* to the relevant *content* for emergencies, the application of *analytical* methods and *filtering* techniques to support the *evaluation* of gathered information. On frontend level, the filtering by charts, keywords, relevance, quality, and grouping of information were conducive to the establishment of *usable configurability*. As the determination of information quality or relevance was enabled by supervised machine learning approaches, *white-box approaches* are required to increase the explainability and transparency of algorithmic decisions. On backend level, our novel approach for relevance classification highlighted that accuracy of machine learning classifiers can be increased by the use of metadata (besides the mere content of social media messages) and the application of active learning (labelling the posts where the classifier has the lowest confidence). Besides these reactive measures of information refinement, *social media guidelines* as non-technical artefacts can be used to increase citizens media literacy and thus the quality of information disseminated before, during, and after emergencies proactively. Furthermore, in contrast to the unstructured use of social media, the proactive increase of information quality can be achieved by the use of crisis apps for *bidirectional communication*, including the structured reporting of observed incidents (e.g. requiring citizens to attach a category, location, description, and multimedia files to a report). However, contextual factors must be considered for the use (barriers regarding law, personnel, time, and connectivity) and interpretation (e.g. distribution of the app as well as the scale and time of the incident) of information gathered by mobile warning apps. More detailed findings are summarised in the following paragraphs:

Design of frontends and filtering techniques. The Emergency Service Interface (ESI), whose design and evaluation was presented in chapter 9, combined several aspects of information refinement (Kaufhold, Rupp, et al., 2020). It was designed as a multi-*channel* application which provides *access* to heterogenous mobile app and social media data in a unified manner. The components for information mining and information quality were not only used as backend algorithms to qualitatively *analyse* data, in case of the EmerGent project *content* related to fires and floods, but also to act as *filters* of information on interface level. The evaluation of the ESI, which is based on 21 semi-structured interviews conducted across field trials, live demonstrations, and workshop exercise, led to implications for the design of social media technologies in emergencies.

With regard to information filtering, *keyword management* was identified as one of the most important requirements to ensure the success of information refinement and mitigate information overload. While predefined keywords used across field trial did only lead to mediocre results, the continuous adaption of keywords according to ongoing events increased the success of finding relevant information in social media data. The *information quality* component, which could be adjusted show low, medium, or only high-quality information on interface level, was perceived as an important filter to focus on the most important results if personnel resources did not allow to scan through all available information. However, the mediocre accuracy led to some obvious misclassifications, calling for the algorithmic improvement of the approach. More importantly, the information gathering, mining, and quality components were implemented as a *black-box* approach, which allowed emergency services no insights on how well their keywords performed and how the relevance and quality scores were computed. This lack of explainability and transparency hindered emergency services to integrate this information confidentially into decision-making. This calls for the implementation of *white-box* approaches, allowing emergency services to understand and configure the system in a way that is more conducive to their needs (Burnett et al., 2017).

Design of backend relevance classification techniques. As already applied in ESI, one measure to mitigate information overload in social media is the use of supervised machine learning to create models for automatic relevance classification to filter out irrelevant information. Using the base technology of the Social Media Observatory (SMO), chapter 10 outlined implications for the improvement of relevance classifiers based on the evaluation of two datasets related to fire and flood scenarios (Kaufhold, Bayer, et al., 2020). The classification results can be improved by not only analysing the content, but also the *metadata*, such as location or time, available. Here, different metadata combinations yield better classification results for different datasets. Following the assumption of uncertainty sampling, i.e. it is reasonable to label the posts where the classifier has the lowest confidence (Lewis & Catlett, 2014), the use of *active* learning is capable of reaching well-performing classifiers with considerably less training effort, i.e. manual annotation of data. Furthermore, feedback classification was used to correct misclassifications of the model. Here, the retraining time of the model cloud be significantly reduced by the application of *feature subset selection.*

Design of guidelines for social media use. Despite the potentials of these tools for the analysis of social media by emergency services, they follow reactive approaches by refining information after they entered the information space. The social media guidelines for citizens (SMG), on the other hand, aimed for reducing the dissemination of chaotic and unorganised information proactively (Kaufhold, Gizikis, et al., 2019). By providing tips on how to act in social media before, during and after emergencies, the guidelines (chapter 8) aim to increase the users' media literacy, reduce misuse, and improve emergency-related behaviour accordingly. The evaluation of the guidelines indicates a high approval among the German population, with 72% declaring them as a useful artefact and each guideline item reaching an agreement of at least 50%. The positive attitude towards guidelines was consistent across all three emergency phases and the agreement of smartphone and social media users towards guidelines was higher. Furthermore, a small effect was visible for citizens with high education and income. The overall positive feedback was strengthened in the qualitative part of the survey, although citizens desired the creation of guidelines for further communication channels since media literacy and technological access varies across the population.

Design of crisis apps for reporting and bidirectional communication. Another way of refining information proactively is the addition of crisis-specific channels tailored to the needs of emergency services. The evaluation of the mobile crisis app 112.social (chapter 11) for bidirectional communication between emergencies and citizens revealed implications for the design of such apps and contextual factors of their use (Kaufhold et al., 2018). In terms of *design* and for the reporting of incidents by citizens, emergency services valued the categorisation of incidents but required their adjustability according to organisational needs, e.g. in terms of available categories. Textual descriptions and multimedia files of a report were perceived as a measure to support situational awareness but could be supplemented by tutorials on how to best formulate text or take pictures. While the transmitted location should be as precise as possible and thus rely on GPS, insufficient connectivity might lead to situations where a user's manual location annotation is more precise or the only available option. A similar trade-off must be considered with regard to timely information: While the user is capable of submitting a multimedia files directly via 112.social, pictures might be taken outside the app and a life-threatening situation can delay the submission of multimedia files or the whole report. Thus, a good threshold for the currency of files must be found to limit misuse of such app, including fake content.

As *contextual factor* and given their limited resources in personnel and time, emergency services indicated that the processing of individual app alerts might be too time-consuming in large-scale emergencies. In contrast, the location (including local population density), scale and time (including day and daytime) of the incident affects the interpretation of information since multiple app reports at the same location sharing similar pictures could be an indicator of the correctness of information. However, although the credibility of 112.social information was estimated to be higher than in social media, emergency services preferred the implementation of credibility check mechanisms and to hand out the app the app to qualified citizens or trusted volunteers only. By employing VOST for the analysis of app alerts and limiting the number of users, the issues of credibility and volume could be addressed, which however comes with the trade-off of potentially missing out information from non-included citizens. As the use of mobile crisis apps is still low across Germany, which is also influenced by the varying technological access and literacy across the population, there is a need to further examine the promotion and user motivations for installing crisis apps. At the time of evaluation, authorities feared barriers regarding law, personnel and time that have to be solved before a crisis app for bidirectional could be used productively.

13.2 Empirical and Theoretical Contributions

Based on the first research question, this dissertation elaborates three contributions of empirical or theoretical nature (C1–C3). Empirical research contributions provide new knowledge through findings based on observation and data gathering (Wobbrock & Kientz, 2016). Furthermore, according to Wobbrock and Kientz (2016, p. 41), theoretical research contributions "consist of new or improved concepts, definitions, models, principles, or frameworks".

Understanding citizens' use and perception of new media, such as crisis apps and social media, to improve crisis response and tailor technology (C1). The dissertation contributed by the conduction of three *quantitative surveys* on *communication infrastructures* (chapter 5), *social media* (chapter 6) and *crisis apps* (chapter 7) that are representative according to the German population with regard to age, gender, region, education, and income. Especially for social media use in emergencies, there is a large body of qualitative case studies and considerably lower number of quantitative surveys (Reuter, Ludwig, Kaufhold, et al., 2016; Reuter, Hughes, et al., 2018). The latter mostly use opportunity-based samples for communication infrastructures (Petersen et al., 2017), social media (Flizikowski et al., 2014; Reuter & Spielhofer, 2017), and crisis apps (Fischer et al., 2019), which limits the generalizability of statements to some extent. In this way, the first survey comprises findings on the media use during crises, expectations towards critical infrastructure operators, measures during an emergency call failure, and anticipated behaviour in crisis situations (Kaufhold, Grinko, et al., 2019). Moreover, the second survey contributes insights on the use, expectations and barriers of social media use in emergencies as well as the current and intended future use of crisis apps (Reuter, Kaufhold, Spielhofer, et al., 2017). Finally, the third survey complements with the analysis of the distribution, attitudes and reasons for crisis apps use (Grinko et al., 2019).

Understanding and systematising usage, role, and perceptions patterns of citizens and emergency services in crises (C2). Chapter 4 researched *usage, role,* and *perception patterns* of social media in emergencies. It updated established work on usage patterns using the crisis communication matrix, comprising interaction types of crisis communication, inter-organizational crisis management, integration of citizen-generated content, and self-help communities (Reuter et al., 2012). It furthermore proposes a novel role typology matrix, which integrates identified roles according to affiliation (authority/organisation or citizen/public) and realm (real or virtual), emphasizing emergent groups, incident management teams, virtual operations support teams, and virtual & technical communities (Reuter & Kaufhold, 2018). Despite identifying potentials for combining volunteers of the real and virtual realm, existing roles and role typologies either focused on organizational or public roles, but did not integrate them (Bergstrand et al., 2013; Detjen et al., 2016; Reuter et al., 2013)

Design of an information refinement framework for emergency services analysis of media in crises (C3). By the integration of results, chapter 12 proposed an *information refinement framework*, which considers event-based, organisational, and societal boundary conditions and comprises the human-technical design space of information channels, access, content, analysis, filtering, and evaluation. The framework was inspired by the information processing pipeline of the EmerGent project (Kaufhold, Rupp, et al., 2020) and established social media analytics frameworks (Fan & Gordon, 2014; Stieglitz, Mirbabaie, Ross, et al., 2018). While the EmerGent information processing pipeline comprised the steps of information gathering, enrichment, mining, quality, alerts, and publication (Kaufhold, Rupp, et al., 2020), it had a focus on technology and did not consider event-based, organisational, and societal boundary conditions. Furthermore, established social media frameworks followed general-purpose approaches and were not designed to capture the mentioned boundary conditions of crisis management as well as the nuances of information refinement in crisis informatics (Fan & Gordon, 2014; Stieglitz, Mirbabaie, Ross, et al., 2018).

13.3   Design and Practical Contributions

The design of artefacts is a prevalent research activity across the domains of CSCW, CS, HCI, and IS. While empirical contributions "arise from descriptive discovery-driven activities (science), artefact contributions arise from generative design-driven activities (invention)" (Wobbrock & Kientz, 2016, p. 40). Based on the second research question, this dissertation comprises four artefact contributions (C4–C7), which were accompanied with empirical evaluations to get insights on current and future desired functionality, practical use, and usability, and to derive implications for further design and research.

Using social media guidelines to enhance media literacy and facilitate the spread of high-quality information (C4). Chapter 8 presented the design and evaluation of social media guidelines for citizens before, during and after emergencies (SMG). Although research indicated a multitude of general or crisis-specific guidelines, they mostly focus on the organisational use of social media (Kaufhold, Gizikis, et al., 2019). Despite three guidelines included recommendations for citizens (Belfo et al., 2015; Emergency 2.0 Wiki, 2015; Helsloot et al., 2015), practitioners from workshops perceived that these were too extensive for reaching a broad audience (Gizikis, Susaeta, et al., 2017; O'Brien et al., 2016). Thus, this chapter did contribute with the design and evaluation of concise citizen guidelines, which were then illustrated as a picture for distribution (by emergency services) in social media.

Providing interfaces that reduce information overload, i.e. the conversion of large volumes of noisy data into a low volume of rich data (C5). Chapter 9 presented the design and evaluation of a cross-platform alerting system for emergency services to mitigate information overload (ESI). Although there is a variety of existing commercial and research tools for social media analytics, most commercial tools are not tailored to the domain and requirements of crisis management, research tools often focus on a single social media platform, and none of the existing tools comprised a sophisticated component for information quality assessment (Kaufhold et al., 2017; Pohl, 2013). In the information gathering component, ESI combined heterogeneous data (Facebook, Google+, Instagram, Twitter, and

YouTube) into the exchange format Activity Streams 2.0 for unified analysis. After information enrichment, the analysis then comprised an information mining component for reducing irrelevant information, a novel information quality component to measure the quality of information, and an alert generator which similar groups social media posts. The results are then presented on a dashboard comprising key indicators, a map and list views, as well as configurable relevance and quality filters. Furthermore, based on requirement analyses and an evaluation (Reuter, Amelunxen, et al., 2016), the tool was tailored to the needs of emergency services.

Using metadata and active learning techniques to establish the rapid creation of accurate relevance classifiers (C6). Chapter 10 comprises a social media collection and analysis tool which integrates a novel concept for rapid relevance classification of social media posts in disasters and emergencies (SMO). Despite ESI integrated an information mining component for relevance classification, research revealed that the used Naïve Bayes classifier was outperformed by other supervised machine learning techniques, especially Random Forest (Habdank et al., 2017). While the former work only analysed the content of social media posts, the evaluation of chapter 10 revealed that the incorporation of metadata of social media posts, such as location or time, improved the quality of relevance classification using Random Forests. The set of metadata, however, varied across different datasets requiring individual optimization for different emergency scenarios, such as fires or floods. Furthermore, the creation of a classifier is a time-consuming task, which requires the manual annotation of training data, which is often not possible due to the time-critical constraints of emergencies (M. Imran et al., 2018). By the use of active learning, our approach was able to reduce the training time of Random Forest classifiers, which required up to only a quarter of labelled data compared to a traditional batch learning approach.

Using crisis apps for bidirectional, organised, and structured information exchange (C7). Chapter 11 comprises a novel mobile app for bidirectional communication between emergency services and citizens (112.social). Existing and established crisis apps, such as Katwarn and NINA in Germany, focus on crisis warning functionality and recommendations of action as measures of unidirectional communication (Reuter, Kaufhold, Leopold, et al., 2017b). In contrast, 112.social facilitates bidirectional communication allowing citizens to report emergencies (by setting a category, subcategory, description, multimedia files and location of the incident) and react to broadcasts of emergency services in a chat-based manner. In this way, emergency services can request further information from potential eyewitnesses. By evaluating 112.social with emergency services and citizens, it furthermore contributed to the lack of scientific studies on the evaluation of crisis (Groneberg et al., 2017).

13.4   Limitations and Future Work

This thesis contains articles published to research domains such as CSCW, CS, HCI, and IS, which emphasises the interdisciplinary nature of crisis informatics (Palen & Anderson, 2016). As such, this dissertation is subject to limitations and offers potentials for future research. When interpreting the crisis communication and role typology matrices in chapter 4, it has to be considered that these models provide structured but limited worldviews on their specific topic. In both matrices, for instance, the role of media organisations is not subject of analysis (Ehnis et al., 2014). However, in future work CSCW research will be useful to further elaborate identified usage and role patterns in new media as a foundation to improve collaboration between emergency services, citizens, and other involved organisation in the ever-changing landscape of mobile technologies and social media.

Although the representative surveys on communication infrastructures (chapter 5), social media (chapter 6), and mobile apps (chapter 7) convey representative results in terms of age, gender, region, income, and education, they were distributed as online surveys which just covers people how are willing to do online surveys. Besides the potential prevalence of social desirability bias, the reported behaviour in the survey might not be congruent with the actual use. Furthermore, the surveys constitute single points of inquiry and insights from disciplines such as psychology and social sciences could be valuable in professionalising and standardising the underlying questionnaires as a foundation for longitudinal research and trend analyses. For instance, while chapter 5 highlighted that German citizens perceived the usefulness of social media higher than crisis apps, this shifted in favour of crisis apps in follow-up studies (Haunschild et al., 2020; Kaufhold, Haunschild, et al., 2020).

Analogously, the social media guidelines for citizens (SMG) were evaluated using a representative sample of German citizens (chapter 8). However, intercultural differences such as *risk cultures* could lead to different perceptions of the relevance of specific recommendations. In contrast to the authority-centred risk culture in Germany, which emphasises the obligation of authorities to solve crises, citizens of individual-centred risk cultures, such as prevalent in the Netherlands, feel themselves more responsible for overcoming crises (Reuter, Kaufhold, et al., 2019). In this way, supported by insights from social sciences and media studies, future social media guidelines could be refined according to risk cultures or educate, for instance, citizens in an authority-centred risk culture on more individual crisis preparation and response as a complement to increase societal resilience. Moreover, as the perception towards social media guidelines was evaluated using a survey, the designed social media guidelines may be distributed in social media before or during emergencies to observe and analyse their *actual* reception by citizens.

From the perspective of HCI, the evaluation of the ESI (chapter 9) revealed the emergency services' need for *usable configurability*, which "demands, on the one hand, easy-to-use and integrated systems and, on the other hand, a configurability of (complex) components regarding the users' and organisations' use cases to achieve, in this case, the goal of a low volume of rich and useful content for emergency services" (Kaufhold, Rupp, et al., 2020, p. 336). Furthermore, to facilitate the integration of citizen-generated content by mobile and social media technologies into decision-making and action, research into *whitebox representations* of algorithms and their decisions are required to enhance the explainability, transparency and trust of emergency services in complex machine learning systems.

Besides interfaces and the representation of algorithms, the computational optimization of algorithms themselves are subject of design. The evaluation of the relevance classification module of the SMO (chapter 10) reveals that CS research could further contribute by examining the varying impact of metadata combinations on the accuracy of relevance classification using supervised machine learning methods such as Random Forests. Since the designed active learning approach showed a varying success in reducing the required number of manually annotated data across both evaluated datasets, future research should examine which characteristics of data led to these different outcomes. Another promising research direction seems the combination of active learning and *domain adaptation*, whereby the latter examines how datasets from pasts events can be combined with data from new emergencies to reduce the amount of required annotated data (M. Imran et al., 2018).

Considering the design of crisis apps for bidirectional communication between emergency services and citizens, such as 112.social (chapter 11), further CS research is required on how to make crisis reporting and warning possible when central infrastructures fail. Infrastructure-less technologies such as off-grid ad-hoc networks could be further explored as an opportunity to move information to devices or into zones with established connectivity (Alvarez et al., 2016). Additionally, infrastructure-*sparse* technologies such as low-power wide-area networks could be used to provide warning messages using a low number of nodes with emergency generators (Höchst et al., 2020). Furthermore, insights from legal theory and organisational strategies and tactics are required to overcome the existing barriers regarding law, personnel, and time. Here, IS research could further contribute by mediating organisational and technological design trade-offs and provide rigour evaluations of established crisis apps (Groneberg et al., 2017).

Finally, the developed information refinement framework (chapter 12) was inspired by the overall findings of this thesis, the EmerGent information processing pipeline, and existing social media analytics frameworks. In this way, the framework is not the result of empirical and theoretical investigations that were designed for the purpose of creating a framework; instead, it is based on high-level reflections upon the accumulated findings. A systematic review of related literature and empirical investigations, such as a Delphi study or interviews with software designers, practitioners, and researchers (Hiltz et al., 2020), explicitly designed for the issue of information refinement are likely to revise and lead to a more mature version of the proposed framework, including the elaboration of the human-technical interplay (HCI) as well as further practical and theoretical implications.

References

Abel, F., Hauff, C., Houben, G.-J., Stronkman, R., Tao, K., & Stronkman, R. (2012). Semantics + Filtering + Search = Twitcident Exploring Information in Social Web Streams Categories and Subject Descriptors. *23rd ACM Conference on Hypertext and Social Media, HT'12*, 285–294. https://doi.org/10.1145/2309996.2310043

Abel, F., Hauff, C., & Stronkman, R. (2012). Twitcident: Fighting Fire with Information from Social Web Streams. *Proceedings of the 21st International Conference Companion on World Wide Web*, 5–8.

Acar, A., & Muraki, Y. (2011). Twitter for crisis communication: lessons learned from Japan's tsunami disaster. *International Journal of Web Based Communities, 7*(3), 392–402. https://doi.org/10.1504/IJWBC.2011.041206

Adam, N., Eledath, J., Mehrotra, S., & Venkatasubramanian, N. (2012). Social Media Alert and Response to Threats to Citizens (SMART-C). *Proceedings Of The 2012 8th International Conference on Collaborative Computing: Networking, Applications and Worksharing (Collaboratecom 2012)*, 181–189. https://doi.org/10.4108/icst.collaboratecom.2012.250713

Agarwal, N., & Yiliyasi, Y. (2010). Information quality challenges in social media. *Proceedings of the 15th International Conference on Information Quality (ICIQ-2010), January 2010*, 234–248.

Agogo, D., & Hess, T. J. (2018). "How does tech make you feel?" a review and examination of negative affective responses to technology use. *European Journal of Information Systems, 27*(5), 570–599. https://doi.org/10.1080/0960085X.2018.1435230

Agogo, D., & Hess, T. J. (2015). Technostress and Technology Induced State Anxiety: Scale Development and Implications. *Thirty Sixth International Conference on Information Systems (ICIS)*, 1–11.

Aha, D. W., Kibler, D., & Albert, M. K. (1991). Instance-based learning algorithms. *Machine Learning*. https://doi.org/10.1007/BF00153759

Akerkar, R., Friberg, T., & Amelunxen, C. (2016). *EmerGent Deliverable 3.5: User Requirements, Version 2*. https://www.fp7-emergent.eu/wp-content/uploads/2016/03/20160229_D3.5_RequirementsVersion2_EmerGent_final2.pdf

Akerkar, R., Friberg, T., & Gizikis, A. (2014). *EmerGent Deliverable 3.4: User Requirements, Version 1.* https://www.fp7-emergent.eu/d3-4-user-requirements-version-1/

Akhgar, B., Fortune, D., Hayes, R. E., Guerra, B., & Manso, M. (2013). Social media in crisis events: Open networks and collaboration supporting disaster response and recovery. *2013 IEEE International Conference on Technologies for Homeland Security (HST)*, 760–765. https://doi.org/10.1109/THS.2013.6699099

Al-Akkad, A., & Raffelsberger, C. (2014). How do I get this app? A discourse on distributing mobile applications despite disrupted infrastructure. In S. R. Hiltz, M. S. Pfaff, L. Plotnick, & P. C. Shih (Eds.), *Proceedings of the International Conference on Information Systems for Crisis Response and Management (ISCRAM)* (pp. 565–569). ISCRAM.

Al-Akkad, A., Ramirez, L., Boden, A., Randall, D., Zimmermann, A., & Augustin, S. (2014). Help Beacons: Design and Evaluation of an Ad-Hoc Lightweight S.O.S. System for Smartphones. In M. Jones, P. Palanque, A. Schmidt, & T. Grossman (Eds.), *Proceedings of the SIGCHI Conference on Human Factors in Computing Systems (CHI '14)* (pp. 1485–1494). ACM. https://doi.org/10.1145/2556288.2557002

Alam, F., Ofli, F., & Imran, M. (2018). Processing Social Media Images by Combining Human and Machine Computing during Crises. *International Journal of Human-Computer Interaction, 34*(4), 311–327. https://doi.org/10.1080/10447318.2018.1427831

Alam, F., Ofli, F., & Imran, M. (2019). Descriptive and visual summaries of disaster events using artificial intelligence techniques: case studies of Hurricanes Harvey, Irma, and Maria. *Behaviour & Information Technology (BIT)*, 1–31. https://doi.org/10.1080/0144929X.2019.1610908

Albris, K. (2017). The switchboard mechanism: How social media connected citizens during the 2013 floods in Dresden. *Journal of Contingencies and Crisis Management.* https://doi.org/10.1111/1468-5973.12201

Alcaidinho, J., Freil, L., Kelly, T., Marland, K., Wu, C., Wittenbrook, B., Valentin, G., & Jackson, M. (2017). Mobile Collaboration for Human and Canine Police Explosive Detection Teams. *Proceedings of the 2017 ACM Conference on Computer Supported Cooperative Work and Social Computing – CSCW '17*, 925–933.

Alexander, D. E. (2014). Social Media in Disaster Risk Reduction and Crisis Management. *Science and Engineering Ethics, 20*(3), 717–733. https://doi.org/10.1007/s11948-013-9502-z

Allen, C. (2004). *Tracing the Evolution of Social Software.* https://www.lifewithalacrity.com/2004/10/tracing_the_evo.html

Alvarez, F., Hollick, M., & Gardner-Stephen, P. (2016). Maintaining both availability and integrity of communications: Challenges and guidelines for data security and privacy during disasters and crises. *GHTC 2016 – IEEE Global Humanitarian Technology Conference: Technology for the Benefit of Humanity, Conference Proceedings*, 62–69. https://doi.org/10.1109/GHTC.2016.7857261

American Red Cross. (2012). *More Americans Using Mobile Apps in Emergencies.* https://www.redcross.org/news/press-release/More-Americans-Using-Mobile-Apps-in-Emergencies

An, J., Kwak, H., Mejova, Y., & Oger, D. (2016). Are you Charlie or Ahmed? Cultural pluralism in Charlie Hebdo response on Twitter. *International AAAI Conference on Web and Social Media 2016*, 1–10.

Andrews, C., Fichet, E., Ding, Y., Spiro, E. S., & Starbird, K. (2016). Keeping Up with the Tweet-dashians: The Impact of "Official" Accounts on Online Rumoring. *Proceedings of the 19th ACM Conference on Computer-Supported Cooperative Work & Social Computing (CSCW '16)*, 452–465. https://doi.org/10.1145/2818048.2819986

Arif, A., Robinson, J. J., Stanek, S. A., Fichet, E. S., Townsend, P., Worku, Z., & Starbird, K. (2017). A Closer Look at the Self-Correcting Crowd: Examining Corrections in Online Rumors. *Proceedings of the 2017 ACM Conference on Computer Supported Cooperative Work and Social Computing*, 155–168. https://doi.org/10.1145/2998181.2998294

Ashktorab, Z., Brown, C., Nandi, M., & Culotta, A. (2014). Tweedr: Mining Twitter to Inform Disaster Response. In S. R. Hiltz, M. S. Pfaff, L. Plotnick, & P. C. Shih (Eds.), *Proceedings of the International Conference on Information Systems for Crisis Response and Management (ISCRAM)* (Issue May, pp. 354–358). https://doi.org/10.1145/1835449.183 5643

Austin, L., Liu, B. F., & Jin, Y. (2012). How audiences seek out crisis information: Exploring the social-mediated crisis communication model. *Journal of Applied Communication Research*, *40*(2), 188–207. https://doi.org/10.1080/00909882.2012.654498

Avvenuti, M., Cresci, S., Marchetti, A., Meletti, C., & Tesconi, M. (2014). EARS (Earthquake Alert and Report System): A Real Time Decision Support System for Earthquake Crisis Management. *Proceedings of the 20th ACM SIGKDD International Conference on Knowledge Discovery and Data Mining*, 1749–1758. https://doi.org/10.1145/2623330. 2623358

Bachmann, D. J., Jamison, N. K., Martin, A., Delgado, J., & Kman, N. E. (2015). Emergency Preparedness and Disaster Response: There's An App for That. *Prehospital and Disaster Medicine*, *30*(5), 486–490. https://doi.org/10.1017/S1049023X15005099

Baird, M. E. (2010). *The Phases of Emergency Management*. https://www.memphis.edu/ifti/ pdfs/cait_phases_of_emergency_mngt.pdf

Bannon, L. J., & Schmidt, K. (1989). CSCW – Four Characters in Search of a Context. *DAIMI Report Series*, *18*(289), 358–372. https://doi.org/10.7146/dpb.v18i289.6667

BBK. (2015). *Warn-App NINA*. https://www.bbk.bund.de/DE/NINA/Warn-App_NINA.html

Becker, J., & Niehaves, B. (2007). Epistemological perspectives on IS research: A framework for analysing and systematizing epistemological assumptions. *Information Systems Journal*, *17*(2), 197–214. https://doi.org/10.1111/j.1365-2575.2007.00234.x

Bekele, D., Eshete, B., Villafiorita, A., & Weldemariam, K. (2010). Context information refinement for pervasive medical systems. *4th International Conference on Digital Society, ICDS 2010, Includes CYBERLAWS 2010: The 1st International Conference on Technical and Legal Aspects of the e-Society*, 210–215. https://doi.org/10.1109/ICDS.2010.42

Belfo, J., Simão, L., Schmidt, S., Rhode, D., Freitag, S., Lück, A., Schönefeld, M., Lemanski, S., Ross, D., Cooke, M., Curristan, S., Gouveia, R., Grilo, P., Delavallade, T. T., Päivinen, N., Jäntti, M., Kurki, T., Hokkanen, L., Honkanen, M., & Villot, E. (2015). *Deliverable D2.271: iSAR+ Guidelines.*

Bergstrand, F., Landgren, J., & Green, V. (2013). Authorities don't tweet, employees do! *Proceedings of the International Conference on Human-Computer Interaction with Mobile Devices and Services (MobileHCI).*

Berliner Feuerwehr. (2012). *Social-Media-Guideline – Empfehlungen für einen sicheren Umgang mit sozialen Medien*. https://www.berliner-feuerwehr.de/fileadmin/bfw/dokume nte/Download/2012/2012_01_SM-Guideline.pdf

Bernard, J., Zeppelzauer, M., Lehmann, M., Müller, M., & Sedlmair, M. (2018). Towards User-Centered Active Learning Algorithms. *Computer Graphics Forum, 37*(3), 121–132. https://doi.org/10.1111/cgf.13406

Birkbak, A. (2012). Crystallizations in the Blizzard: Contrasting Informal Emergency Collaboration In Facebook Groups. *Proceedings of the Nordic Conference on Human-Computer Interaction (NordiCHI), 428–437.*

Blei, D. M., Ng, A. Y., & Jordan, M. I. (2003). Latent Dirichlet Allocation. *Journal of Machine Learning Research, 3*(4–5), 993–1022. https://doi.org/10.1016/b978-0-12-411519-4.00006-9

Blum, J., Kefalidou, G., Houghton, R., Flintham, M., Arunachalam, U., & Goulden, M. (2014). Majority report: Citizen empowerment through collaborative sensemaking. *Proceedings of the International Conference on Information Systems for Crisis Response and Management (ISCRAM), May,* 767–771.

Boin, A., & McConnell, A. (2007). Preparing for critical infrastructure breakdowns: The limits of crisis management and the need for resilience. *Journal of Contingencies and Crisis Management, 15*(1), 50–59. https://doi.org/10.1111/j.1468-5973.2007.00504.x

Boin, A., Stern, E., & Sundelius, B. (2005). *The Politics of Crisis Management: Public Leadership Under Pressure.* Cambridge University Press.

Borlund, P. (2003). The Concept of Relevance in Information Retrieval. *Journal of the American Society for Information Science and Technology, 54*(10), 913–925. https://doi.org/10.1002/asi.10286

Botega, L. C., Ferreira, L. C., Oliveira, N. P., Oliveira, A., Berti, C. B., De Neris, V. P., & De Araújo, R. B. (2015). SAW-oriented user interfaces for emergency dispatch systems. *Lecture Notes in Computer Science (Including Subseries Lecture Notes in Artificial Intelligence and Lecture Notes in Bioinformatics), 9173,* 537–548. https://doi.org/10.1007/978-3-319-20618-9_53

Botega, L. C., Valdir, A. P., Oliveira, A. C. M., Saran, J. F., Villas, L. A., & De Araujo, R. B. (2017). Quality-aware human-driven information fusion model. *International Conference on Information Fusion.* https://doi.org/10.23919/ICIF.2017.8009851

Bruns, A. (2014). Social media and journalism during times of crisis. In J. Hunsinger & T. Senft (Eds.), *The Social Media Handbook.* Routledge.

Brynielsson, J., Granåsen, M., Lindquist, S., Narganes Quijano, M., Nilsson, S., & Trnka, J. (2018). Informing crisis alerts using social media: Best practices and proof of concept. *Journal of Contingencies and Crisis Management (JCCM), 26*(1), 28–40. https://doi.org/10.1111/1468-5973.12195

Brynielsson, J., Johansson, F., Jonsson, C., & Westling, A. (2014). Emotion classification of social media posts for estimating people's reactions to communicated alert messages during crises. *Security Informatics, 3*(1), 7. https://doi.org/10.1186/s13388-014-0007-3

Bundesamt für Sicherheit in der Informationstechnik. (2008). *Notfallmanagement – BSI-Standard 100-4.* https://www.bsi.bund.de/SharedDocs/Downloads/DE/BSI/Publikationen/ITGrundschutzstandards/BSI-Standard_1004.pdf

Bundesanstalt Technisches Hilfswerk. (2011). *Verhalten in sozialen Netzwerken.* https://thw-eisenach.de/uploads/media/social_Media_guidelines_final_02_03_2012.pdf

Bundesministerium des Inneren. (2009). *Nationale Strategie zum Schutz Kritischer Infrastrukturen.* https://www.bmi.bund.de/SharedDocs/downloads/DE/publikationen/themen/bevoelkerungsschutz/kritis.html

Bundesministerium des Inneren. (2014). *Leitfaden Krisenkommunikation.* https://www.bmi. bund.de/SharedDocs/downloads/DE/publikationen/themen/bevoelkerungsschutz/leitfa den-krisenkommunikation.pdf

Bundesnetzagentur und Bundeskartellamt. (2016). *Monitoringbericht 2016.* https://www.bun desnetzagentur.de/

Bundeszentrale für politische Bildung (bpb). (2013). *Hochwasser in Deutschland 2013.* https://www.bpb.de/politik/hintergrund-aktuell/163064/hochwasser-in-deutschland-12-06-2013

Bundeszentrale für politische Bildung (bpb). (2016). *Datenreport 2016. Ein Sozialbericht für die Bundesrepublik Deutschland.* Statistisches Bundesamt. https://www.destatis.de/

Burnap, P., Williams, M. L., Sloan, L., Rana, O., Housley, W., Edwards, A., Knight, V., Procter, R., & Voss, A. (2014). Tweeting the terror: modelling the social media reaction to the Woolwich terrorist attack [Article]. *Social Network Analysis and Mining, 4*(1), 1–14. https://doi.org/10.1007/s13278-014-0206-4

Burnett, M., Kulesza, T., Oleson, A., Ernst, S., Beckwith, L., Cao, J., Jernigan, W., & Grigoreanu, V. (2017). Toward Theory-Based End-User Software Engineering. In F. Paternò & V. Wulf (Eds.), *New Perspectives in End-User Development* (pp. 231–268). Springer International Publishing. https://doi.org/10.1007/978-3-319-60291-2_10

Cambridge Dictionary. (2020). *Definition of Refinement.* https://dictionary.cambridge.org/dic tionary/english/refinement

Cameron, M. A., Power, R., Robinson, B., & Yin, J. (2012). Emergency Situation Awareness from Twitter for Crisis Management. *Proceedings of the 21st International Conference Companion on World Wide Web,* 695–698. https://doi.org/10.1145/2187980.2188183

Canadian Red Cross. (2012). *Social Media during Emergencies.* https://www.redcross.ca/cms lib/general/pub_social_media_in_emergencies_survey_oct2012_en.pdf

Caragea, C., Mcneese, N., Jaiswal, A., Traylor, G., Kim, H., Mitra, P., Wu, D., Tapia, A. H., Giles, L., Jansen, B. J., & Yen, J. (2011). Classifying Text Messages for the Haiti Earthquake. *Proceedings of the International Conference on Information Systems for Crisis Response and Management (ISCRAM), May,* 1–10. https://doi.org/10.1.1.370.6804

Caragea, C., Silvescu, A., & Tapia, A. H. (2016). Identifying Informative Messages in Disasters using Convolutional Neural Networks. In A. H. Tapia, P. Antunes, V. A. Bañuls, K. Moore, J. Porto, & 13th International Conference on Information Systems for Crisis Response and Management (Eds.), *Proceedings of the International Conference on Information Systems for Crisis Response and Management (ISCRAM).* Federal University of Rio de Janeiro.

Caragea, C., Squicciarini, A., Stehle, S., Neppalli, K., & Tapia, A. H. (2014). Mapping Moods: Geo-Mapped Sentiment Analysis During Hurricane Sandy. In S. R. Hiltz, M. S. Pfaff, L. Plotnick, & A. C. Robinson (Eds.), *Proceedings of the International Conference on Information Systems for Crisis Response and Management (ISCRAM)* (Issue May, pp. 642–651). ISCRAM.

Castillo, C. (2016). *Big Crisis Data: Social Media in Disasters and Time-Critical Situations.* Cambridge University Press.

Chaturvedi, A., Simha, A., & Wang, Z. (2015). ICT infrastructure and social media tools usage in disaster/crisis management. *2015 Regional Conference of the International Telecommunications Society (ITS), Los Angeles, CA, 25–28 October, 2015.*

Chauhan, A., & Hughes, A. L. (2017). Providing Online Crisis Information: An Analysis of Official Sources during the 2014 Carlton Complex Wildfire. *Proceedings of the 35th International Conference on Human Factors in Computing Systems (CHI 2017)*. https://doi.org/10.1145/3025453.3025627

Chen, C., Carolina, N., & Ractham, P. (2011). Lessons Learned from the Use of Social Media in Combating a Crisis: A Case Study of 2011 Thailand Flooding Disaster. *Proceedings of the International Conference on Information Systems (ICIS)*, 1–17.

Cheong, M., & Lee, V. C. S. (2011). A microblogging-based approach to terrorism informatics: Exploration and chronicling civilian sentiment and response to terrorism events via Twitter. *Information Systems Frontiers, 13*(1), 45–59. https://doi.org/10.1007/s10796-010-9273-x

Choudrie, J., & Dwivedi, Y. K. (2005). Investigating the Research Approaches for Examining Technology Adoption Issues. *Journal of Research Practice, 1*(1), 1–12.

Choudrie, J., & Dwivedi, Y. K. (2006). A comparative study to examine the socio-economic characteristics of broadband adopters and non-adopters. *Electronic Government, an International Journal, 3*(3), 272–288. https://doi.org/10.1504/EG.2006.009599

Choudrie, J., Weerakkody, V., & Jones, S. (2005). Realising e-government in the UK: rural and urban challenges. *The Journal of Enterprise Information Management, 18*(5), 568–585. https://doi.org/10.1108/17410390510624016

Chrpa, L., & Thórisson, K. R. (2013). On Applicability of Automated Planning for Incident Management. *The International Scheduling and Planning Applications WoRKshop*, 1–7.

Clarivate Analytics. (1994). *The Clarivate Analytics Impact Factor*. https://clarivate.com/web ofsciencegroup/essays/impact-factor/

Cobb, C., McCarthy, T., Perkins, A., Bharadwaj, A., Comis, J., Do, B., & Starbird, K. (2014). Designing for the Deluge: Understanding & Supporting the Distributed, Collaborative Work of Crisis Volunteers. *Proceedings of the Conference on Computer Supported Cooperative Work (CSCW)*, 888–899. https://doi.org/10.1145/2531602.2531712

Coleman, L. (2006). Frequency of Man-Made Disasters in the 20th Century. *Journal of Contingencies and Crisis Management (JCCM), 14*(1), 3–11.

Computing Research and Education Association of Australasia. (2018). *CORE Conference and Journal Rankings*. https://www.core.edu.au/conference-portal

Convertino, G., Mentis, H. M., Slavkovic, A., Rosson, M. B., & Carroll, J. M. (2011). Supporting common ground and awareness in emergency management planning. *ACM Transactions on Computer-Human Interaction (TOCHI), 18*(4), 1–34.

Coombs, W. T. (2009). Conceptualizing Crisis Communication. In R. L. Heath & D. O'Hair (Eds.), *Handbook of Risk and Crisis Communication* (pp. 99–118). Routledge.

Coombs, W. T. (2010). Parameters for Crisis Communication. In W. T. Coombs & S. J. Holladay (Eds.), *The Handbook of Crisis Communication* (pp. 17–53). Wiley-Blackwell.

Coombs, W. T. (2014). *Ongoing crisis communication: planning, managing, and responding*. Sage Publications Ltd.

Cornia, A., Dressel, K., & Pfeil, P. (2016). Risk cultures and dominant approaches towards disasters in seven European countries. *Journal of Risk Research, 19*(3), 288–304. https://doi.org/10.1080/13669877.2014.961520

Dailey, D., & Starbird, K. (2017). Social Media Seamsters: Stitching Platforms & Audiences into Local Crisis Infrastructure. *Proceedings of the ACM Conference on Computer-Supported Cooperative Work & Social Computing (CSCW)*.

Daimler AG. (2012). *Social Media Leitfaden – 10 Tipps zum Umgang mit Social Media.* https:// www.daimler.com/dokumente/konzern/sonstiges/daimler-socialmedialeitfaden-de.pdf

Davis, F. D. (1993). User acceptance of information technology: system characteristics, user perceptions and behavioral impacts. *International Journal of Man-Machine Studies, 38*(3), 475–487.

Davis, F. D., & Venkatesh, V. (2004). Toward preprototype user acceptance testing of new information systems: Implications for software project management. *IEEE Transactions on Engineering Management, 51*(1), 31–46. https://doi.org/10.1109/TEM.2003.822468

Dawes, S. S., Cresswell, A. M., & Cahan, B. B. (2004). Learning from Crisis: Lessons in Human and Information Infrastructure from the World Trade Center Response. *Social Science Computer Review, 22*(1), 52–66. https://doi.org/10.1177/0894439303259887

de Albuquerque, J. P., Fonte, C. C., De Almeida, J.-P., & Cardoso, A. (2016). How volunteered geographic information can be integrated into emergency management practice? First lessons learned from an urban fire simulation in the city of Coimbra. *Proceedings of the International Conference on Information Systems for Crisis Response and Management (ISCRAM).*

de Albuquerque, J. P., Herfort, B., Brenning, A., & Zipf, A. (2015). A geographic approach for combining social media and authoritative data towards identifying useful information for disaster management. *International Journal of Geographical Information Science, 29*(4), 667–689. https://doi.org/10.1080/13658816.2014.996567

Delibaši, B., Vuki, M., Jovanovi, M., & Suknovi, M. (2013). White-Box or Black-Box Decision Tree Algorithms: Which to Use in Education? *IEEE Transactions on Education, 56*(3), 287–291. https://doi.org/10.1109/TE.2012.2217342

Denef, S., Bayerl, P. S., & Kaptein, N. (2013). Social Media and the Police — Tweeting Practices of British Police Forces during the August 2011 Riots. *Proceedings of the Conference on Human Factors in Computing Systems (CHI),* 3471–3480.

Detjen, H., Volkert, S., & Geisler, S. (2016). Categorization of Volunteers and their Motivation in Catastrophic Events. *Mensch & Computer 2016: Sozial Digital – Gemeinsam Auf Neuen Wegen, September,* 1–8.

Deutscher Caritasverband. (2014). *Das Soziale ins Netz bringen – die Caritas und soziale Medien.* https://www.caritas.de/cms/contents/caritasde/medien/dokumente/zentraledoku mente/socialmedialeitlinie/social_media_leitlinien_caritas_mitarbeiter.pdf

Deutsches Rotes Kreuz. (2012). *Ein Leitfaden zum Umgang mit Social Media im DRK.* https://www.drk-baden-wuerttemberg.de/fileadmin/Eigene_Bilder_und_Vid eos/Publikationen/Mediathek/Social_Media_Leitfaden_DRK.pdf

Deutsches Rotes Kreuz. (2013). *DRK-Untersuchung zur Rolle von ungebundenen HelferInnen und Sozialen Netzwerken bei der Bewältigung des Jahrhunderthochwassers im Juni 2013.* https://www.b-b-e.de/fileadmin/inhalte/aktuelles/2013/10/NL22_DRK_Definition.pdf

Deutsches Rotes Kreuz. (2018). *Rotkreuz-App "Mein DRK."* https://mobil.drk-intern.de/

Dittus, M., Quattrone, G., & Capra, L. (2017). Mass Participation During Emergency Response: Event-centric Crowdsourcing in Humanitarian Mapping. *Proceedings of the 2017 ACM Conference on Computer-Supported Cooperative Work and Social Computing (CSCW '17),* 1290–1303. https://doi.org/10.1145/2998181.2998216

Dressel, K. (2015). Risk culture and crisis communication. *International Journal of Risk Assessment and Management, 18*(2), 115–124. https://doi.org/10.1504/IJRAM.2015.069020

Dressel, K., & Pfeil, P. (2017). Socio-cultural factors of risk and crisis communication: Crisis communication or what civil protection agencies should be aware of when communicating with the public in crisis situations. In M. Klafft (Ed.), *Risk and Crisis Communication for Disaster Prevention and Management* (pp. 64–76).

Ehnis, C., Mirbabaie, M., Bunker, D., & Stieglitz, S. (2014). The role of social media network participants in extreme events. *25th Australian Conference of Information Systems, Hasselmann.* https://doi.org/10.1109/HICSS.2016.33

Eisenberg, M. B. (1988). Measuring Relevance Judgments. *Information Processing and Management, 24*(4), 373–389. https://doi.org/10.1016/0306-4573(88)90042-8

Eismann, K., Posegga, O., & Fischbach, K. (2016). Collective Behaviour, Social Media, and Disasters: a Systematic Literature Review. *European Conference on Information Systems (ECIS).*

Eismann, K., Posegga, O., & Fischbach, K. (2018). Decision Making in Emergency Management: The Role of Social Media. *Proceedings of the 26th European Conference on Information Systems (ECIS).*

Emergency 2.0 Wiki. (2015). *Emergency 2.0 Wiki.* https://emergency20wiki.org/

EmerGent. (2017). *Emergency Management in Social Media Generation.* https://www.fp7-emergent.eu/

Endsley, T., Wu, Y., Eep, J., & Reep, J. (2014). The Source of the Story: Evaluating the Credibility of Crisis Information Sources. *Proceedings of the International Conference on Information Systems for Crisis Response and Management (ISCRAM), 1*(1), 158–162.

Eriksson, M., & Olsson, E. K. (2016). Facebook and Twitter in Crisis Communication: A Comparative Study of Crisis Communication Professionals and Citizens. *Journal of Contingencies and Crisis Management, 24*(4), 198–208. https://doi.org/10.1111/1468-5973.12116

Eshghi, K., & Larson, R. C. (2008). Disasters: lessons from the past 105 years. *Disaster Prevention and Management: An International Journal, 17*(1), 62–82. https://doi.org/10.1108/09653560810855883

Facebook. (2018). *Crisis Response.* https://www.facebook.com/crisisresponse/

Fahad, A., Alshatri, N., Tari, Z., Alamri, A., Khalil, I., Zomaya, A. Y., Foufou, S., & Bouras, A. (2014). A survey of clustering algorithms for big data: Taxonomy and empirical analysis. *IEEE Transactions on Emerging Topics in Computing, 2*(3), 267–279. https://doi.org/10.1109/TETC.2014.2330519

Fan, W., & Gordon, M. D. (2014). The Power of Social Media Analytics. *Communications of the ACM, 57*(6), 74–81. https://doi.org/10.1145/2602574

Fathi, R., Thom, D., Koch, S., Ertl, T., & Fiedrich, F. (2020). VOST: A case study in voluntary digital participation for collaborative emergency management. *Information Processing and Management, 57*(4), 102174. https://doi.org/10.1016/j.ipm.2019.102174

Federal Emergency Management Agency. (2006). *Select Emergency Management-Related Terms and Definitions.* https://training.fema.gov/hiedu/docs/hazdem/appendix - select em-related terms and definitions.doc

Ferrara, E., Varol, O., Davis, C., Menczer, F., & Flammini, A. (2016). The Rise of Social Bots. *Communications of the ACM, 59*(7), 96–104. https://dl.acm.org/citation.cfm?id=2818717

Fichet, E., Robinson, J., & Starbird, K. (2015). Eyes on the Ground: Emerging Practices in Periscope Use during Crisis Events. *Proceedings of the International Conference on Information Systems for Crisis Response and Management (ISCRAM).*

Fischer, D., Posegga, O., & Fischbach, K. (2016). Communication Barriers in Crisis Management: A Literature Review. *European Conference on Information Systems (ECIS)*, 1–18.

Fischer, D., Putzke-Hattori, J., & Fischbach, K. (2019). Crisis Warning Apps: Investigating the Factors Influencing Usage and Compliance with Recommendations for Action. *Proceedings of the 52nd Hawaii International Conference on System Sciences*, 639–648.

Flizikowski, A., Hołubowicz, W., Stachowicz, A., Hokkanen, L., & Delavallade, T. (2014). Social Media in Crisis Management – the iSAR + Project Survey. *Proceedings of the International Conference on Information Systems for Crisis Response and Management (ISCRAM)*, 707–711.

Fraunhofer FOKUS. (2018). *Katwarn*. https://katwarn.de/

Fuchs, G., Andrienko, N., Andrienko, G., Bothe, S., & Stange, H. (2013). Tracing the German Centennial Flood in the Stream of Tweets: First Lessons Learned. *SIGSPATIAL International Workshop on Crowdsourced and Volunteered Geographic Information*, 2–10.

Fung, I. C.-H., Tse, Z. T. H., Cheung, C.-N., Miu, A. S., & Fu, K.-W. (2014). Ebola and the social media. *The Lancet, 384*(9961), 2207. https://doi.org/10.1016/S0140-6736(14)6241 8-1

Fürnkranz, J. (2018). *Introduction to Machine Learning, TU-Darmstadt Data Mining und Maschinelles Lernen 2018–2019, Präsentation.*

Gao, H., Barbier, G., & Goolsby, R. (2011). Harnessing the Crowdsourcing Power of Social Media for Disaster Relief. *IEEE Intelligent Systems, 26*(3), 10–14. https://doi.org/10.1109/MIS.2011.52

Geenen, E. M. (2009). Warnung der Bevölkerung. In Schutzkommission beim Bundesminister des Inneren (Ed.), *Gefahren und Warnung* (pp. 61–102).

German Federal Foreign Office. (2018). *Sicher Reisen – Ihre Reise-App.* https://www.auswae rtiges-amt.de/de/ReiseUndSicherheit/app-sicher-reisen/350382

Giuliani, G. (2016). Monstrosity, Abjection and Europe in the War on Terror [JOUR]. *Capitalism Nature Socialism, 27*(4), 96–114. https://doi.org/10.1080/10455752.2016.119 2212

Gizikis, A., O'Brien, T., Susaeta, I. G., Moi, M., Schubert, A., Reuter, C., Kaufhold, M.-A., Cullen, J., Muddiman, A., Perruzza, M., & Delprato, U. (2017). *EmerGent Deliverable 7.3: Guidelines to increase the benefit of social media in emergencies.* https://www.fp7-emergent.eu/publications/20170529_D7.3_Guidelines_to_increase_the_benefit_of_social_media_EmerGent.pdf

Gizikis, A., Susaeta, I. G., Spielhofer, T., & Bizjak, G. (2017). *EmerGent Deliverable 2.7: Workshop III.* https://www.fp7-emergent.eu/wp-content/uploads/2017/04/20170330_D2.7_Workshop_III_EmerGent_pub.pdf

Glaser, B. G., & Strauss, A. L. (1967). *The Discovery of Grounded Theory: Strategies for Qualitative Research.* Aldine de Gruyter.

Gonzalez, R. A., & Bharosa, N. (2009). A framework linking information quality dimensions and coordination challenges during interagency crisis response. *Proceedings of the 42nd Annual Hawaii International Conference on System Sciences, HICSS*, 1–10. https://doi.org/10.1109/HICSS.2009.15

Goolsby, R. (2010). Social media as crisis platform: The future of community maps/crisis maps. *ACM Transactions on Intelligent Systems and Technology, 1*(1), 1–11. https://doi.org/10.1145/1858848.1858955

Gorrell, G., & Bontcheva, K. (2016). Classifying Twitter favorites: Like, bookmark, or Thanks? *Journal of the Association for Information Science and Technology*, *67*(1), 17–25. https://doi.org/10.1002/asi.23352

Grinko, M., Kaufhold, M.-A., & Reuter, C. (2019). Adoption, Use and Diffusion of Crisis Apps in Germany: A Representative Survey. In F. Alt, A. Bulling, & T. Döring (Eds.), *Mensch und Computer 2019* (pp. 263–274). ACM.

Groneberg, C., Heidt, V., Knoch, T., & Helmerichs, J. (2017). *Analyse internationaler Bevölkerungsschutz-Apps*. https://smarter-projekt.de/wp-content/uploads/2017/10/Analyse_internationaler_Bevoelkerungsschutz-Apps.pdf

Gründer-Fahrer, S., Schlaf, A., Wiedemann, G., & Heyer, G. (2018). Topics and topical phases in German social media communication during a disaster. In *Natural Language Engineering* (Vol. 24, Issue 2). https://doi.org/10.1017/S1351324918000025

Guha-Sapir, D., Hargitt, D., & Hoyois, P. (2004). *Thirty Years of Natural Disasters 1974–2003: The Numbers*. Presses universitaires de Louvain.

Guha-Sapir, D., Hoyois, P., Wallemacq, P., & Below, R. (2017). *Annual Disaster Statistical Review 2010: The numbers and trends*. https://doi.org/10.1093/rof/rfs003

GWS Production. (2018). *The Safeture App*. https://globalwarningsystem.com/safeturesolutions/

Habdank, M., Rodehutskors, N., & Koch, R. (2017). Relevancy Assessment of Tweets using Supervised Learning Techniques Mining emergency related Tweets for automated relevancy classification. *2017 4th International Conference on Information and Communication Technologies for Disaster Management (ICT-DM)*.

Hagar, C. (2007). The information needs of farmers and use of ICTs. In B. Nerlich & M. Doring (Eds.), *From Mayhem to Meaning: Assessing the social and cultural impact of the 2001 foot and mouth outbreak in the UK*. Manchester University Press.

Hagar, C. (2010). Crisis Informatics: Introduction. *Bulletin of the American Society for Information Science and Technology*, *36*(5), 10–12. https://doi.org/10.1002/bult.2010.1720360504

Hall, M. A., Frank, E., Holmes, G., Pfahringer, B., Reutemann, P., & Witten, I. H. (2009). The WEKA data mining software: an update. *SIGKDD Explorations*, *11*(1), 10–18. https://doi.org/10.1145/1656274.1656278

Harrald, J. R., Egan, D. M., & Jefferson, T. (2002). Web Enabled Disaster and Crisis Response: What Have We Learned from the September 11th. *Proceedings of the Bled EConference*, 69–83.

Hartwig, K., & Reuter, C. (2019). TrustyTweet: An Indicator-based Browser-Plugin to Assist Users in Dealing with Fake News on Twitter. *Proceedings of the International Conference on Wirtschaftsinformatik (WI)*.

Hassenzahl, M., Burmester, M., & Koller, F. (2003). AttrakDiff: Ein Fragebogen zur Messung wahrgenommener hedonischer und pragmatischer Qualität. *Mensch & Computer 2003: Interaktion in Bewegung*, 187–196.

Hastie, T., Tibshirani, R., & Friedman, J. (2009). The Elements of Statistical Learning. *Elements*, *1*, 337–387. https://doi.org/10.1007/b94608

Haunschild, J., Kaufhold, M.-A., & Reuter, C. (2020). Sticking with Landlines? Citizens' and Police Social Media Use and Expectation During Emergencies. *Proceedings of the International Conference on Wirtschaftsinformatik (WI)*.

Helsloot, I., de Vries, D., Groenendaal, J., Scholtens, A., in 't Veld, M., van Melick, G., Baruh, L., Salvatore, S., Günel, Z., Watson, H., Wadhwa, K., Hagen, K., Kalemaki, E., Papadimitriou, A., & Vontas, A. M. (2015). *Deliverable D6.1 & D6.2: Guidelines for the use of new media in crisis situations; Project "Cosmic – The COntribution of Social Media In Crisis management."* https://doi.org/10.5281/ZENODO.16235

Helsloot, I., & Groenendaal, J. (2013). Twitter: An Underutilized Potential during Sudden Crises? *Journal of Contingencies and Crisis Management, 21*(3), 178–183. https://doi.org/10.1111/1468-5973.12023

Helsloot, I., & Ruitenberg, A. (2004). Citizen Response to Disasters: a Survey of Literature and Some Practical Implications. *Journal of Contingencies and Crisis Management, 12*(3), 98–111. https://doi.org/10.1111/j.0966-0879.2004.00440.x

here. (2019). *HERE Geocoder API.* https://developer.here.com/c/geocoding

Hermann, S. D., Wolisz, A., & Sortais, M. (2007). Enhancing the accuracy of position information through superposition of location server data. *IEEE International Conference on Communications*, 2030–2037. https://doi.org/10.1109/ICC.2007.337

Heverin, T., & Zach, L. (2010). Microblogging for Crisis Communication: Examination of Twitter Use in Response to a 2009 Violent Crisis in the Seattle-Tacoma, Washington Area. *Proceedings of the International Conference on Information Systems for Crisis Response and Management (ISCRAM).*

Hevner, A. R. (2007). A Three Cycle View of Design Science Research. *Scandinavian Journal of Information Systems, 19*(2), 87–92. https://aisel.aisnet.org/sjis/vol19/iss2/4

Hevner, A. R., & Chatterjee, S. (2010). *Design Research in Information Systems: Theory and Practice.* Springer.

Hevner, A. R., March, S. T., Park, J., & Ram, S. (2004). Design science in information systems research. *MIS Quarterly, 28*(1), 75–105. https://doi.org/10.2307/25148625

Hewett, T., Baecker, R., Card, S., Carey, T., Gasen, J., Mantei, M., Perlman, G., Strong, G., & Verplank, W. (1992). ACM SIGCHI Curricula for Human-Computer Interaction. In *ACM SIGCHI Curricula for Human-Computer Interaction.* https://doi.org/10.1145/2594128

Hilligoss, B., & Rieh, S. Y. (2008). Developing a unifying framework of credibility assessment: Construct, heuristics, and interaction in context. *Information Processing and Management, 44*(4), 1467–1484. https://doi.org/10.1016/j.ipm.2007.10.001

Hiltz, S. R., Diaz, P., & Mark, G. (2011). Introduction: Social Media and Collaborative Systems for Crisis Management. *ACM Transactions on Computer-Human Interaction (ToCHI), 18*(4), 1–6. https://doi.acm.org/10.1145/2063231.2063232

Hiltz, S. R., Hughes, A. L., Imran, M., Plotnick, L., Power, R., & Turoff, M. (2020). Exploring the usefulness and feasibility of software requirements for social media use in emergency management. *International Journal of Disaster Risk Reduction (IJDDR), 42*(January), 101367. https://doi.org/10.1016/j.ijdrr.2019.101367

Hiltz, S. R., Kushma, J., & Plotnick, L. (2014). Use of Social Media by US Public Sector Emergency Managers: Barriers and Wish Lists. *Proceedings of the International Conference on Information Systems for Crisis Response and Management (ISCRAM)*, 600–609.

Hiltz, S. R., & Plotnick, L. (2013). Dealing with Information Overload When Using Social Media for Emergency Management: Emerging Solutions. In T. Comes, F. Fiedrich, S. Fortier, J. Geldermann, & T. Müller (Eds.), *Proceedings of the International Conference on Information Systems for Crisis Response and Management (ISCRAM)* (pp. 823–827). ISCRAM Digital Library.

Hiltz, S. R., van de Walle, B., & Turoff, M. (2011). The Domain of Emergency Management Information. In B. Van De Walle, M. Turoff, & S. R. Hiltz (Eds.), *Information Systems for Emergency Management* (pp. 3–20). M.E. Sharpe.

Höchst, J., Baumgärtner, L., Kuntke, F., Penning, A., Sterz, A., & Freisleben, B. (2020). LoRa-based Device-to-Device Smartphone Communication for Crisis Scenarios. *Proceedings of the International Conference on Information Systems for Crisis Response and Management (ISCRAM)*.

Holenstein, M., & Küng, L. (2008). Stromausfall – was denkt die Bevölkerung. *Sicherheit, 3*, 61.

Homeland Security. (2013). *Lessons Learned: Social Media and Hurricane Sandy – Virtual Social Media Working Group and Group DHS First Responders* (Issue June).

Höppe, P. (2015). *Naturkatastrophen – immer häufiger, heftiger, tödlicher, teurer?* https://www.munichre-foundation.org/dms/MRS/Documents/20150303_DF2015_Hoeppe_Disasters/DF2015_March_Handout_Hoeppe.pdf

Huang, Y. L., Starbird, K., Orand, M., Stanek, S. A., & Pedersen, H. T. (2015). Connected through crisis: emotional proximity and the spread of misinformation online. *Proceedings of the 18th ACM Conference on Computer Supported Cooperative Work & Social Computing (CSCW '15)*, 969–980. https://doi.org/10.1145/2675133.2675202

Hughes, A. L., & Chauhan, A. (2015). Online Media as a Means to Affect Public Trust in Emergency Responders. *Proceedings of the International Conference on Information Systems for Crisis Response and Management (ISCRAM)*.

Hughes, A. L., & Palen, L. (2012). The Evolving Role of the Public Information Officer: An Examination of Social Media in Emergency Management. *Journal of Homeland Security and Emergency Management (JHSEM), 9*(1), Article 22. https://doi.org/10.1515/1547-7355.1976

Hughes, A. L., & Palen, L. (2014). Social Media in Emergency Management: Academic Perspective. In J. E. Trainor & T. Subbio (Eds.), *Critical Issues in Disaster Science and Management: A Dialogue Between Scientists and Emergency Managers*. Federal Emergency Management Agency.

Hughes, A. L., & Palen, L. (2009). Twitter adoption and use in mass convergence and emergency events. In J. Landgren & S. Jul (Eds.), *Proceedings of the International Conference on Information Systems for Crisis Response and Management (ISCRAM)* (Vol. 6, Issue 3/4). https://doi.org/10.1504/IJEM.2009.031564

Hughes, A. L., Palen, L., Sutton, J., Liu, S. B., & Vieweg, S. (2008). "Site-Seeing" in Disaster: An Examination of On-Line Social Convergence. *Proceedings of the International Conference on Information Systems for Crisis Response and Management (ISCRAM)*, 1–10.

Hughes, A. L., St. Denis, L. A., Palen, L., & Anderson, K. M. (2014). Online Public Communications by Police & Fire Services during the 2012 Hurricane Sandy. *Proceedings of the Conference on Human Factors in Computing Systems (CHI)*, 1505–1514. https://doi.org/10.1145/2556288.2557227

Hughes, A. L., Starbird, K., Leavitt, A., Keegan, B., & Semaan, B. (2016). Information Movement Across Social Media Platforms During Crisis Events. *CHI'16 Extended Abstracts.*, 1–5. https://doi.org/10.1145/2851581.2856500

Hughes, A. L., & Tapia, A. H. (2015). Social Media in Crisis: When Professional Responders Meet Digital Volunteers. *Journal of Homeland Security and Emergency Management, 12*(3), 679–706. https://doi.org/10.1515/jhsem-2014-0080

Hulten, G., Spencer, L., & Domingos, P. (2001). Mining time-changing data streams. *Seventh ACM SIGKDD International Conference on Knowledge Discovery and Data Mining,* 97–106. https://doi.org/10.1145/502512.502529

IBM. (2014). *Statistical Package for the Social Sciences.* https://www.ibm.com/products/spss-statistics

IFRC. (2015). *World Disaster Report 2015: Focus on local actors, the key to humanitarian effectiveness.* https://doi.org/10.1017/CBO9781107415324.004

Imran, M., Castillo, C., Diaz, F., & Vieweg, S. (2015). Processing Social Media Messages in Mass Emergency: A Survey. In *ACM Computing Surveys* (Vol. 47, Issue 4). ACM. https://doi.org/10.1145/2771588

Imran, M., Castillo, C., Diaz, F., & Vieweg, S. (2018). Processing Social Media Messages in Mass Emergency: Survey Summary. *Companion Proceedings of the The Web Conference 2018 (WWW '18),* 507–511. https://doi.org/10.1145/3184558.3186242

Imran, M., Castillo, C., Lucas, J., Meier, P., & Vieweg, S. (2014). AIDR: Artificial Intelligence for Disaster Response. *Proceedings of the Companion Publication of the 23rd International Conference on World Wide Web Companion,* 159–162. https://doi.org/10.1145/2567948.2577034

Imran, M., Elbassuoni, S., Castillo, C., Diaz, F., & Meier, P. (2013a). Extracting Information Nuggets from Disaster-Related Messages in Social Media. In T. Comes, F. Fiedrich, S. Fortier, J. Geldermann, & L. Yang (Eds.), *Proceedings of the International Conference on Information Systems for Crisis Response and Management (ISCRAM)* (pp. 791–800). https://doi.org/10.1145/2534732.2534741

Imran, M., Elbassuoni, S., Castillo, C., Diaz, F., & Meier, P. (2013b). Practical extraction of disaster-relevant information from social media. *Proceedings of the 22nd International Conference on World Wide Web – WWW '13 Companion,* 1021–1024. https://doi.org/10.1145/2487788.2488109

Imran, M., Mitra, P., & Castillo, C. (2016). Twitter as a Lifeline: Human-annotated Twitter Corpora for NLP of Crisis-related Messages. *ArXiv Preprint* ArXiv:1605.05894. https://arxiv.org/abs/1605.05894

Imran, M., Mitra, P., & Srivastava, J. (2016). Cross-Language Domain Adaptation for Classifying Crisis-Related Short Messages. *ArXiv Preprint* ArXiv:1602.05388, *May.*

Imran, M., Mitra, P., & Srivastava, J. (2017). Enabling Rapid Classification of Social Media Communications During Crises. *International Journal of Information Systems for Crisis Response and Management.* https://doi.org/10.4018/ijiscram.2016070101

Imran, N., Seet, B. C., & Fong, A. C. M. (2015). Distributed video coding for wireless video sensor networks: a review of the state-of-the-art architectures. *SpringerPlus, 4*(513). https://doi.org/10.1186/s40064-015-1300-4

Institute for Crisis Disaster and Risk Management. (2009). *ICDRM/GWU Emergency Management Glossary of Terms.* The George Washington University. https://www.gwu.edu/~icdrm/publications/PDF/EM_Glossary_ICDRM.pdf

International Association of Chiefs of Police. (2010). *2010 Social Media Survey Results.* https://www.iacpsocialmedia.org/Portals/1/documents/Survey Results Document.pdf

International Association of Chiefs of Police. (2015). *2015 Social Media Survey Results.* https://www.iacpsocialmedia.org/Portals/1/documents/FULL 2015 Social Media Survey Results.pdf

Jagtman, H. M. (2010). Cell broadcast trials in The Netherlands: Using mobile phone technology for citizens' alarming. *Reliability Engineering and System Safety, 95*(1), 18–28. https://doi.org/10.1016/j.ress.2009.07.005

Jennex, M. E. (2012). Social Media – Truly Viable For Crisis Response? In L. Rothkrantz, J. Ristvej, & Z. Franco (Eds.), *Proceedings of the International Conference on Information Systems for Crisis Response and Management (ISCRAM)* (Issue April, pp. 1–5).

Jensen, G. E. (2012). *Key criteria for information quality in the use of online social media for emergency management in New Zealand. October.*

Jin, Y., Liu, B. F., & Austin, L. L. (2014). Examining the Role of Social Media in Effective Crisis Management. *Communication Research, 41*(1), 74–94. https://doi.org/10.1177/009 3650211423918

Jochimsen, R. (1966). *Theorie der Infrastruktur, Grundlagen der marktwirtschaftlichen Entwicklung.* J. C. B. Mohr.

Johansson, F., Brynielsson, J., & Quijano, M. N. (2012). Estimating Citizen Alertness in Crises Using Social Media Monitoring and Analysis. *2012 European Intelligence and Security Informatics Conference,* 189–196. https://doi.org/10.1109/EISIC.2012.23

John, G. H., Kohavi, R., & Pfleger, K. (1994). Irrelevant Features and the Subset Selection Problem. In *Machine Learning Proceedings 1994* (pp. 121–129). https://doi.org/10.1016/ B978-1-55860-335-6.50023-4

Jurgens, M., & Helsloot, I. (2017). The effect of social media on the dynamics of (self) resilience during disasters: A literature review. *Journal of Contingencies and Crisis Management (JCCM), 26,* 79–88. https://doi.org/10.1111/1468-5973.12212

Kaewkitipong, L., Chen, C., & Ractham, P. (2012). Lessons Learned from the Use of Social Media in Combating a Crisis: A Case Study of 2011 Thailand Flooding Disaster. *Proceedings of the International Conference on Information Systems (ICIS),* 1–17.

Kandari, J. (2010). Information quality on the World Wide Web: A User Perspective. *International Journal of Information Quality, 2*(4), 324–343. https://doi.org/10.1504/IJIQ.2011. 043784

Kaplan, A. M., & Haenlein, M. (2010). Users of the world, unite! The challenges and opportunities of Social Media. *Business Horizons, 53*(1), 59–68. https://doi.org/10.1016/j.bus hor.2009.09.003

Karl, I., Rother, K., & Nestler, S. (2015). Crisis-related Apps: Assistance for Critical and Emergency Situations. *International Journal of Information Systems for Crisis Response and Management (IJISCRAM), 7*(2), 19–35. https://services.igi-global.com/resolvedoi/ resolve.aspx?doi=10.4018/IJISCRAM.2015040102

Kaufhold, M.-A., Bayer, M., & Reuter, C. (2020). Rapid relevance classification of social media posts in disasters and emergencies: A system and evaluation featuring active, incremental and online learning. *Information Processing and Management, 57*(1), 1–32. https:// doi.org/10.1016/j.ipm.2019.102132

Kaufhold, M.-A., Gizikis, A., Reuter, C., Habdank, M., & Grinko, M. (2019). Avoiding Chaotic Use of Social Media before, during, and after Emergencies: Design and Evaluation of Citizens' Guidelines. *Journal of Contingencies and Crisis Management (JCCM), 27*(3), 198–213. https://doi.org/10.1111/1468-5973.12249

Kaufhold, M.-A., Grinko, M., Reuter, C., Schorch, M., Langer, A., Skudelny, S., & Hollick, M. (2019). Potentiale von IKT beim Ausfall kritischer Infrastrukturen: Erwartungen, Informationsgewinnung und Mediennutzung der Zivilbevölkerung in Deutschland. *Proceedings of the International Conference on Wirtschaftsinformatik*. https://www.peasec.de/paper/ 2019/2019_KaufholdGrinkoReuteretal_KritischeInfrastrukturenErwartungen_WI.pdf

Kaufhold, M.-A., Haunschild, J., & Reuter, C. (2020). Warning the Public: A Survey on Attitudes, Expectations and Use of Mobile Crisis Apps in Germany. *Proceedings of the European Conference on Information Systems (ECIS)*.

Kaufhold, M.-A., & Reuter, C. (2014). Vernetzte Selbsthilfe in Sozialen Medien am Beispiel des Hochwassers 2013 / Linked Self-Help in Social Media using the example of the Floods 2013 in Germany. *I-Com – Zeitschrift Für Interaktive Und Kooperative Medien*, *13*(1), 20–28.

Kaufhold, M.-A., & Reuter, C. (2016). The Self-Organization of Digital Volunteers across Social Media: The Case of the 2013 European Floods in Germany. *Journal of Homeland Security and Emergency Management (JHSEM)*, *13*(1), 137–166. https://doi.org/10.1515/ jhsem-2015-0063

Kaufhold, M.-A., & Reuter, C. (2019). Cultural Violence and Peace in Social Media. In C. Reuter (Ed.), *Information Technology for Peace and Security – IT-Applications and Infrastructures in Conflicts, Crises, War, and Peace* (pp. 361–381). Springer Vieweg. https://doi.org/10.1007/978-3-658-25652-4_17

Kaufhold, M.-A., & Reuter, C. (2017). The Impact of Social Media in Emergencies: A Case Study with the Fire Department of Frankfurt. In Tina Comes, F. Bénaben, C. Hanachi, & M. Lauras (Eds.), *Proceedings of the International Conference on Information Systems for Crisis Response and Management (ISCRAM)* (pp. 603–612). ISCRAM. https://idl.iscram.org/files/marc-andrekaufhold/2017/1494_Marc-AndreKauf hold+ChristianReuter2017.pdf

Kaufhold, M.-A., Reuter, C., & Ludwig, T. (2019). Cross-Media Usage of Social Big Data for Emergency Services and Volunteer Communities: Approaches, Development and Challenges of Multi-Platform Social Media Services. ArXiv:1907.07725 *[Cs.SI]*, 1–11. https:// arxiv.org/pdf/1907.07725.pdf

Kaufhold, M.-A., Reuter, C., Ludwig, T., & Scholl, S. (2017). Social Media Analytics: Eine Marktstudie im Krisenmanagement. In M. Eibl & M. Gaedke (Eds.), *INFORMATIK 2017, Lecture Notes in Informatics (LNI), Gesellschaft für Informatik* (pp. 1325–1338).

Kaufhold, M.-A., Rupp, N., Reuter, C., Amelunxen, C., & Cristaldi, M. (2018). 112.social: Design and Evaluation of a Mobile Crisis App for Bidirectional Communication between Emergency Services and Citizens. *European Conference on Information Systems (ECIS)*.

Kaufhold, M.-A., Rupp, N., Reuter, C., & Habdank, M. (2020). Mitigating Information Overload in Social Media during Conflicts and Crises: Design and Evaluation of a Cross-Platform Alerting System. *Behaviour & Information Technology (BIT)*, *39*(3), 319–342.

Keim, D., Andrienko, G., Fekete, J., Carsten, G., & Melan, G. (2008). Visual Analytics: Definition, Process and Challenges. *Information Visualization – Human-Centered Issues and Perspectives*, 154–175. https://doi.org/10.1007/978-3-540-70956-5_7

Khouzam, B. (2009). *Incremental decision trees*. https://www.vincentlemaire-labs.fr/stu dents/master_thesis_bassem_khouzam.pdf

Kietzmann, J. H., Hermkens, K., Mccarthy, I. P., & Silvestre, B. S. (2011). Social media? Get serious! Understanding the functional building blocks of social media. *Business History*, *54*(3), 241–251. https://doi.org/10.1016/j.bushor.2011.01.005

Kim, M., Sharman, R., Cook-Cottone, C. P., Rao, H. R., & Upadhyaya, S. J. (2012). Assessing roles of people, technology and structure in emergency management systems: a public sector perspective. *Behaviour & Information Technology (BIT)*, *31*(12), 1147–1160. https://doi.org/10.1080/0144929X.2010.510209

Kircher, F. (2014). Ungebundene Helfer im Katastrophenschutz – Die Sicht der Behörden und Organisationen mit Sicherheitsaufgaben. *BRANDSchutz – Deutsche Feuerwehr-Zeitung*, 593–597.

Klafft, M. (2013). Diffusion of emergency warnings via multi-channel communication systems an empirical analysis. *Eleventh International Symposium on Autonomous Decentralized Systems (ISADS)*, 1–5. https://doi.org/10.1109/ISADS.2013.6513437

Klafft, M., & Reinhardt, N. (2016). Information and interaction needs of vulnerable groups with regard to disaster alert apps. In B. Weyers & A. Dittmar (Eds.), *Mensch & Computer 2016: Workshopband* (pp. 1–7).

Koch, Michael. (2008). CSCW and Enterprise 2.0 – towards an integrated perspective. *Proceedings of the Bled EConference*.

Koch, W., & Frees, B. (2016). Dynamische Entwicklung bei mobiler Internetnutzung sowie Audios und Videos. *Media Perspektiven*, *9*, 418–437.

Kogan, M., Anderson, J., Palen, L., Anderson, K. M., & Soden, R. (2016). Finding the Way to OSM Mapping Practices: Bounding Large Crisis Datasets for Qualitative Investigation. *Proceedings of the 2016 CHI Conference on Human Factors in Computing Systems*, 2783–2795. https://doi.org/10.1145/2858036.2858371

Kogan, M., Palen, L., & Anderson, K. M. (2015). Think Local, Retweet Global: Retweeting by the Geographically-Vulnerable during Hurricane Sandy. *Proceedings of the 18th ACM Conference on Computer Supported Cooperative Work & Social Computing (CSCW '15)*, 981–993. https://doi.org/10.1145/2675133.2675218

Kotthaus, C., Ludwig, T., & Pipek, V. (2016). Persuasive System Design Analysis of Mobile Warning Apps for Citizens. *Adjunct Proceedings of the 11th International Conference on Persuasive Technology*.

Krafft, P., Zhou, K., Edwards, I., Starbird, K., & Spiro, E. S. (2017). Centralized, Parallel, and Distributed Information Processing during Collective Sensemaking. *ACM CHI Conference on Human Factors in Computing Systems (CHI)*, 2976–2987. https://doi.org/10.1145/302 5453.3026012

Kroll, L. (2016). *Infografik mit den Nutzerzahlen der wichtigsten Social Networks weltweit*. Social Media Institute. https://socialmedia-institute.com/uebersicht-aktueller-social-media-nutzerzahlen/

Kulessa, M. (2015). *Online-Lernen von zufälligen Entscheidungsbäumen*. https://www.ke.tu-darmstadt.de/lehre/arbeiten/bachelor/2015/Kulessa_Moritz.pdf

Kuttschreuter, M., Rutsaert, P., Hilverda, F., Regan, Á., Barnett, J., & Verbeke, W. (2014). Seeking information about food-related risks: The contribution of social media. *Food Quality and Preference*, *37*, 10–18. https://doi.org/10.1016/j.foodqual.2014.04.006

Landesberger, T. Von, Kuijper, A., Schreck, T., Kohlhammer, J., Wijk, J. Van, Fekete, J.-D., & Fellner, D. (2011). Visual Analysis of Large Graphs: State-of-the-Art and Future

Research Challenges. *Computer Graphics Forum, 30*(6), 1719–1749. https://doi.org/10.1111/j.1467-8659.2011.01898.x

LanguageTool. (2019). *LanguageTool.* https://languagetool.org/

Latonero, M., & Shklovski, I. (2011). Emergency Management, Twitter, and Social Media Evangelism. *International Journal of Information Systems for Crisis Response and Management (IJISCRAM), 3*(4), 1–16.

Leitungsgremium der GI-Fachgruppe CSCW. (2009). *CSCW-Orientierungsliste.* https://fg-cscw.gi.de/fileadmin/FG/CSCW/dokumente/Journal_und_Konferenzratings-FGCSCW-2009-02.pdf

Lewin, K. (1948). *Resolving Social Conflicts.* Harper and Row.

Lewin, K. (1958). *Group Decision and Social Change.* Holt; Rinehart and Winston.

Lewis, D. D., & Catlett, J. (2014). Heterogeneous Uncertainty Sampling for Supervised Learning. In *Machine Learning Proceedings 1994.* https://doi.org/10.1016/b978-1-55860-335-6.50026-x

Ley, B., Ludwig, T., Pipek, V., Randall, D., Reuter, C., & Wiedenhoefer, T. (2014). Information and Expertise Sharing in Inter-Organizational Crisis Management. *Computer Supported Cooperative Work: The Journal of Collaborative Computing (JCSCW), 23*(4–6), 347–387.

Li, H., Caragea, D., Caragea, C., & Herndon, N. (2017). Disaster response aided by tweet classification with a domain adaptation approach. *Journal of Contingencies and Crisis Management,* 16–27. https://doi.org/10.1111/1468-5973.12194

Li, H., Guevara, N., Herndon, N., Caragea, D., Neppalli, K., Caragea, C., Squicciarini, A., & Tapia, A. H. (2015). Twitter Mining for Disaster Response: A Domain Adaptation Approach. *Proceedings of the International Conference on Information Systems for Crisis Response and Management (ISCRAM),* [-7]. https://doi.org/10.1503/cmaj.082001

Li, J., & Rao, H. R. (2010). Twitter As a Rapid Response News Service: an Exploration in the Context of the 2008 China Earthquake. *The Electronic Journal on Information Systems in Developing Countries, 42,* 1–22. https://doi.org/10.1002/j.1681-4835.2010.tb00300.x

Li, Y., & Manoharan, S. (2013). A performance comparison of SQL and NoSQL databases. *IEEE Pacific RIM Conference on Communications, Computers, and Signal Processing – Proceedings,* 15–19. https://doi.org/10.1109/PACRIM.2013.6625441

Lieberman, H., Paternò, F., Klann, M., & Wulf, V. (2006). End-user development: An emerging paradigm. *End User Development, 9,* 1–8. https://doi.org/10.1007/1-4020-5386-X_1

Lin, X., Spence, P. R., Sellnow, T. L., & Lachlan, K. A. (2016). Crisis communication, learning and responding: Best practices in social media. *Computers in Human Behavior, 65,* 601–605. https://doi.org/10.1016/j.chb.2016.05.080

Link, D., Meesters, K., Hellingrath, B., & Van De Walle, B. (2014). Reference task-based design of crisis management games. *Proceedings of the International Conference on Information Systems for Crisis Response and Management (ISCRAM), May,* 592–596.

Liu, B. F., Austin, L., & Jin, Y. (2011). How publics respond to crisis communication strategies: The interplay of information form and source. *Public Relations Review, 37*(4), 345–353. https://doi.org/10.1016/j.pubrev.2011.08.004

Liu, B. F., Lin, Y., & Austin, L. L. (2013). The Tendency To Tell: Understanding Publics' Communicative Responses To Crisis Information Form and Source. *Journal of Public Relations Research, 25,* 51–67. https://doi.org/10.1080/1062726X.2013.739101

Liu, S., Palen, L., & Sutton, J. (2008). In search of the bigger picture: The emergent role of on-line photo sharing in times of disaster. *Proceedings of the International Conference on Information Systems for Crisis Response and Management (ISCRAM), May.*

Ludwig, T. (2017). *Researching Complex Information Infrastructures: Design Characteristics of ICT Tools for Examining Modern Technology Usage.* Springer.

Ludwig, T., Kotthaus, C., Reuter, C., Dongen, S. Van, Pipek, V., van Dongen, S., & Pipek, V. (2017). Situated crowdsourcing during disasters: Managing the tasks of spontaneous volunteers through public displays. *International Journal on Human-Computer Studies (IJHCS), 102*(C), 103–121. https://doi.org/j.ijhcs.2016.09.008

Ludwig, T., Reuter, C., Friberg, T., Cristaldi, M., Sangiorgio, F., Toscano, F., Muddiman, A., & Markham, D. (2014). *EmerGent Deliverable 5.3: Design of Social Apps.* https://www.fp7-emergent.eu/d5-3-design-of-social-apps/

Ludwig, T., Reuter, C., & Pipek, V. (2015). Social Haystack: Dynamic Quality Assessment of Citizen-Generated Content during Emergencies. *Transactions on Human Computer Interaction (ToCHI), 21*(4), 17:1–17:27.

Ludwig, T., Reuter, C., & Pipek, V. (2013). What You See Is What I Need: Mobile Reporting Practices in Emergencies. In O. W. Bertelsen, L. Ciolfi, A. Grasso, & G. A. Papadopoulos (Eds.), *Proceedings of the 13th European Conference on Computer Supported Cooperative Work (ECSCW)* (pp. 181–206). Springer. https://doi.org/10.1007/978-1-4471-5346-7_10

Ludwig, T., Reuter, C., Siebigteroth, T., & Pipek, V. (2015). CrowdMonitor: Mobile Crowd Sensing for Assessing Physical and Digital Activities of Citizens during Emergencies. In B. Begole, J. Kim, K. Inkpen, & W. Woo (Eds.), *Proceedings of the Conference on Human Factors in Computing Systems (CHI)* (pp. 4083–4092). ACM Press.

Ma, J., Saul, L. K., Savage, S., & Voelker, G. M. (2009). Identifying Suspicious URLs: An Application of Large-Scale Online Learning. *ICML '09 Proceedings of the 26th Annual International Conference on Machine Learning,* 681–688. https://doi.org/10.1145/155 3374.1553462

Maier, C., Laumer, S., Eckhardt, A., & Weitzel, T. (2012). When Social Networking Turns To Social Overload: Explaining the Stress, Emotional Exhaustion, and Quitting Behavior From Social Network Sites' Users. *Twentieth European Conference on Information Systems (ECIS),* 71–82.

Maier, C., Laumer, S., Weinert, C., & Weitzel, T. (2015). The effects of technostress and switching stress on discontinued use of social networking services: A study of Facebook use. *Information Systems Journal, 25*(3), 275–308. https://doi.org/10.1111/isj.12068

Manning, C., Surdeanu, M., Bauer, J., Finkel, J., Bethard, S., & McClosky, D. (2014). The Stanford CoreNLP Natural Language Processing Toolkit. *Proceedings of 52nd Annual Meeting of the Association for Computational Linguistics: System Demonstrations,* 55–60. https://doi.org/10.3115/v1/P14-5010

Marcus, A., Bernstein, M., Badar, O., Karger, D. R., Madden, S., & Miller, R. C. (2011). Twitinfo: aggregating and visualizing microblogs for event exploration. *Proceedings of the Conference on Human Factors in Computing Systems (CHI),* 227–236.

Markham, D., & Muddiman, A. (2016). *EmerGent Deliverable 4.4: Specification of Mining methods to develop, Version 2.* https://www.fp7-emergent.eu/d4-4-specification-of-min ing-methods-to-develop-version-2/

Marktplatz GmbH. (2018). *BIWAPP – BÜRGER INFO & WARN APP.* https://www.biw app.de/

McAfee, A., & Brynjolfsson, E. (2012). Big Data: The Management Revolution. *Harvard Business Review, 90*(10), 61–67.

McKinney, E. H. (2011). Crisis IT design implications for high risk systems: systems, control and information propositions. *Behaviour & Information Technology (BIT), 30*(3), 339–352. https://doi.org/10.1080/0144929X.2010.535854

McMahon, K. (2011). *The Psychology of Disaster*. Blog. https://carolynbaker.net/2011/03/17/the-psychology-of-disaster-by-kathy-mcmahon/

Medina, R. Z., & Diaz, J. C. L. (2016). Social Media Use in Crisis Communication Management: An Opportunity for Local Communities? In M. Z. Sobaci (Ed.), *Social Media and Local Governments* (pp. 321–335). Springer International Publishing. https://doi.org/10.1007/978-3-319-17722-9_17

Meissen, U., Hardt, M., & Voisard, A. (2014). Towards a general system design for community-centered crisis and emergency warning systems. *Proceedings of the International Conference on Information Systems for Crisis Response and Management (ISCRAM)*, 155–159.

Mendoza, M., Poblete, B., & Castillo, C. (2010). Twitter Under Crisis: Can we trust what we RT? *Proceedings of the First Workshop on Social Media Analytics*, 71–79.

Menski, U., & Gardemann, J. (2008). *Auswirkungen des Ausfalls Kritischer Infrastrukturen auf den Ernährungssektor am Beispiel des Stromausfalls im Münsterland im Herbst 2005*. Fachhochschule Münster.

Meurisch, C., Hamza, Z., Bayrak, B., & Muhlhauser, M. (2019). Enhanced detection of crisis-related microblogs by spatiotemporal feedback loops. *Proceedings – International Computer Software and Applications Conference, 1*, 507–512. https://doi.org/10.1109/COMPSAC.2019.00078

Miller, G. A. (1956). The magical number seven, plus or minus two: some limits on our capacity for processing information. In *Psychological Review* (Vol. 63, Issue 2, pp. 81–97). American Psychological Association. https://doi.org/10.1037/h0043158

Miller, R. (2017). *Dataminr announces new tool to assist first responders*. Tech Crunch. https://techcrunch.com/2017/05/15/dataminr-announces-new-tool-to-assist-first-responders-at-techcrunch-disrupt

Mirbabaie, M., & Zapatka, E. (2017). Sensemaking in Social Media Crisis Communication – A Case Study on the Brussels Bombings in 2016. *Twenty-Fifth European Conference on Information Systems (ECIS)*, 2169–2186.

Moi, M., Friberg, T., Marterer, R., Reuter, C., Ludwig, T., Markham, D., Hewlett, M., & Muddiman, A. (2015). Strategy for Processing and Analyzing Social Media Data Streams in Emergencies. *Proceedings of the International Conference on Information and Communication Technologies for Disaster Management (ICT-DM)*, 1–7.

Moi, M., Habig, T., Schubert, A., Brune, M., Witter, F., & Kiel, M. (2017). *EmerGent Deliverable 4.5: Information Quality Criteria and Indicators*. https://www.fp7-emergent.eu/d4-5-information-quality-criteria-and-indicators/

Moore, A. W. (1991). An intoductory tutorial on kd-trees. *Efficient Memory-Based Learning for Robot Control*. https://doi.org/10.1016/j.matcom.2008.01.003

Munich Re. (2017). *Topics Geo: Natural catastrophes 2017*. https://www.munichre.com/topics-online/en/2018/topics-geo/topics-geo-2017

Muralidharan, S., Dillistone, K., & Shin, J.-H. (2011). The Gulf Coast oil spill: Extending the theory of image restoration discourse to the realm of social media and beyond petroleum. *Public Relations Review*, *37*(3), 226–232. https://doi.org/10.1016/j.pubrev.2011.04.006

Murphy, T., & Jennex, M. E. (2006). Knowledge Management, Emergency Response, and Hurricane Katrina. *International Journal of Intelligent Control Systems*, *11*(4), 199–208. https://doi.org/10.4018/978-1-59904-916-8.ch021

Nagy, A., Valley, C., & Stamberger, J. (2012). Crowd Sentiment Detection during Disasters and Crises. *Proceedings of the International Conference on Information Systems for Crisis Response and Management (ISCRAM), April*, 1–9.

Naumann, F., & Rolker, C. (2000). Assessment Methods for Information Quality Criteria. *Proceedings of the International Conference on Information Quality (IQ)*, 148–162.

Nestler, S. (2017). Flächendeckende Kommunikation im Stromausfall durch regionale IKT Krisenszenario: Längerfristiger Stromausfall. In M. Burghardt, R. Wimmer, C. Wolff, & C. Womser-Hacker (Eds.), *Mensch und Computer 2017 – Workshopband* (pp. 9–16). Gesellschaft für Informatik e. V.

Nguyen, D. T., Mannai, K. A. Al, Joty, S., Sajjad, H., Imran, M., & Mitra, P. (2016). Rapid Classification of Crisis-Related Data on Social Networks using Convolutional Neural Networks. *International AAAI Conference on Web and Social Media*.

Nguyen, D. T., Ofli, F., Imran, M., & Mitra, P. (2017). *Damage Assessment from Social Media Imagery Data During Disasters*. 569–576. https://doi.org/10.1145/3110025.3110109

Nguyen, M.-T., Kitamoto, A., & Nguyen, T.-T. (2015). TSum4act: A Framework for Retrieving and Summarizing Actionable Tweets during a Disaster for Reaction. In T. Cao, E.-P. Lim, Z.-H. Zhou, T.-B. Ho, D. Cheung, & H. Motoda (Eds.), *Advances in Knowledge Discovery and Data Mining. PAKDD 2015. Lecture Notes in Computer Science* (pp. 64–75). Springer. https://doi.org/10.1007/978-3-319-18032-8

Nielsen, J. (1993). *Usability Engineering*. Academic Press.

Nilges, J., Balduin, N., & Dierich, B. (2009). Information and Communication Platform for Crisis Management (IKK). *Proceedings of the International Conference and Exhibition on Electricity Distribution (CIRED)*.

Norris, F. H., Stevens, S. P., Pfefferbaum, B., Wyche, K. F., & Pfefferbaum, R. L. (2008). Community resilience as a metaphor, theory, set of capacities, and strategy for disaster readiness. *American Journal of Community Psychology*, *41*(1–2), 127–150. https://doi.org/10.1007/s10464-007-9156-6

North, K. (2016). Die Wissenstreppe. In *Wissensorientierte Unternehmensführung: Wissensmanagement gestalten* (pp. 33–65). Springer Fachmedien Wiesbaden. https://doi.org/10.1007/978-3-658-11643-9_3

O'Brien, T., Gizikis, A., Brugghemans, B., Spielhofer, T., Moi, M., & Friberg, T. (2016). *EmerGent Deliverable 2.6: Workshops I and II*. https://www.fp7-emergent.eu/d2-6-workshops-i-and-ii/

O'Reilly, T. (2006). *Web 2.0 Compact Definition: Trying Again*. https://radar.oreilly.com/2006/12/web-20-compact-definition-tryi.html

O'Reilly, T. (2007). What Is Web 2.0: Design Patterns and Business Models for the Next Generation of Software. *International Journal of Digital Economics*, *65*(March), 17–37.

O'Sullivan, T. L., Kuziemsky, C. E., Toal-Sullivan, D., & Corneil, W. (2013). Unraveling the complexities of disaster management: A framework for critical social infrastructure to promote population health and resilience. *Social Science and Medicine, 93*, 238–246. https://doi.org/10.1016/j.socscimed.2012.07.040

OASIS. (2010). *Common Alerting Protocol*. OASIS Standard. https://docs.oasis-open.org/emergency/cap/v1.2/CAP-v1.2-os.html

Oh, O., Agrawal, M., & Rao, R. (2013). Community Intelligence and Social Media Services: A Rumor Theoretic Analysis of Tweets during Social Crises. *Management Information Systems Quarterly, 37*(2), 407–426.

Okolloh, O. (2009). Ushahidi, or "testimony": Web 2.0 tools for crowdsourcing crisis information. *Participatory Learning and Action, 59*(1), 65–70.

Oliveira, M., & Gama, J. (2012). An overview of social network analysis. *WIREs Data Mining and Knowledge Discovery, 2*, 99–115. https://doi.org/10.1002/widm.1048

Olshannikova, E., Olsson, T., Huhtamäki, J., & Kärkkäinen, H. (2017). Conceptualizing Big Social Data. *Journal of Big Data, 4*(1), 1–19. https://doi.org/10.1186/s40537-017-0063-x

Olteanu, A., Vieweg, S., & Castillo, C. (2015). What to Expect When the Unexpected Happens: Social Media Communications Across Crises. *Proceedings of the 18th ACM Conference on Computer Supported Cooperative Work & Social Computing*, 994–1009. https://doi.org/10.1145/2675133.2675242

Onorati, T., Díaz, P., & Carrion, B. (2018). From social networks to emergency operation centers: A semantic visualization approach. *Future Generation Computer Systems*. https://doi.org/10.1016/j.future.2018.01.052

Organisation for Economic Co-operation and Development (OECD). (2019). *Virtual Operations Support Team (VOST)*. https://www.oecd.org/governance/observatory-public-sector-innovation/innovations/page/virtualoperationssupportteamvost.htm

Oxford Learner's Dictionaries. (2020). *Definition of Refinement*. Oxford University Press. https://www.oxfordlearnersdictionaries.com/definition/english/refinement?q=refinement

Pacific Disaster Center. (2018). *Disaster Alert* [TM] *App*. https://www.pdc.org/apps/disaster-alert/

Palen, L. (2008). Online Social Media in Crisis Events. *Educause Quarterly, 3*, 76–78.

Palen, L., & Anderson, K. M. (2016). Crisis informatics: New data for extraordinary times. *Science, 353*(6296), 224–225. https://doi.org/10.1126/science.aag2579

Palen, L., Anderson, K. M., Mark, G., Martin, J., Sicker, D., Palmer, M., & Grunwald, D. (2010). A vision for technology-mediated support for public participation & assistance in mass emergencies & disasters. *Proceedings of the 2010 ACMBCS Visions of Computer Science Conference*, 1–12.

Palen, L., & Hughes, A. L. (2018). Social Media in Disaster Communication. In H. Rodríguez, W. Donner, & J. E. Trainor (Eds.), *Handbook of Disaster Research* (pp. 497–518). Springer International Publishing. https://doi.org/10.1007/978-3-319-63254-4_24

Palen, L., & Liu, S. B. (2007). Citizen communications in crisis: anticipating a future of ICT-supported public participation. *Proceedings of the Conference on Human Factors in Computing Systems (CHI)*, 727–736.

Palen, L., Vieweg, S., Liu, S. B., & Hughes, A. L. (2009). Crisis in a Networked World: Features of Computer-Mediated Communication in the April 16, 2007, Virginia Tech Event. *Social Science Computer Review, 27*(4), 467–480. https://doi.org/10.1177/0894439309332302

Palen, L., Vieweg, S., Sutton, J., Liu, S. B., & Hughes, A. L. (2007). Crisis Informatics: Studying Crisis in a Networked World. *Proceedings of the International Conference on E-Social Science.*

Pang, B., & Lee, L. (2008). Opinion Mining and Sentiment Analysis. *Foundations and Trends in Information Retrieval, 2*(1–2), 1–135. https://doi.org/10.1561/1500000011

Párraga Niebla, C., Weber, T., Skoutaridis, P., Hirst, P., Ramírez, J., Rego, D., Gil, G., Engelbach, W., Brynielsson, J., Wigro, H., Grazzini, S., & Dosch, C. (2011). Alert4All: An Integrated Concept for Effective Population Alerting in Crisis Situations. *Proceedings of the International Conference on Information Systems for Crisis Response and Management (ISCRAM).*

Paternò, F., & Wulf, V. (2017). New Perspectives in End-User Development. In *New Perspectives in End-User Development.* Springer. https://doi.org/10.1007/978-3-319-60291-2

Pereira, V. A., Sanches, M. F., Botega, L. C., Souza, J., Coneglian, C. S., Fusco, E., & de Campos, M. R. (2015). Multi-criteria Fusion of Heterogeneous Information for Improving Situation Awareness on Emergency Management Systems. In S. Yamamoto (Ed.), *Lecture Notes in Computer Science* (pp. 3–14). Springer International Publishing.

Perng, S.-Y., Büscher, M., Wood, L., Halvorsrud, R., Stiso, M., Ramirez, L., & Al-Akkad, A. (2012). Peripheral response: Microblogging during the 22/7/2011 Norway attacks. In L. Rothkrantz, J. Ristvej, & Z. Franco (Eds.), *Proceedings of the International Conference on Information Systems for Crisis Response and Management (ISCRAM)* (pp. 1–11).

Perry, K. (2017). *As I #prayforlasvegas I pray for us all. Find each other out there....* https://www.instagram.com/p/BZwx8oVle7s/ *[Tweet].*

Petersen, L., Fallou, L., Reilly, P., & Serafinelli, E. (2017). Public expectations of social media use by critical infrastructure operators in crisis communication. *Proceedings of the International Conference on Information Systems for Crisis Response and Management (ISCRAM),* 522–531.

Petter, S., DeLone, W., & McLean, E. R. (2013). Information Systems Success: The Quest for the Independent Variables. *Journal of Management Information Systems, 29*(4), 7–62. https://doi.org/10.2753/MIS0742-1222290401

Pipek, V., Liu, S. B., & Kerne, A. (2014). Special Issue: Crisis Informatics and Collaboration. *Computer Supported Cooperative Work (CSCW), 23*(4–6).

Pipek, V., Reuter, C., Ley, B., Ludwig, T., & Wiedenhoefer, T. (2013). Sicherheitsarena – Ein Ansatz zur Verbesserung des Krisenmanagements durch Kooperation und Vernetzung. *Crisis Prevention – Fachmagazin Für Innere Sicherheit, Bevölkerungsschutz Und Katastrophenhilfe, 3*(1), 58–59.

Pipek, V., & Wulf, V. (2009). Infrastructuring: Towards an Integrated Perspective on the Design and Use of Information Technology. *Journal of the Association for Information Systems, 10*(5), 447–473. https://doi.org/10.17705/1jais.00195

Plotnick, L., & Hiltz, S. R. (2016). Barriers to Use of Social Media by Emergency Managers. *Journal of Homeland Security and Emergency Management, 13*(2), 247–277. https://doi.org/10.1515/jhsem-2015-0068

Plotnick, L., & Hiltz, S. R. (2018). Software Innovations to Support the Use of Social Media by Emergency Managers. *International Journal of Human-Computer Interaction, 34*(4), 367–381. https://doi.org/10.1080/10447318.2018.1427825

Plotnick, L., Hiltz, S. R., Kushma, J. A., & Tapia, A. (2015). Red Tape: Attitudes and Issues Related to Use of Social Media by U.S. County-Level Emergency Managers. *Proceedings of the International Conference on Information Systems for Crisis Response and Management (ISCRAM)*.

Pohl, D. (2013). *Social Media Analysis for Crisis Management: A Brief Survey.* https://stcsn. ieee.net/e-letter/vol-2-no-1/social-media-analysis-for-crisis-management-a-brief-survey

Pohl, D., Bouchachia, A., & Hellwagner, H. (2015). Social media for crisis management: clustering approaches for sub-event detection. *Multimedia Tools and Applications, 74*(11), 3901–3932. https://doi.org/10.1007/s11042-013-1804-2

Porter, M. (2018). *Snowball: An English stop word list.* https://snowball.tartarus.org/algori thms/english/stop.txt

Powers, D. M. W. (2011). Evaluation: From Precision, Recall and F-Measure To Roc, Informedness, Markedness & Correlation. *Journal of Machine Learning Technologies, 2*(1), 37–63. https://doi.org/10.1.1.214.9232

Purohit, H., Castillo, C., Diaz, F., Sheth, A., & Meier, P. (2014). Emergency-relief coordination on social media: Automatically matching resource requests and offers. *First Monday, 19*(1), 1–7.

Purohit, H., Castillo, C., Imran, M., & Pandey, R. (2018). Ranking of Social Media Alerts with Workload Bounds in Emergency Operation Centers. *2018 IEEE/WIC/ACM International Conference on Web Intelligence (WI)*, 206–213. https://doi.org/10.1109/WI.2018.00-88

Purohit, H., Hampton, A., Bhatt, S., Shalin, V. L., Sheth, A. P., & Flach, J. M. (2014). Identifying Seekers and Suppliers in Social Media Communities to Support Crisis Coordination. *Computer Supported Cooperative Work: The Journal of Collaborative Computing (JCSCW), 23*(4–6), 513–545.

Qu, Y., Huang, C., Zhang, P., & Zhang, J. (2011). Microblogging after a Major Disaster in China: A Case Study of the 2010 Yushu Earthquake. *Proceedings of the Conference on Computer Supported Cooperative Work (CSCW)*, 25–34.

Qu, Y., Wu, P. F., & Wang, X. (2009). Online Community Response to Major Disaster: A Study of Tianya Forum in the 2008 Sichuan Earthquake. *Proceedings of the Hawaii International Conference on System Sciences (HICSS), January.*

Quarantelli, E. L. (1984). Emergent Citizen Groups in Disaster Preparedness and Recovery Activities. In *Final Project Report.* University of Delaware.

Quarantelli, E. L. (1985). What is a Disaster? The Need for Clarification in Definition and Conceptualization in Research. In B. Sowder (Ed.), *Disasters and Mental Health: Selected Contemporary Perspectives.* US Government Printing Office.

Quarantelli, E. L. (1987). What Should We Study? Questions and Suggestions for Researchers About the Concept of Disasters. *International Journal of Mass Emergencies and Disasters, 5*(1), 7–32.

Quarantelli, E. L. (1988). Disaster Crisis Management: A summary of research findings. *Journal of Management Studies, 25*(4), 373–385. https://doi.org/10.1111/j.1467-6486.1988. tb00043.x

Quarantelli, E. L., & Dynes, R. R. (1977). Response to Social Crisis and Disaster. *Annual Review of Sociology, 3*(1), 23–49. https://doi.org/10.1146/annurev.so.03.080177.000323

Rao, R., Plotnick, L., & Hiltz, S. R. (2017). Supporting the Use of Social Media by Emergency Managers: Software Tools to Overcome Information Overload. In *Proceedings of the 50th Hawaii International Conference on System Sciences (HICSS)*.

Ren, S., Lian, Y., & Zou, X. (2014). Incremental Naïve Bayesian Learning Algorithm based on Classification Contribution Degree. *Journal of Computers, 9*(8), 1967–1974. https://doi.org/10.4304/jcp.9.8.1967-1974

Reuter, C. (2014a). Communication between Power Blackout and Mobile Network Overload. *International Journal of Information Systems for Crisis Response and Management (IJISCRAM), 6*(2), 38–53. https://doi.org/10.4018/ijiscram.2014040103

Reuter, C. (2014b). *Emergent Collaboration Infrastructures: Technology Design for Inter-Organizational Crisis Management (Ph.D. Thesis).* Springer Gabler. https://doi.org/10.1007/978-3-658-08586-5

Reuter, C. (2018). *Sicherheitskritische Mensch-Computer-Interaktion: Interaktive Technologien und Soziale Medien im Krisen- und Sicherheitsmanagement.* Springer Vieweg (Lehrbuch/Fachbuch). https://doi.org/10.1007/978-3-658-19523-6

Reuter, C. (2019). *Information Technology for Peace and Security – IT-Applications and Infrastructures in Conflicts, Crises, War, and Peace.* Springer Vieweg. https://doi.org/10.1007/978-3-658-25652-4

Reuter, C., & Amelunxen, C. (2016). *EmerGent Deliverable 3.7: Potentials of Social Media Usage by EMS and citizens' involvement in the EMC enabled by Emer-Gent.* https://www.fp7-emergent.eu/wp-content/uploads/2017/07/20170630_D7.3_Guidelines_to_increase_the_benefit_of_social_media_EmerGent.pdf

Reuter, C., Amelunxen, C., & Moi, M. (2016). Semi-Automatic Alerts and Notifications for Emergency Services based on Cross-Platform Social Media Data – Evaluation of a Prototype. In H. C. Mayr & M. Pinzger (Eds.), *Informatik 2016: von Menschen für Menschen.* GI-Edition-Lecture Notes in Informatics (LNI).

Reuter, C., Backfried, G., Kaufhold, M.-A., & Spahr, F. (2018). ISCRAM turns 15: A Trend Analysis of Social Media Papers 2004–2017. In K. Boersma & B. Tomaszewski (Eds.), *Proceedings of the International Conference on Information Systems for Crisis Response and Management (ISCRAM).* ISCRAM. https://idl.iscram.org/files/christianreuter/2018/1570_ChristianReuter_etal2018.pdf

Reuter, C., Hartwig, K., Kirchner, J., & Schlegel, N. (2019). Fake News Perception in Germany: A Representative Study of People's Attitudes and Approaches to Counteract Disinformation. *Proceedings of the International Conference on Wirtschaftsinformatik (WI)*, 1069–1083.

Reuter, C., Heger, O., & Pipek, V. (2013). Combining Real and Virtual Volunteers through Social Media. In T. Comes, F. Fiedrich, S. Fortier, J. Geldermann, & T. Müller (Eds.), *Proceedings of the International Conference on Information Systems for Crisis Response and Management (ISCRAM)* (pp. 780–790). https://doi.org/10.1126/science.1060143

Reuter, C., Hughes, A. L., & Kaufhold, M.-A. (2018). Social Media in Crisis Management: An Evaluation and Analysis of Crisis Informatics Research. *International Journal on Human-Computer Interaction (IJHCI), 34*(4), 280–294. https://doi.org/10.1080/10447318.2018.1427832

Reuter, C., & Kaufhold, M.-A. (2018). Fifteen Years of Social Media in Emergencies: A Retrospective Review and Future Directions for Crisis Informatics. *Journal of Contingencies and Crisis Management (JCCM), 26*(1), 41–57. https://doi.org/10.1111/1468-5973.12196

Reuter, C., Kaufhold, M.-A., Leopold, I., & Knipp, H. (2017a). Informing the Population: Mobile Warning Apps. In M. Klafft (Ed.), *Risk and Crisis Communication in Disaster Prevention and Management* (pp. 31–41).

Reuter, C., Kaufhold, M.-A., Leopold, I., & Knipp, H. (2017b). Katwarn, NINA, or FEMA? Multi-Method Study on Distribution, Use and Public Views on Crisis Apps. *European Conference on Information Systems (ECIS)*, 2187–2201.

Reuter, C., Kaufhold, M.-A., Schmid, S., Hahne, A. S., & Spielhofer, T. (2019). The Impact of Risk Cultures: Citizens' Perception of Social Media Use in Emergencies across Europe. *Technological Forecasting and Social Change (TFSC)*, *148*(119724). https://doi.org/10.1016/j.techfore.2019.119724

Reuter, C., Kaufhold, M.-A., & Spielhofer, T. (2017). *EmerGent Deliverable 2.5: Continuous Citizens and EMS Involvement by Social Media.* https://www.fp7-emergent.eu/wp-content/uploads/2017/09/20170427_D2-5_CitizenIntegration_EmerGent_incl_Appendix_small.pdf

Reuter, C., Kaufhold, M.-A., Spielhofer, T., & Hahne, A. S. (2017). Social Media in Emergencies: A Representative Study on Citizens' Perception in Germany. *Proceedings of the ACM: Human Computer Interaction (PACM): Computer-Supported Cooperative Work and Social Computing*, *1*(2), 1–19. https://doi.org/10.1145/3134725

Reuter, C., Kaufhold, M. A., Spahr, F., Spielhofer, T., & Hahne, A. S. (2020). Emergency service staff and social media – A comparative empirical study of the attitude by emergency services staff in Europe in 2014 and 2017. *International Journal of Disaster Risk Reduction (IJDRR)*, *46*(101516). https://doi.org/10.1016/j.ijdrr.2020.101516

Reuter, C., & Ludwig, T. (2013). Anforderungen und technische Konzepte der Krisenkommunikation bei Stromausfall. In M. Hornbach (Ed.), *Informatik 2013 – Informatik angepasst an Mensch, Organisation und Umwelt, GI-Edition-Lecture Notes in Informatics (LNI)* (pp. 1604–1618). GI.

Reuter, C., Ludwig, T., Friberg, T., Moi, M., Akerkar, R., Wanczura, S. P., Gizikis, A., & Brien, T. O. (2014). *EmerGent Deliverable 3.1: Usage Patterns of Social Media in Emergencies.* https://www.fp7-emergent.eu/wp-content/uploads/2014/09/D3.1_UsagePatternsOfSocialMediaInEmergencies.pdf

Reuter, C., Ludwig, T., Friberg, T., Pratzler-Wanczura, S., & Gizikis, A. (2015). Social Media and Emergency Services? Interview Study on Current and Potential Use in 7 European Countries. *International Journal of Information Systems for Crisis Response and Management (IJISCRAM)*, *7*(2), 36–58.

Reuter, C., Ludwig, T., Kaufhold, M.-A., & Hupertz, J. (2017a). Social Media Resilience during Infrastructure Breakdowns using Mobile Ad-Hoc Networks. In V. Wohlgemuth, F. Fuchs-Kittowski, & J. Wittmann (Eds.), *Advances and New Trends in Environmental Informatics – Proceedings of the 30th EnviroInfo Conference* (pp. 75–88). Springer. https://doi.org/10.1007/978-3-319-44711-7_7

Reuter, C., Ludwig, T., Kaufhold, M.-A., & Pipek, V. (2015). XHELP: Design of a Cross-Platform Social-Media Application to Support Volunteer Moderators in Disasters. *Proceedings of the Conference on Human Factors in Computing Systems (CHI)*, 4093–4102. https://doi.org/10.1145/2702123.2702171

Reuter, C., Ludwig, T., Kaufhold, M.-A., & Spielhofer, T. (2016). Emergency Services Attitudes towards Social Media: A Quantitative and Qualitative Survey across Europe.

International Journal on Human-Computer Studies (IJHCS), 95, 96–111. https://doi.org/ 10.1016/j.ijhcs.2016.03.005

Reuter, C., Ludwig, T., Kotthaus, C., Kaufhold, M.-A., von Radziewski, E., & Pipek, V. (2016). Big Data in a Crisis? Creating Social Media Datasets for Emergency Management Research. *I-Com: Journal of Interactive Media, 15*(3), 249–264. https://doi.org/10.1515/ icom-2016-0036

Reuter, C., Ludwig, T., & Pipek, V. (2014). Ad Hoc Participation in Situation Assessment: Supporting Mobile Collaboration in Emergencies. *ACM Transactions on Computer-Human Interaction (ToCHI), 21*(5). https://doi.org/10.1145/2651365

Reuter, C., Ludwig, T., & Pipek, V. (2016). Kooperative Resilienz – ein soziotechnischer Ansatz durch Kooperationstechnologien im Krisenmanagement. *Gruppe. Interaktion. Organisation. Zeitschrift Für Angewandte Organisationspsychologie (GIO).*

Reuter, C., Ludwig, T., Ritzkatis, M., & Pipek, V. (2015). Social-QAS: Tailorable Quality Assessment Service for Social Media Content. In P. Díaz, V. Pipek, C. Ardito, C. Jensen, I. Aedo, & A. Boden (Eds.), *Proceedings of the International Symposium on End-User Development (IS-EUD)* (pp. 156–170). Lecture Notes in Computer Science.

Reuter, C., Marx, A., & Pipek, V. (2012). Crisis Management 2.0: Towards a Systematization of Social Software Use in Crisis Situations. *International Journal of Information Systems for Crisis Response and Management (IJISCRAM), 4*(1), 1–16. https://doi.org/10.4018/jis crm.2012010101

Reuter, C., Marx, A., & Pipek, V. (2011). Social Software as an Infrastructure for Crisis Management – a Case Study about Current Practice and Potential Usage. In M. A. Santos, L. Sousa, & E. Portela (Eds.), *Proceedings of the International Conference on Information Systems for Crisis Response and Management (ISCRAM).*

Reuter, C., Mentler, T., & Geisler, S. (2015). Special Issue on Human Computer Interaction in Critical Systems I: Citizen and Volunteers. *International Journal of Information Systems for Crisis Response and Management (IJISCRAM), 7*(2).

Reuter, C., Pätsch, K., & Runft, E. (2017a). IT for Peace? Fighting Against Terrorism in Social Media – An Explorative Twitter Study. *I-Com: Journal of Interactive Media, 16*(2).

Reuter, C., Pätsch, K., & Runft, E. (2017b). Terrorbekämpfung mithilfe sozialer Medien – ein explorativer Einblick am Beispiel von Twitter. *Proceedings of the International Conference on Wirtschaftsinformatik (WI).*

Reuter, C., Ritzkatis, M., & Ludwig, T. (2014). Entwicklung eines SOA – basierten und anpassbaren Bewertungsdienstes für Inhalte aus sozialen Medien. *Informatik 2014 – Big Data – Komplexität Meistern,* 977–988.

Reuter, C., & Scholl, S. (2014). Technical Limitations for Designing Applications for Social Media. In M. Koch, A. Butz, & J. Schlichter (Eds.), *Mensch & Computer: Workshopband* (pp. 131–140). Oldenbourg-Verlag.

Reuter, C., & Schröter, J. (2015). Microblogging during the European Floods 2013: What Twitter May Contribute in German Emergencies. *International Journal of Information Systems for Crisis Response and Management (IJISCRAM), 7*(1), 22–41.

Reuter, C., & Spielhofer, T. (2017). Towards Social Resilience: A Quantitative and Qualitative Survey on Citizens' Perception of Social Media in Emergencies in Europe. *Journal Technological Forecasting and Social Change (TFSC), 121,* 168–180.

Reuter, C., Stieglitz, S., & Imran, M. (2020). Social media in conflicts and crises. *Behaviour and Information Technology, 39*(3), 241–251. https://doi.org/10.1080/0144929X.2019. 1629025

Reynolds, B., & W Seeger, M. (2005). Crisis and emergency risk communication as an integrative model. *Journal of Health Communication, 10*(1), 43–55. https://doi.org/10. 1080/10810730590904571

Ritter, A., Clark, S., Mausam, & O, E. (2011). Named Entity Recognition in Tweets: An Experimental Study. In P. Merlo, R. Barzilay, & M. Johnson (Eds.), *Proceedings of the Conference on Empirical Methods in Natural Language Processing (EMNLP)* (pp. 1524– 1534). Association for Computational Linguistics.

Robinson, T., Callahan, C., Boyle, K., Rivera, E., & Cho, J. K. (2017). I ♥ FB: A Q-Methodology Analysis of Why People 'Like' Facebook.' *International Journal of Virtual Communities and Social Networking (IJVCSN), 9*(2), 46–61. https://doi.org/10.4018/IJV CSN.2017040103

Rohde, M., Brödner, P., Stevens, G., Betz, M., & Wulf, V. (2017). Grounded Design-a praxeological IS research perspective. *Journal of Information Technology, 32*(2), 163–179. https://doi.org/10.1057/jit.2016.5

Rohde, M., Stevens, G., Brödner, P., & Wulf, V. (2009). Towards a Paradigmatic Shift in IS: Designing for Social Practice. *Proceedings of the International Conference on Design Science Research in Information Systems and Technology (DESRIST)*. https://doi.org/10. 1145/1555619.1555639

Rohweder, J. P., Kasten, G., Malzahn, D., Piro, A., & Schmid, J. (2011). Informationsqualität – Definitionen, Dimensionen und Begriffe. In *Daten und Informationsqualität – Auf dem Weg zur Information Excellence* (pp. 25–45). https://doi.org/10.1007/978-3-8348-9953-8

Roidl, E., & Rüsing, O. (2014). Erfolgsfaktoren zur Verhaltensveränderung durch mobile Apps. *UP14 – Kurzvorträge*, 1–15.

Romero, C., Olmo, J. L., & Ventura, S. (2013). A meta-learning approach for recommending a subset of white-box classification algorithms for Moodle datasets. *Proceedings of the International Conference on Educational Data Mining (EDM)*, 268–271.

Rother, K., Karl, I., & Nestler, S. (2015). Virtual Reality Crisis Simulation for Usability Testing of Mobile Apps. *Mensch Und Computer 2015 – Workshopband*, 69–76. https:// doi.org/10.1515/9783110443905-010

Rudra, K., Ganguly, N., Goyal, P., & Ghosh, S. (2018). Extracting and Summarizing Situational Information from the Twitter Social Media during Disasters. *ACM Transactions on the Web, 12*(3), 1–35. https://doi.org/10.1145/3178541

Rudra, K., Ghosh, S., Ganguly, N., Goyal, P., & Ghosh, S. (2015). Extracting Situational Information from Microblogs During Disaster Events: A Classification-Summarization Approach. *Proceedings of the 24th ACM International on Conference on Information and Knowledge Management*, 583–592. https://doi.org/10.1145/2806416.2806485

Rudra, K., Goyal, P., Ganguly, N., Mitra, P., & Imran, M. (2018). Identifying Sub-events and Summarizing Disaster-Related Information from Microblogs. *SIGIR '18 The 41st International ACM SIGIR Conference on Research & Development in Information Retrieval*, 265–274. https://doi.org/10.1145/3209978.3210030

Ryan, B. (2011). How people seek information when their community is in a disaster. *Mergency Media and Public Affairs: Partnering with the Media*.

Sackmann, S., Hofmann, M., & Betke, H. J. (2014). Organizing On-Site Volunteers: An App-Based Approach. *Proceedings of the 2014 Ninth International Conference on Availability, Reliability and Security*, 438–439. https://doi.org/10.1109/ARES.2014.66

safeREACH. (2018). *Krisenalarmierung für Unternehmen*. https://www.safereach.net/

Sagar, V. C. (2016). As the Water Recedes: Sri Lanka Rebuilds. *RSIS Commentaries, 141*.

Sakaki, T., Okazaki, M., & Matsuo, Y. (2010). Earthquake shakes Twitter users: real-time event detection by social sensors. *WWW '10: Proceedings of the 19th International Conference on World Wide Web*, 851. https://doi.org/10.1145/1772690.1772777

San, Y., Wardell III, C., & Thorkildsen, Z. (2013). *Social Media in the Emergency Management Field: 2012 Survey Results*. https://www.cna.org/sites/default/files/research/SocialMedia_EmergencyManagement.pdf

Saracevic, T. (1975). Relevance: A Review of and a Framework for the Thinking on the Notion in Information Science. *Journal of the American Society for Information Science, 26*(6), 321–343. https://doi.org/10.1002/asi.4630260604

Saracevic, T. (2007). Relevance: A Review of the Literature and a Framework for Thinking on the Notion in Information Science. Part II: Nature and Manifestations of Relevance. *Journal of the American Society for Information Science and Technology, 58*(13), 1915–1933. https://doi.org/10.1002/asi.20682

Schamber, L., & Eisenberg, M. B. (1988). Relevance: The Search for a Definition. *Annual Meeting of the American Society for Information Science*, 17.

Schamber, L., Eisenberg, M. B., & Nilan, M. S. (1990). A Re-Examination of Relevance: Toward a Dynamic, Situational Definition. *Information Processing and Management, 26*(6), 755–776. https://doi.org/10.1016/0306-4573(90)90050-C

Schmidt, A., Wolbers, J., Ferguson, J., & Boersma, K. (2017). Are you Ready2Help? Conceptualizing the management of online and onsite volunteer convergence. *Journal of Contingencies and Crisis Management*. https://doi.org/10.1111/1468-5973.12200

Schulz, A., Hadjakos, A., Paulheim, H., Nachtwey, J., & Mühlhäuser, M. (2013). A multi-indicator approach for geolocalization of tweets. *Proceedings of the 7th International Conference on Weblogs and Social Media, ICWSM 2013*, 573–582.

Sebastiani, F. (2002). Machine {Learning} in {Automated} {Text} {Categorization}. *ACM Comput. Surv., 34*(1), 1–47. https://doi.org/10.1145/505282.505283

Seeger, M. W. (2006). Best practices in crisis communication: An expert panel process. *Journal of Applied Communication Research, 34*(3), 232–244. https://doi.org/10.1080/00909880600769944

Semaan, B., & Mark, G. (2011). Technology-mediated social arrangements to resolve breakdowns in infrastructure during ongoing disruption. *ACM Transactions on Computer-Human Interaction (TOCHI), 18*(4, Article 21), 1–21. https://doi.org/10.1145/2063231.2063235

Semaan, B., & Mark, G. (2012). "Facebooking" towards crisis recovery and beyond: disruption as an opportunity. *Proceedings of the ACM 2012 Conference on Computer Supported Cooperative Work (CSCW '12)*, 27–36. https://doi.org/10.1145/2145204.2145214

Sen, A., Rudra, K., & Ghosh, S. (2015). Extracting Situational Awareness from Microblogs during Disaster Events. *2015 7th International Conference on Communication Systems and Networks (COMSNETS)*, 1–6.

Settles, B. (2010). Active learning literature survey. *University of Wisconsin, Madison, 15*(2), 201–221. https://doi.org/10.1.1.167.4245

Shankaranarayanan, G., & Blake, R. (2017). From Content to Context: The Evolution and Growth of Data Quality Research. *ACM Journal of Data and Information Quality, 8*(2), 9:1–9:28. https://doi.org/10.1145/2996198

Shankaranarayanan, G., Iyer, B., & Stoddard, D. (2012). Quality of Social Media Data and Implications of Social Media for Data Quality. *Proceedings of the 17th International Conference on Information Quality (ICIQ-2012)*, 311–325.

Shaw, G. L., & Harrald, J. R. (2004). The core competencies required of executive level business crisis and continuity managers. *Journal of Homeland Security and Emergency Management, 1*(1), 1–15. https://doi.org/10.2202/1547-7355.1003

Shih, F., Seneviratne, O., Liccardi, I., Patton, E., Meier, P., & Castillo, C. (2013). Democratizing mobile app development for disaster management. *Joint Proceedings of the Workshop on AI Problems and Approaches for Intelligent Environments and Workshop on Semantic Cities – AIIP '13*, 39–42. https://doi.org/10.1145/2516911.2516915

Shklovski, I., Palen, L., & Sutton, J. (2008). Finding Community Through Information and Communication Technology During Disaster Events. *Proceedings of the Conference on Computer Supported Cooperative Work (CSCW)*.

Shneiderman, B. (1996). The Eyes Have It: A Task by Data Type Taxonomy for Information Visualizations. *Proceedings 1996 IEEE Symposium on Visual Languages*, 1–8.

Simon, T., Goldberg, A., Aharonson-Daniel, L., Leykin, D., & Adini, B. (2014). Twitter in the Cross Fire—The Use of Social Media in the Westgate Mall Terror Attack in Kenya. *PLOS ONE, 9*(8), e104136.

Soden, R., & Palen, L. (2016). Infrastructure in the Wild: What Mapping in Post-Earthquake Nepal Reveals About Infrastructural Emergence. *Proceedings of the 2016 CHI Conference on Human Factors in Computing Systems*, 2796–2807. https://doi.org/10.1145/2858036.2858545

Soden, R., & Palen, L. (2018). Informating Crisis: Expanding Critical Perspectives in Crisis Informatics. *Proceedings of the ACM on Human-Computer Interaction*. https://doi.org/10.1145/3274431

Spence, P. R., Lachlan, K. A., Lin, X., & del Greco, M. (2015). Variability in Twitter Content Across the Stages of a Natural Disaster: Implications for Crisis Communication. *Communication Quarterly, 63*(2), 171–186. https://doi.org/10.1080/01463373.2015.1012219

Spielhofer, T., Greenlaw, R., Markham, D., & Hahne, A. (2016). Data mining Twitter during the UK floods: Investigating the potential use of social media in emergency management. *2016 3rd International Conference on Information and Communication Technologies for Disaster Management (ICT-DM)*, 1–6. https://doi.org/10.1109/ICT-DM.2016.7857213

Sreenivasan, N. D., Lee, C. S., & Goh, D. H. L. (2011). Tweet me home: Exploring information use on Twitter in crisis situations. *Proceedings of the 14th International Conference on Human-Computer Interaction*, 120–129. https://doi.org/10.1007/978-3-642-21796-8_13

Sriram, B., Fuhry, D., Demir, E., Ferhatosmanoglu, H., & Demirbas, M. (2010). Short Text Classification in Twitter to Improve Information Filtering. *Proceedings of the 33rd International ACM SIGIR Conference on Research and Development in Information Retrieval SE – SIGIR '10, July*, 841–842. doi: https://doi.org/10.1145/1835449.1835643

St. Denis, L. A., Hughes, A. L., & Palen, L. (2012). Trial by Fire: The Deployment of Trusted Digital Volunteers in the 2011 Shadow Lake Fire. In L. Rothkrantz, J. Ristvej, & Z.

Franco (Eds.), *Proceedings of the International Conference on Information Systems for Crisis Response and Management (ISCRAM)* (pp. 1–10). ISCRAM.

St. Denis, L. A., Palen, L., & Anderson, K. M. (2014). Mastering social media: An analysis of Jefferson County's communications during the 2013 Colorado floods. *Proceedings of the International Conference on Information Systems for Crisis Response and Management (ISCRAM), May,* 737–746.

Stallings, R. A., & Quarantelli, E. L. (1985). Emergent Citizen Groups and Emergency Management. *Public Administration Review, 45*(Special Issue), 93–100. https://doi.org/10.2307/3135003

Starbird, K. (2013). Delivering Patients to Sacré Coeur: Collective Intelligence in Digital Volunteer Communities. In W. E. Mackay, S. Brewster, & S. Bødker (Eds.), *Proceedings of the Conference on Human Factors in Computer Systems (CHI)* (pp. 801–810).

Starbird, K., & Palen, L. (2012). (How) will the revolution be retweeted?: information diffusion and the 2011 Egyptian uprising. *Proceedings of the Conference on Computer Supported Cooperative Work (CSCW),* 7–16.

Starbird, K., & Palen, L. (2010). Pass It On?: Retweeting in Mass Emergency. In S. French, B. Tomaszewski, & C. Zobel (Eds.), *Proceedings of the International Conference on Information Systems for Crisis Response and Management (ISCRAM)* (Issue December 2004, pp. 1–10).

Starbird, K., & Palen, L. (2011). Voluntweeters: Self-Organizing by Digital Volunteers in Times of Crisis. *Proceedings of the Conference on Human Factors in Computing Systems (CHI),* 1071–1080.

Starbird, K., Palen, L., Hughes, A. L., & Vieweg, S. (2010). Chatter on the red: what hazards threat reveals about the social life of microblogged information. *Proceedings of the Conference on Computer Supported Cooperative Work (CSCW),* 241–250. https://doi.org/10.1145/1718918.1718965

Starbird, K., Spiro, E., Edwards, I., Zhou, K., Maddock, J., & Narasimhan, S. (2016). Could This Be True?: I Think So! Expressed Uncertainty in Online Rumoring. *Proceedings of the 2016 CHI Conference on Human Factors in Computing Systems,* 360–371. https://doi.org/10.1145/2858036.2858551

Starbird, K., & Stamberger, J. (2010). Tweak the Tweet: Leveraging Microblogging Proliferation with a Prescriptive Syntax to Support Citizen Reporting. In S. French, B. Tomaszewski, & C. Zobel (Eds.), *Proceedings of the International Conference on Information Systems for Crisis Response and Management (ISCRAM)* (pp. 1–5).

Statista. (2015). *Bevölkerung Deutschlands nach Altersgruppen 2015.* https://de.statista.com/statistik/daten/studie/1365/umfrage/bevoelkerung-deutschlands-nach-altersgruppen/

Statista. (2016). *Bevölkerung Deutschlands nach Altersgruppen 2016.* https://de.statista.com/statistik/daten/studie/1365/umfrage/bevoelkerung-deutschlands-nach-altersgruppen/

Statista. (2017a). *Häufigkeit der Nutzung von Social Media in Deutschland 2015.* https://de.statista.com/statistik/daten/studie/168920/umfrage/haeufigkeit-der-nutzung-von-community-plattformen/

Statista. (2017b). *Most famous social network sites worldwide as of September 2017, ranked by number of active users (in millions).* https://www.statista.com/statistics/272014/global-social-networks-ranked-by-number-of-users/

Statista. (2017c). *Number of smartphone users worldwide from 2014 to 2020 (in billions).* https://www.statista.com/statistics/330695/number-of-smartphone-users-worldwide/

Statista. (2018). *Ranking der größten sozialen Netzwerke und Messenger nach der Anzahl der monatlich aktiven Nutzer (MAU) im Januar 2018 (in Millionen)*. https://de.statista.com/statistik/daten/studie/181086/umfrage/die-weltweit-groessten-social-networks-nach-anzahl-der-user/

Statista. (2020a). *Most popular social networks worldwide as of April 2020, ranked by number of active users*. https://www.statista.com/statistics/272014/global-social-networks-ranked-by-number-of-users/

Statista. (2020b). *Number of smartphone users worldwide from 2016 to 2021*. https://www.statista.com/statistics/330695/number-of-smartphone-users-worldwide/

Statistisches Bundesamt. (2016). *Bildungsstand der Bevölkerung 2016*. https://www.destatis.de/DE/Publikationen/Thematisch/BildungForschungKultur/Bildungsstand/Bildungsstandbevoelkerung5210002167004.pdf

Stephen, D. G., Dixon, J. A., & Isenhower, R. W. (2009). Dynamics of Representational Change: Entropy, Action, and Cognition. *Journal of Experimental Psychology: Human Perception and Performance, 35*(6), 1811–1832. https://doi.org/10.1037/a0014510

Stevens, G., & Pipek, V. (2018). Making Use: Understanding, Studying, and Supporting Appropriation. In *Socio-Informatics: A Practice-Based Perspective on the Design and Use of IT Artifacts* (pp. 139–176). Oxford University Press.

Stevens, G., Rohde, M., Korn, M., & Wulf, V. (2018). Grounded Design: A Research Paradigm inPractice-based Computing. In *Socio-Informatics: A Practice-Based Perspective on the Design and Use of IT Artifacts* (pp. 23–46).

Stieglitz, S., Bunker, D., Mirbabaie, M., & Ehnis, C. (2017). Sense-Making in Social Media During Extreme Events. *Journal of Contingencies and Crisis Management (JCCM)*. https://doi.org/10.1111/1468-5973.12193

Stieglitz, S., Dang-Xuan, L., Bruns, A., & Neuberger, C. (2014). Social Media Analytics: An Interdisciplinary Approach and Its Implications for Information Systems. *Business and Information Systems Engineering, 6*(2), 89–96. https://doi.org/10.1007/s12599-014-0315-7

Stieglitz, S., Mirbabaie, M., Fromm, J., & Melzer, S. (2018). The Adoption of Social Media Analytics for Crisis Management – Challenges and Opportunities. *Proceedings of the 26th European Conference on Information Systems (ECIS)*.

Stieglitz, S., Mirbabaie, M., & Milde, M. (2018). Social Positions and Collective Sense-Making in Crisis Communication. *International Journal of Human-Computer Interaction, 34*(4), 328–355. https://doi.org/10.1080/10447318.2018.1427830

Stieglitz, S., Mirbabaie, M., Ross, B., & Neuberger, C. (2018). Social media analytics – Challenges in topic discovery, data collection, and data preparation. *International Journal of Information Management, 39*, 156–168. https://doi.org/10.1016/j.ijinfomgt.2017.12.002

Strauss, A. L. (1987). *Qualitative Analysis for Social Scientists*. Cambridge Press.

Strauss, A. L., & Corbin, J. (1998). *Basics of qualitative research: Techniques and procedures for developing grounded theory*. Sage Publications.

Sutton, J. (2010). Twittering Tennessee: Distributed networks and Collaboration Following a Technological Disaster. In S. French, B. Tomaszewski, & C. Zobel (Eds.), *Proceedings of the International Conference on Information Systems for Crisis Response and Management (ISCRAM)*.

Sutton, J., Woods, C., & Vos, S. C. (2017). Willingness to click: Risk information seeking during imminent threats. *Journal of Contingencies and Crisis Management (JCCM)*, 1–12. https://doi.org/10.1111/1468-5973.12197

Taieb, M. H., Chouinard, J.-Y., & Wang, D. (2013). Spatial correlation-based side information refinement for distributed video coding. *EURASIP Journal on Advances in Signal Processing, 2013*(168), 1–19. https://doi.org/10.1186/1687-6180-2013-168

Tan, M. L., Prasanna, R., Stock, K., Hudson-Doyle, E., Leonard, G., & Johnston, D. (2017). Mobile applications in crisis informatics literature: A systematic review. *International Journal of Disaster Risk Reduction (IJDRR)*, 24, 297–311. https://doi.org/10.1016/j.ijdrr. 2017.06.009

Techno et Control GmbH & Co. KG. (2018). *SoftAngel App.* https://getsoftangel.com/

The Council of the European Union. (2008). Council Directive 2008/114/EC of 8 December 2008 on the indentification and designation of European critical infrastructures and the assessment of the need to improve their protection. *Official Journal of the European Union*, 75–82.

Thieken, A. H., Kreibich, H., Müller, M., & Merz, B. (2007). Coping with floods: Preparedness, response and recovery of flood-affected residents in Germany in 2002. *Hydrological Sciences Journal, 52*(5), 1016–1037. https://doi.org/10.1623/hysj.52.5.1016

Tjong Kim Sang, E. F., & De Meulder, F. (2003). Introduction to the CoNLL-2003 shared task. *Proceedings of the Seventh Conference on Natural Language Learning at HLT-NAACL 2003 –*, 4, 142–147. https://doi.org/10.3115/1119176.1119195

Tomé, A., & Pereira, F. (2011). Low delay distributed video coding with refined side information. *Signal Processing: Image Communication, 26*(4–5), 220–235. https://doi.org/10.1016/j.image.2011.01.005

Trilateral Research. (2015). *Comparative Review of Social Media Analysis Tools for Preparedness* (Issue July). https://trilateralresearch.com/tenders/#comparative-review-of-social-media-analysis-tools-for-preparedness

Tucker, A., Deek, F., Jones, J., Mccowan, D., Stephenson, C., & Verno, A. (2006). A Model Curriculum for K-12 Computer Science. In *Computer Science Association*.

Tucker, S., Ireson, N., Lanfranchi, V., & Ciravegna, F. (2012). "Straight to the information I need": Assessing collational interfaces for emergency response. *Proceedings of the International Conference on Information Systems for Crisis Response and Management (ISCRAM)*.

Twidale, M., Randall, D., & Bentley, R. (1994). Situated evaluation for cooperative systems. *Proceedings of the Conference on Computer Supported Cooperative Work (CSCW)*, 441–452.

United Nations. (2006). *Global Survey of Early Warning Systems.* https://www.undrr.org/publication/global-survey-early-warning-systems

United Nations Department of Humanitarian Affairs. (2000). *Internationally agreed glossary of basic terms related to Disaster Management.* United Nations.

Utz, S., Schultz, F., & Glocka, S. (2013). Crisis communication online: How medium, crisis type and emotions affected public reactions in the Fukushima Daiichi nuclear disaster. *Public Relations Review, 39*(1), 40–46. https://doi.org/10.1016/j.pubrev.2012.09.010

Uysal, I., & Croft, W. B. (2011). User oriented tweet ranking. *Proceedings of the 20th ACM International Conference on Information and Knowledge Management – CIKM '11*, 2261. https://doi.org/10.1145/2063576.2063941

Valecha, R., Oh, O., & Rao, R. (2013). An Exploration of Collaboration over Time in Collective Crisis Response during the Haiti 2010 Earthquake. In R. Baskerville & M. Chau (Eds.), *Proceedings of the International Conference on Information Systems (ICIS)* (pp. 1–10). AISeL.

van Gorp, A. F. (2014). Integration of Volunteer and Technical Communities into the Humanitarian Aid Sector: Barriers to Collaboration. *Proceedings of the International Conference on Information Systems for Crisis Response and Management (ISCRAM)*, May, 620–629.

Veil, S. R., Buehner, T., & Palenchar, M. J. (2011). A Work-In-Process Literature Review: Incorporating Social Media in Risk and Crisis Communication. *Journal of Contingencies and Crisis Management (JCCM)*, *19*(2), 110–122. https://doi.org/10.1111/j.1468-5973.2011.00639.x

Venkatesh, V., & Davis, F. D. (2000). A Theoretical Extension of the Technology Acceptance Model: Four Longitudinal Field Studies. *Management Science*, *46*(2), 186–204. https://doi.org/10.1287/mnsc.46.2.186.11926

Venkatesh, V., Morris, M. G., Davis, G. B., & Davis, F. D. (2003). User Acceptance of Information Technology: Toward a Unified View. *MIS Quarterly*, *27*(3), 425–478. https://doi.org/10.2307/30036540

Verma, S., Vieweg, S., Corvey, W. J., Palen, L., Martin, J. H., Palmer, M., Schram, A., & Anderson, K. M. (2011). Natural Language Processing to the Rescue? Extracting "Situational Awareness" Tweets During Mass Emergency. *Proceedings of the 5th International AAAI Conference on Weblogs and Social Media*, 385–392.

Vieweg, S. (2012a). *Situational Awareness in Mass Emergency: A Behavioral and Linguistic Analysis of Microblogged Communications*. 1–300.

Vieweg, S. (2012b). Twitter communications in mass emergency. *Proceedings of the ACM 2012 Conference on Computer Supported Cooperative Work Companion – CSCW '12*, 227. https://doi.org/10.1145/2141512.2141584

Vieweg, S., Hughes, A. L., Starbird, K., & Palen, L. (2010). Microblogging During Two Natural Hazards Events: What Twitter May Contribute to Situational Awareness. *Proceedings of the Conference on Human Factors in Computing Systems (CHI)*, 1079–1088.

Viviani, M., & Pasi, G. (2017). Credibility in social media: opinions, news, and health information—a survey. *Wiley Interdisciplinary Reviews: Data Mining and Knowledge Discovery*, *7*(5), e1209–n/a. https://doi.org/10.1002/widm.1209

Volgger, S., Walch, S., Kumnig, M., & Penz, B. (2006). *Kommunikation vor, während und nach der Krise – Leitfaden für Kommunikationsmanagement anhand der Erfahrungen des Hochwasserereignisses Tirol 2005.*

vom Brocke, J., Simons, A., Riemer, K., Niehaves, B., Plattfaut, R., & Cleven, A. (2015). Standing on the Shoulders of Giants: Challenges and Recommendations of Literature Search in Information Systems Research. *Communications of the Association for Information Systems*, *37*, 205–224. https://doi.org/10.17705/1CAIS.03709

VOSG. (2017). *Virtual Operations Support Teams (VOST)*. https://vosg.us/

Wade, J. (2012). Using mobile apps in disasters. *Risk Management*, *59*(9), 6–8.

Wan, S., & Paris, C. (2015). Understanding Public Emotional Reactions on Twitter. *Proceedings of the Ninth International AAAI Conference on Web and Social Media*, 715–716.

Wang, A., Wan, G., Cheng, Z., & Li, S. (2009). An incremental extremely random forest classifier for online learning and tracking. *Proceedings – International Conference on Image Processing, ICIP*. https://doi.org/10.1109/ICIP.2009.5414559

Wang, R. Y., & Strong, D. M. (1996). Beyond Accuracy: What Data Quality Means to Data Consumers. *Journal of Management Information Systems, 12*(4), 5–33. https://doi.org/10.1080/07421222.1996.11518099

Weißweiler, L., & Fraser, A. (2017). Developing a Stemmer for German Based on a Comparative Analysis of Publicly Available Stemmers. *International Conference of the German Society for Computational Linguistics and Language Technology.*

Werbos, P. J. (1994). *The roots of backpropagation: from ordered derivatives to neural networks and political forecasting.* Wiley-Interscience.

White, C., Plotnick, L., Kushma, J., Hiltz, S. R., & Turoff, M. (2009). An online social network for emergency management. *International Journal of Emergency Management (IJEM), 6*(3/4), 369–382. https://doi.org/10.1504/IJEM.2009.031572

White, J. I., & Palen, L. (2015). Expertise in the Wired Wild West. *Proceedings of the ACM Conference on Computer-Supported-Cooperative Work and Social Computing (CSCW),* 662–675. https://doi.org/10.1145/2675133.2675167

White, J. I., Palen, L., & Anderson, K. M. (2014). Digital Mobilization in Disaster Response: The Work & Self – Organization of On-Line Pet Advocates in Response to Hurricane Sandy. In S. R. Fussell, W. G. Lutters, M. R. Morris, & M. Reddy (Eds.), *Proceedings of the Conference on Computer Supported Cooperative Work (CSCW)* (Issue October, pp. 866–876). ACM.

Wiegand, S., & Middleton, S. E. (2016). Veracity and Velocity of Social Media Content during Breaking News: Analysis of November 2015 Paris Shootings. *Proceedings of the 25th International Conference Companion on World Wide Web, November 2015,* 751–756. https://doi.org/10.1145/2872518.2890095

Wikipedia. (2018). *Cell Broadcast.* https://en.wikipedia.org/wiki/Cell_Broadcast

Wilensky, H. (2014). Twitter as a Navigator for Stranded Commuters during the Great East Japan Earthquake. *Proceedings of the International Conference on Information Systems for Crisis Response and Management (ISCRAM), May,* 695–704.

Wilson, T., Stanek, S. A., Spiro, E. S., & Starbird, K. (2017). Language Limitations in Rumor Research? Comparing French and English Tweets Sent During the 2015 Paris Attacks. *Proceedings of the International Conference on Information Systems for Crisis Response and Management (ISCRAM),* 546–553.

Wise Bitch. (2009). *Country residents outside of Fargo are surrounded by flood waters. Some R being rescued [Tweet].*

Wissenschtliche Kommission Wirtschaftsinformatik. (2008). WI-Orientierungslisten. *Wirtschaftsinformatik, 50*(2). https://doi.org/10.1365/s11576-008-0040-2

Wobbrock, J. O., & Kientz, J. A. (2016). Research contribution in human-computer interaction. *Interactions, 23*(3), 38–44. https://doi.org/10.1145/2907069

World Wide Web Consortium. (2016). *Activity Streams 2.0.* W3C Recommendation. https://www.w3.org/TR/activitystreams-core/

Wu, S., Hofman, J. M., Mason, W. a., & Watts, D. J. (2011). Who says what to whom on twitter. In S. Sadagopan & K. Ramamritham (Eds.), *Proceedings of the 20th International Conference on World Wide Web* (pp. 705–714). https://doi.org/10.1145/1963405.1963504

Wukich, C. (2015). Social media use in emergency management. *Journal of Emergency Management*, *13*(4), 281–294. https://doi.org/10.5055/jem.2015.0242

Wulf, V., Misaki, K., Atam, M., Randall, D., & Rohde, M. (2013). 'On the Ground' in Sidi Bouzid: Investigating Social Media Use during the Tunisian Revolution. *Proceedings of the Conference on Computer Supported Cooperative Work (CSCW)*, 1409–1418.

Wulf, V., Müller, C., Pipek, V., Randall, D., & Rohde, M. (2015). Practice-Based Computing: Empirically Grounded Conceptualizations Derived from Design Case Studies. In V. Wulf, K. Schmidt, & D. Randall (Eds.), *Designing Socially Embedded Technologies in the Real-World* (pp. 111–150). Springer. https://doi.org/10.1007/978-1-4471-6720-4

Wulf, V., Rohde, M., Pipek, V., & Stevens, G. (2011). Engaging with Practices: Design Case Studies as a Research Framework in CSCW. *Proceedings of the Conference on Computer Supported Cooperative Work (CSCW)*, 505–512.

Xie, B., He, D., Mercer, T., Wang, Y., Wu, D., Fleischmann, K. R., Zhang, Y., Yoder, L. H., Stephens, K. K., Mackert, M., & Lee, M. K. (2020). Global healtah crises are also information crises: A call to action. *Journal of the Association for Information Science and Technology*, 1–5. https://doi.org/10.1002/asi.24357

Yang, S., Chung, H., Lin, X., Lee, S., & Chen, L. (2013). PhaseVis: What, When, Where, and Who in Visualizing the Four Phases of Emergency Management Through the Lens of Social Media. *Proceedings of the International Conference on Information Systems for Crisis Response and Management (ISCRAM)*, 912–917.

Yang, Y., & Loog, M. (2017). Active learning using uncertainty information. *Proceedings – International Conference on Pattern Recognition.* https://doi.org/10.1109/ICPR.2016.7900034

Zade, H., Shah, K., Rangarajan, V., Kshirsagar, P., Imran, M., & Starbird, K. (2018). From Situational Awareness to Actionability: Towards Improving the Utility of Social Media Data for Crisis Response. *Proceedings of the ACM on Human-Computer Interaction.*

Zeng, J., Chan, C., & Fu, K. (2016). What is social media platforms' role in constructing 'truth' around crisis events? A case study of Weibo's rumour management strategies after the 2015 Tianjin blasts. *The Internet, Policy & Politics Conferences.*

Zhou, Z. H., & Liu, X. Y. (2006). Training cost-sensitive neural networks with methods addressing the class imbalance problem. *IEEE Transactions on Knowledge and Data Engineering*, *18*(1), 63–77. https://doi.org/10.1109/TKDE.2006.17

Zipf, A. (2016). Mit Netz und Geodaten. Katastrophen-Management Online. *Ruperto Carola Forschungsmagazin*, *8*, 42–49.

Printed in the United States
by Baker & Taylor Publisher Services